Reminiscences of a Life Well Spent

by an Old Redleg (Artilleryman)

Reminiscences of a Life Well Spent

by an Old Redleg (Artilleryman)

BUILDING A FOUNDATION FOR SUCCESS

Col(Ret) Gerald W. "Jerry" Sharpe

Husband, Father, Son, Brother, Uncle, Cousin, Nephew, Partner, Artilleryman,
Kiwanian, Rotarian, and American living the good life in the Philippines

Printed in The United States of America
Cover and interior design by Deeds Creative

ISBN 978-1-961505-16-2

Books are available in quantity for promotional or premium use.
For information, email info@deedspublishing.com.

First Edition, 2024

10 9 8 7 6 5 4 3 2 1

DEDICATION

TO MY MOTHER, WHO I CONSIDER TO BE A SAINT, WHO LOVED GOD AND LIVED AN EXEMPLARY life. She raised me with love; taught me to appreciate reading at a very early age; taught me how to do many things like bake cookies, paint, sew, and keep my room clean; and did it with patience, understanding, and a willingness to show me how to do things rather than just telling me how.

To my father, who taught me by example. He would say, "Watch, this is how you do it." I learned the value of hard work as dad rose at dawn and worked a full day, usually seven days a week. He had a multitude of projects ongoing at any given time and I was expected to help. He taught me the value of money as I had to earn what I received. He had a "Midas Touch" when it came to real estate and I learned many other things from him that I describe in my early chapters.

To my four brothers and sister and their families who have significantly affected my life.

To my four children who encouraged me to put down on paper the events in my life.

To my high school classmates and friends who played a role in my early development.

To a number of ministers of the Pillar of Fire Church who had a profound effect on my life and contributed directly to my later successes.

To the many supervisors and commanders I had in my early Army days who taught me both positive and negative lessons I describe more fully in this book. They were truly responsible for, as my subtitle suggests, "Building a Foundation for Success.

To my Partner, Myla "Mae" Figueroa who has supported my efforts, encouraged me daily, and has been my "Rock of Gibraltar."

CONTENTS

TESTIMONIALS

I AM PLEASED TO HAVE THE OPPORTUNITY TO RECOMMEND MY HIGH SCHOOL CLASSMATE'S book as it has brought back many memories of earlier times. I have known Jerry (or Gerald, as we all knew him back then) since my early childhood. We first met when our family went on vacation to California in the 1950's. Our families are closely intertwined by marriage as my mother's sister, Gertrude married Jerry's father's brother, Edward Sharpe, in New Jersey

My mother, Grace, grew up in New Jersey and attended Pillar of Fire Schools in Zarephath, NJ. She later taught in church schools in Cincinnati, OH and at Belleview where she met my father, Carl Jacobson, who grew up in Westminster, and attended Belleview schools. They were married in July of 1935 and we lived in a home next to the campus.

Once Jerry's family moved to Belleview, we attended school in the 8th grade and then through four years of high school. Jerry was always enthusiastic, bright, outgoing, and slightly mischievous! We enjoyed intermural sports and school programs. I remember one of our teachers asked each of us where we wanted to go to college, and I recall Jerry replied, "M.I.T."

All nine of us graduated from Belleview in 1959. As Jerry relates, he went off to Western State College while I decided to work for a year. Not sure what happened to his "M.I.T" dream. I joined Jerry at Western State in 1960 and we were classmates for a year. Jerry would occasionally ride with me back to Westminster for the weekend and he relates the story of our trip on December 16, 1960, when as we were heading toward Monarch Pass on Highway 50, we heard on the radio that two planes had collided over New York City and one had crashed into a church property in New Jersey, owned by the Pillar of Fire. Jerry's 90-year-old great grandfather, Wallace Lewis, was killed.

I am not sure I knew the story of why Jerry joined the Army instead of returning to Western State, but in reading his stories it seems he found his calling. I did not see him much, if at all, over his years in the Army, but I received Christmas newsletters, showing the beautiful homes he built and where he raised his family.

It is commendable that Jerry was able to assemble so many photos and recall so many things about his childhood. I, too, enjoy the connection with my past, as well as the history between the Sunday Sharpe and Edward Sharpe families, of which I am a part. Thanks, Jerry, for taking the time to assemble such a fine publication!

—**Arthur "Art" Jacobson,** High School Classmate

READING JERRY'S BOOK CERTAINLY EVOKES MANY MEMORIES, BOTH GOOD AND BAD OF OUR families' early years in the Pillar of Fire Church. Jerry has been able to use diaries, letters, and recollections from others to talk about growing up without being separated from his parents. Readers will get some sense of what growing up in the Pillar of Fire Church was like. As for me, the 95-year-old brother of Jerry's mother, I recall Bishop Alma White often breaking up families as she did with ours, depositing me at age 8 in a church orphanage for two years. Jerry tells the story of how he used his experiences growing up in the church as a basis for a successful Army career. I highly recommend reading this book which gives many examples of how to be successful by sheer hard work and determination.

—**Ray Ritchie,** Jerry's Uncle

WITH "REMINISCENCES," MY COUSIN JERRY SHARPE HAS PRODUCED SOMETHING REMARK-able! He lays out the story of a person actively, inextricably engaged in life, ever in a "take charge" way. His impressive compilation of bits and pieces creates a uniquely styled mosaic of his life. The result is an overarching high-resolution chronicling of a man and his times; and in its form and result unlike any other memoir I'm familiar with. Further, for anyone who might want to compare today's world with the mid-20th century origins of "silent generation" folks, plain-spoken, everyday details of some of those times are included. They might appeal to anyone seeking to contrast life lived without social media with "the now," and they evoke times when much of today's world would have been unfamiliar or even incomprehensible. This book stands completely on its own as a catalog of Col Sharpe's lifetime of noteworthy accomplishments, credentials, and adventures. Additionally on the personal side, it has provided me a peek at some of his, and our families developments that I had only been peripherally aware of. I really had trouble putting it down, for *all* those reasons!

—**Burton L. "Burt" Sharpe,** Jerry's eldest cousin,
Retired NASA Engineer and Author

I AM HONORED TO RECOMMEND MY BROTHER'S BOOK FOR MANY REASONS. GERALD, AS WE learn in the book, was the only name acceptable to our mother. He is now the patriarch of our family and has dedicated the last several years to writing his life's story and, in the process, has also recounted the history of our family. I thank him for this gift to all of us. This book highlights Gerald's great determination and perseverance which enabled him to overcome obstacles, take charge, and produce outstanding results in his Army career. It makes for fascinating reading. Several of his superiors ranked him the best officer they had ever known. Never one to blow his own horn, he did not talk much about these stories to my knowledge. I, for one, was not fully aware of his spectacular rise in rank and results. His story is inspiring and challenging. Gerald is a talented writer, and his narrative is compelling. I've thoroughly enjoyed even the second and third readings of his book. He has invested a magnificent amount of work and an amazing organization of details to produce this first volume of his life's story. I can't wait to read the next two books.

— Eileen Sharpe Fournier,
Ph.D., Jerry's only sister

I HAVE KNOWN COL (RET) JERRY SHARPE SINCE HE WAS MY BATTERY COMMANDER WHILE I AT-tended Artillery OCS in 1966 at Fort Sill, OK. I was somewhat familiar with his accomplishments and experiences, but "Reminiscences of a Life Well Spent: by an Old Redleg (Artilleryman)" goes "above and beyond". When building a house, the foundation is critical, and Jerry has meticulously constructed the foundation of his life in this first installment of his memoir. During his early years, Jerry described personal experiences, coupled with entries from his parents' diaries, and demonstrated a considerable amount of patience in assembling a cohesive timeline of his life. The attention to detail that Jerry exhibited while writing this autobiography is one of the most important skills for an artilleryman. When firing a 100 lb. artillery projectile over friendly troops at a target that is 10 miles away requires exacting calculations and measurements – attention to detail. His growing up years included relocating to numerous states, which added a considerable amount of personal development and challenges, which he was able to handle relatively well. There is a pattern throughout this part of his life that Jerry relies on heavily. He is very observant and catalogs his experiences, as well as those of others, into what works and what doesn't work for future reference. He demonstrates that he has been able to learn from both good and bad leadership. As a former enlisted soldier and NCO, he also brings a common-sense approach to both minimizing failures and resolving problems.

— George A. Bannon, COL, USA Ret)

CAPTAIN JERRY SHARPE, WHO I KNEW AND WORKED FOR IN VIETNAM, WAS THE MOST PROFES-sional officer that l had the privilege of serving under. He was pleasant, showed interest in his men, and yet was determined to have the most responsible, dedicated, and competent S2 section possible. When l was invited to read his first volume, l was amazed at the quality of the story being told. Details were interwoven, yet it did not become boring, dooming the narrative to monotonous reading or worse, but instead illuminated and highlighted the storyline.

The details are such, especially in the telling of his Vietnam experiences that l was a part of, l just know he had to be journaling. His first work is amazing, telling the story of a boy, adolescent, and young man growing up in the United States and then hearing and responding to the call to serve for his adult life. He did all of this during, probably, the greatest time in the history of this country. I look forward to reading the rest of his saga. I know it will be as exceptional.

— Gary Enzfelder, Sergeant (E5) S2,
Divarty, 1st Cavalry Division (Airmobile)
South Vietnam (69-70)

JERRY'S BIOGRAPHICAL INFORMATION IS INTENSELY PERSONAL, YET FRANK AND UNVARNISHED.
He is justifiably proud of his successful military career, but he also shows personal humility, as appropriate, when acknowledging the contributions of others to his life and career. The road he traveled was challenging, from entering the US Army as an enlisted man with some college, to his retirement as a full colonel.

This path required extraordinary capability and dedication to excellence. I appreciated the extensive glossary he placed at the beginning of his book. Also, I found his Vietnam experiences to be especially compelling, including his firsthand account of Nixon's Cambodian Incursion initiative. Speaking as a friend of Jerry and his family, I must say that I was previously unaware of much of his life. This book answered many questions for me and I am sure it will for others.

— **Frederic N. "Fred" Schissler,**
Brother of Jerry's stepmother, Elaine

REMEMBERING JERRY SHARPE

I first met Jerry Sharpe during our time at Artillery Officers Candidate School in Lawton, Oklahoma. We were both young, eager candidates, ready to take on the grueling six months of schooling, physical training, inspections, and the inevitable harassment that comes with being a newbie. From the very beginning, Jerry stood out. He was an exceptional student, and his attention to detail was unmatched. Whether it was his polished brass, gleaming, mirror-finished shoes, or the razor-sharp creases in his starched uniforms, Jerry consistently set a standard that the rest of us could only aspire to. Competing with him was futile, so instead, I watched with admiration as he continued to receive squad and company accolades. His dedication, discipline, and unwavering commitment to excellence were evident even back then, and it's no surprise that he went on to have such a distinguished career.

After we received our officer's gold bars upon graduation we decided to room together in a small two-bedroom home off base, that I truly got to know the best side of Jerry. He was the ideal roommate—generous, kind, and thoughtful. We never h ad a disagreement while living together, which is a testament to his easygoing yet disciplined nature. As I recall he was a pretty good cook, too, though we took most of our meals on base in the mess hall. While we both enjoyed our time off duty, Jerry was always more business minded than I was, constantly preparing his uniforms and studying for his duties on post.

We faced the decision of choosing our career paths. At that time, the Army offered a six-month active-duty option followed by seven years in the reserves, which I chose. During my six months I was assigned as a motor pool officer at one of the artillery battalions at Fort Sill. Jerry, as you'll learn reading his memoir, chose a more extended and committed path with the Army.

His dedication was a great influence on my early life, and it's clear now that all his hard work and focus foreshadowed the remarkable career, he would go on to have. It was an honor to share those formative years with Jerry, and I'm proud to call him a friend.

— **Edmund J. "Ed" McCormick,** Jr., Artillery OCS Class 5-64,
classmate and author of ten books

I FOUND THAT READING JERRY'S FIRST VOLUME OF HIS AUTOBIOGRAPHY BROUGHT BACK MANY fond memories. Although Jerry and I were about the same age and grew up in the same church organization, we never really knew each other until his family moved to the Belleview Campus in Colorado. He was born in our family's historic Staats farmhouse at Zarephath, NJ. His family later moved into the building where my brother and I were born and where we lived in New Jersey in 1943 for a few months, but we were to young to really know each other.

I believe I first got to know Jerry when he was in the 8th grade in 1954, and I was a freshman in high school. Once again, the Sharpe's followed our family. We moved out of the Hopewell residence into a house two blocks away, and they moved in right behind us. The next year, when we were both in high school, we spent hours and hours together and we became great friends. I had a growth spurt as a freshman and was about 6 ft tall; Jerry's didn't come until much later - so as he relates, we became known as Mutt and Jeff.

Jerry writes about how, just before his junior year in college, he and I were having family problems and we decided that joining the Army was the perfect solution. We both signed up, took our entrance exams, and selected jobs. We then went home and talked to our parents. My parents caved to my wishes, and I chickened out, but Jerry, as he so clearly describes, went in the service and apparently never looked back.

A couple of years later, in 1963, I got drafted. For advanced training, I was assigned to artillery AIT at Fort Sill, OK. At about the same time, Jerry had made Sergeant E-5 in Germany and then entered Officer Candidate School at Fort Sill. We were able to visit several times. While I was still at Fort Sill, Jerry graduated from OCS as a 2nd Lieutenant. He had met and proposed to a girl who lived next door whose father was a retired officer at Ft. Bliss, TX. The wedding was to be at a church in El Paso. I am not sure that Jerry ever knew how I got there, but I was fortunate to hitch rides on Army aircraft and was able to attend.

I often tell people that one of my best friends in high school went from private to full colonel in the Army, a rare feat. I remember many of the full-colonels I was associated with were West Point graduates. The way Jerry has described his life certainly fits his sub-title, "Building a Foundation for Success" and I believe his book can be used as a guide on how to be successful. Not only can I highly recommend Jerry's book to others, but can say that my children, particularly my daughter Cindy, have been reading it with great interest. I am looking forward to following his activities and adventures in Volume 2 scheduled for release next year.

—**Willard Staats,**
Jerry's friend in high school and beyond.

PREFACE

THIS IS THE FIRST OF THREE VOLUMES THAT WILL TELL MY STORY. A STORY OF A LIFE WELL LIVED, of a life in service to my country and the communities where I resided. And of the family that raised me and later of my own family raised with love and affection. My extended family continues to be a blessing to me as I enjoy my remaining years on this earth.

In 2021 at the age of 79, while in my seventh year of living the good life retired in the Philippines, I began to note signs of memory loss that seemed to be more than just old age. I found that I could not remember reading books on my Kindle that I had read less than a year previously; I struggled to recall names of friends, and I would forget things when I went around my house looking for something. I had several War College Classmates who had been institutionalized and died because of Alzheimer's, and the stories their wives told me about how they finally recognized their husband's symptoms sounded eerily familiar. I decided that I wanted to create a history of my life so that my four children, 12 grandchildren, and — at the time — seven great grandchildren would have something to remember me by before — heaven forbid — I was incapacitated by this dreadful disease. So in June of 2021, I began spending a couple hours a day writing an outline with notes of what I wanted to include in what would become this autobiography.

We scheduled a family reunion in Orlando, FL for the last week of November 2021 to celebrate my 80th birthday on the 28th. As had become my custom since moving to the Philippines, I scheduled a week's visit with each of my four children. My partner of five years, Myla "Mae" Figueroa, and I laid out a two-month schedule. Our first stop was at the home of my younger daughter (second child), Barbara Michelle "Shelley" Vonderharr in Colorado Springs, CO. She and her husband, Pete, welcomed the two of us for our second visit there together.

One evening I told Shelley about my fears of possibly being in the early stages of Alzheimer's. She has a PhD in counseling and owns a company that counsels individuals on a wide range of subjects, including alcoholism, drug addiction, and anger management. She told me she had also created a counseling program for early-stage Alzheimer's patients. We spent a couple of hours reviewing actions I could take to keep my brain active and at least delay the onset of more serious symptoms if I, in fact, were developing the disease. She gave me coloring books and puzzle books, and recommended

I go online to download additional memory aid activities to my iPad. I did as she instructed and have continued to do these activities daily.

Upon returning to the Philippines in early December 2021, I began expanding my notes for what was becoming an interesting writing project. Also in December, I lost two of my four brothers. My next younger brother, Stanley, died of Covid on December 14 and my next to youngest brother, Robert, died in the terrible Boulder, CO fire of December 29 that destroyed more than 1,000 homes.

In February of 2022, I received a message from one of my War College Classmates, Major General (Retired) Lawson W. Magruder, that he had just published an autobiography that he had written at the urging of his children entitled *A Soldiers Journey Living His "Why."* I purchased the book and read it cover to cover. In reading about Lawson's life in the US Army, I noted many similarities in his life to that which was in my notes.

I began spending a few hours each day writing down my memories and often something would come to mind, and I would fast forward and write about later happenings. By June of 2022 I had written about 100 pages, and I was only up to the point where I was promoted to captain, perhaps a third of the way through my army career. On the last Sunday of June, I received word that one of my good friends in the Army as well as a War College Classmate, Colonel (Retired) Johnny Lawrence, had passed away. We taught together as Artillerymen in the Infantry School at Fort Benning, GA from 1972 to 1975, and he retired in the area. One of the Infantry Instructors, Colonel (Retired) Larry Redmond, sent me a note saying that he had just published his autobiography, *A Dusty Boot Soldier Remembers*, written at the urging of his three children. Once again, I purchased the book and read it cover to cover. We communicated a few times as I sent him items that I had written that were almost identical to his experiences.

Starting in July 2022 I redoubled my writing effort and tried to spend a minimum of two hours every day putting words down on paper. I continued to color, do puzzles, read, and try to keep my brain active. Fortunately, I have not noted any further deterioration in my memory.

I have depended on my only sister, Eileen, to help with story ideas and photos. She and her husband, Ron Fournier, live in Wendell, NC, a bit east of where my son Sean lives in north Raleigh. Ron has been kind enough to spend hours reading and editing this manuscript. We have been working on getting our mother and father's handwritten diaries typed and put into electronic format with pictures from the family albums. Most of the pictures from my early life have been provided by Eileen.

In late September 2022, Mae and I took a trip to America. We would attend the US Army War College Class of 1986 reunion in Memphis in mid-October that I had arranged and then visit family members. My oldest daughter, Sherrin Louise Patterson, had been battling stage 4 rectal/colon cancer for nearly two years, and we spent ten days with Sherrin and her family in Gaithersburg, MD. While her 57th birthday was actually on October 12 while we were still at our Reunion, we held a surprise birthday party for her on Saturday, October 22. We returned to the Philippines on October 31, 2022. I became immersed in assisting my sister with her work in settling our brother Robert's estate and I did not get back to working on these stories at all for the rest of 2022.

In mid-February of 2023, Sherrin took a very rapid turn for the worse and was admitted to the hospital on February 10. At first it was thought she would go home, but she continued to get worse, and Mae I took a flight to Washington Dulles Airport arriving on the afternoon of February 26 and went straight to Holy Cross Hospital in Germantown, MD. She had received sedation and was sleeping, so we were not able to visit. Fortunately, the next morning she was awake and greeted us with a "Hi Dad" and "Hi Mae," and we were able to carry on a short conversation.

That afternoon Sherrin slipped into a coma and passed away in the early morning of March 1. After her funeral on March 9 and a week- long visit with my son Sean and family in Raleigh, NC and another week with my sister Eileen and husband Ron in Wendell, NC we took a ten-day trip through Savannah, GA; Charleston, SC; and Wilmington, NC.

We returned to the Philippines on April 11, 2023. I received word from my daughter Shelley that the mother of my four children passed away on April 21. My son Terry, who lives in Houston, handled all the arrangements and she was laid to rest in Nacogdoches, TX on Friday, April 25 with her children and most of her extended family in attendance.

Upon arriving back in the Philippines, I returned to writing and sent this first volume to the publisher the first week of December 2023. I hope in reading this you will experience some laughter and perhaps pick up on a life lesson or two I learned along the way.

GLOSSARY

ADC – Assistant Division Commander
AIT – Advanced Individual Training/2nd 8 wks
AK – Bishop Arthur Kent White/ Son of Alma
AO – Area of Operations-space given to a CMD
ARVN – Army, Republic of Vietnam
AFB – Air Force Base
ARTY – Artillery
A&R – Athletics & Recreation
AWOL – Absent Without Leave
BA – Bachelor's Degree
BDE – Brigade -=Above BN/below DIV
BOQ – Bachelors Officers Quarters
BN – Battalion — Above CO/below BDE
BTRY – Battery- Above PLT/below BN
CID – Criminal Investigation Division
CG – Commanding General
CO – Company - Above PLT/below BN
CORPS – Above DIV/below Army
CPL – Corporal — E4, first non-com rank
COF S – Chief of Staff — at DIV and CORP
DC 3 – Douglas Commercial plane/1935 1st Flt
DIVARTY – Division Arty-Above BN/below DIV
DRO – Dining Room Orderly
ENLISTED RANKS
    E1 or PVT - Private
    E2 or PVT - Private
    E3 or PFC- Private First Class
    E4 or SP4 - Specialist 4th Class/Corporal
    E5 or SP5/SGT-Specialist 5th Class /Sergeant
    E6 or SSG — Staff Sergeant

    E7 or SFC — Sergeant First Class
    E8 or MSG or Master Sergeant/1st Sgt
    E9 or SGM or Sergeant Major/CMD SGM
FDC – Fire Direction Center
GPS – Global Positioning System
GRAF – Grafenwoehr Training Area, Germany
HHB – Headquarters Battery of an arty unit
KASERNE – Military Installation in Germany
JARK – An OCS run up a mountain
JFK – John F. Kennedy
KBA – Killed by Air
KPOF – Radio Call sign/POF, CO station
KP – Kitchen Police
KW – Kilowatts, power of a generator
LORAN – Long Range Navigation–629 Ft Tower
MOS – Military Occupation Specialty
MP – Military Police
MPH – Miles Per Hour
NATO – North Atlantic Treaty Organization
NCOIC – Noncommissioned Officer in Charge
NVA – North Vietnamese Army
OCS – Officer Candidate School
OMG – Slang term, Oh My God
OU – OK University
OFFICER RANKS
    2LT — 2nd Lieutenant — Gold Bar
    1LT — 1st Lieutenant — Silver Bar
    CPT — Captain — Two Silver Bars
    MAJ — Major — Gold Leaf
    LTC — Lieutenant Colonel — Silver Leaf

COL — Colonel - Eagle
BG — Brigadier General — One Star
MG — Major General — Two Stars
LTG - Lieutenant General — Three Stars
GEN — General — Four Stars
PHD – Degree — Doctor of Philosophy
PLT – Platoon
POF – Pillar of Fire — Religious Organization
PM – Provost Martial — Head Cop in area
PT – Physical Training — performed daily
PX – Army shopping store
RAF – Royal Air Force — British
ROTC – Reserve Officer Training Corps
RPM – Revolutions per Minute
SASCOM – Special Ammo Support Command
SIG – Signal
SOP – Standard Operating Procedure
STAFF POSITIONS IN A HEADQUARTERS
S1 or G1 — Administration/Personnel Mgt
S2 or G2 — Intelligence collection/ dissemination
S3 or G3 — Operations/Planning
S4 or G4 — Logistics — resupply
TOC – Tactical Operations Center
TWA – Trans World Airlines (bought by American)
UCMJ – Uniform Code of Military Justice
USAWC _ US ARMY WAR COLLEGE
VA – Veterans Administration, Gov unit
VW - VOLKSWAGEN

WAWZ – Radio Call Sign, POF ZA
WD1 – A simple type communications wire
WW II – World War Two
XO – Executive Office — second in command
YORSA – Youth of Red Springs Area
ZA - ZAREPHATH – National HQ of POF Church

1. CHILDHOOD

NOVEMBER 1941 TO SEPTEMBER 1955

I WAS BORN AT SUNSET FARM A FEW MILES FROM ZAREPHATH, NJ, THE NATIONAL HEADQUAR-
ters of the Pillar of Fire Church on November 28, 1941, at 11:00 pm. Mom and Dad were both missionaries in the church. Mother had a difficult pregnancy and had spent much time in bed both before and after I was born. In early June of 1942, six months after my birth, my father graduated from college with four others at Zarephath. Shortly thereafter, Mom and Dad were assigned to Providence, Rhode Island, departing in late June.

My parents spent the last two weeks of August 1942 at "Camp Meeting," the annual church assembly, at Zarephath. They returned to Providence on September 1, and on September 15, Dad took a short trip by train to Wentworth, NC to visit his mother and family. On October 13 Dad went on a missionary trip to Lowell, Massachusetts, about an hour away from Providence.

He discovered that someone from the Pillar of Fire had worked there even though he had been told it was his territory. He fired off a letter of complaint to the church headquarters. On the 15th he received an unsympathetic reply telling him he just needed to find other places to work. Whether or not other things happened between his return from North Carolina and October 20, the result was that Mom and Dad left the Pillar of Fire Church and moved to NC. There were a number of letters exchanged with the bishop threatening to revoke Dad's draft exemption status, demanding the return of Dad's credentials as a minister, and expressing outrage. The bishop even branded Dad a "traitor," saying that Mom and Dad had availed themselves of a free education and this was the thanks the church was receiving. In early November, Dad took it upon himself to notify his draft board in San Francisco that he was no longer eligible for draft deferment.

On my first birthday in Wentworth, NC, Mom and Dad had professional photos taken of the three of us. On December 18, 1942, Dad had an attack of appendicitis and had to have his appendix removed. He spent one week in the hospital and nearly two weeks at home in bed. Perhaps because of his health crisis, a 40-dollar hospital bill, and the distinct possibility of being drafted, Dad sent a letter of apology to the bishop and asked to be allowed to return to the church.

On January 13, 1943, Mom and Dad, with the bishop's permission, drove to Zarephath, arriving on January 14. They were assigned a room on the third floor in what was called the Frame Building. Dad worked in the book bindery as he was still too weak for hard labor. Mother was assigned to teach school. The issue of Dad's draft status came to a head on January 21 when they received a request from the draft board for a copy of my birth certificate. Dad was also requested to send a notarized statement certifying his return to full-time ministry, which he did. On February 11, Reverend Barkman, at the San Francisco church, notified Dad that he had been placed back on "Deferred draft status."

On February 23, 1943, Mom and Dad were assigned to the Pillar of Fire Church at 13624 Stout Avenue in Detroit, MI, and arrived two days later. This location had a large Sunday School of over 100 children plus a small congregation of locals who attended worship services. Also serving there were two other missionaries. One was Miss Phyllis Hoffman, age 21, who had graduated with Dad. She and Mother became lifelong friends. The other was Miss Harriett McCormick, a 35-year-old Irish bundle of energy who also became a life-long friend and cared for our mother in her final years of battling cancer.

Just four months later, in June 1943, my parents were reassigned to Pillar of Fire Church and School at College View Place, College Hill, Cincinnati, OH, to take the place of Fred and Hazel Schissler who were reassigned to Florida. Their seven-year-old daughter, Elaine Schissler, would grow up, marry, have children, lose her husband to cancer, and later marry our father after my mother's death, even though he was 20 years her senior.

Dad was appointed the school headmaster and he not only taught 13 different subjects in elementary and high school but also did missionary work and maintenance on the church grounds. Mother often taught school for Dad when he had work to do around the property. One of Dad's projects was the construction of two stone gates by the entrance to the property — gates that survive to this day. Dad saw fit to purchase a pony and there are a number of pictures of me riding it. Also, if the pictures from the time are any indication, I was a favorite of the female students often being included in pictures with groups of them.

There are no pictures we can find of my second birthday, but Mother describes it as being fun to watch me blow out my two candles while many of the school children gathered around to watch. My Uncle Woodrow, Dad's brother, was visiting from Colorado at the time. Woody, as we all came to know him, had married a Belleview classmate, Fern Brown, on Dad's birthday in 1942.

Dad wrote in his diary that he marveled that I "took to him almost at once and we had a great time." He went on to note that Woodrow seemed to enjoy children so much and wrote, "Gerald sure likes to play with him." Mom reported in her diary that Woodrow wanted to take me home with him. I must note that this was in direct contrast to Dad who seldom if ever "played with us" as he was always too busy with other things. As my sister Eileen notes in the introduction to one of Dad's diaries, "He was simply not playful by nature."

On February 23, 1944, I came down with a case of chicken pox. Dad reported in his diary that I did not have a very serious case, and he seemed to be more concerned with my cold than I was about my chicken pox. Mother wrote, "He is as full of pep as ever and does not seem sick."

Apparently, I was a joy to my mother as she often wrote in her diary about my increasing vocabulary, various new facial expressions, and the many new things I learned on my own, such as how to climb in and out of my carriage. She mentions taking me swimming for the first time and notes that not only was I "not a bit afraid," but that I did not want to get out of the water. Mother talks about my "picky eating habits" (that exist to this day) and she mentions taking me to a Chinese restaurant where I would eat nothing but crackers and a bit of custard. She later mentioned that I liked string beans and noted that she canned ten pints specifically for me.

On May 1, Mother wrote, "Gerald has been having a wonderful time outdoors lately. Today he was feeding grass to the chickens, and one pecked him. He has been talking about it ever since." In my baby book, Mother entered that we took a trip to New York City with Granddad and Grandma Ritchie in May of 1944. Mom and Dad took me on my first visit to a zoo in Cincinnati a month later in June.

On November 20, Dad tells of writing a poem to celebrate my 3rd birthday. He sent it to Zarephath to be published in one of the church papers. He wrote:

TODAY I'M THREE

We have within our home so fair
A little lad with auburn hair;
His eyes are dark and shiny bright–
They glow like diamonds in the light.

Not very small or large is he,
This is the day he says, "I'm three."
Oh, yes, quite like a man he'll stand–
But still goes out to play in sand.

'Tis then we say, "He's in his glory"
Just like the times we read a story
Of Peter Rabbit or Goldilocks,
Of English queens and dainty frocks.

At times he likes to make a zoo,
With camels, bears, and kangaroo;
These all must have a nice cool drink;
'Tis quite a task at three, I think.

Then, when comes time to go inside,

You'd think his life had been denied.
"Please let me stay outside and play!
Now will you, Mother?" I hear him say.

And so it is, when time for bed;
With heavy sighs, he hangs his head.
Oh me! What must this poor world be
For little lads who've just turned three?

—*Wm. S. Sharpe*—

In June 1945, after two years in Cincinnati, Mom and Dad were reassigned. We arrived in Chicago, IL on June 27, 1945. We first lived in the church home at 622 Oakdale Avenue, then in September of 1945 moved to a home at 1115 Barry Avenue. My brother Stanley Marshall was born on April 22, 1946, and Charles Nathan was born on October 19, 1948, both in Chicago.

I do not have many memories of my childhood in Chicago, but one incident is well documented in my mother's diary. On Wednesday December 12, 1945, about two weeks after my 4th birthday, the school children were in the chapel singing, and as I was too young to be in school, I was simply left to my own devices. Dad was on a missionary trip, and Grandmother Bessie Sharpe was visiting. The chapel had a raised platform or stage in front and there was a short altar rail with spindles that ran from one side of the platform to the middle, then there was a free-standing pulpit, and the railing continued on the other side to the end of the platform. In front of the platform was a series of prayer benches that ran the full length of the stage where people could kneel and pray. There was perhaps a foot and a half space between the prayer benches and the platform. I obviously thought it was a good idea to try and walk on the altar rail from one end to the other doing a balancing act on the top of it.

When I reached the pulpit, I apparently reached around to the other side of it and tried to step around it. Since the pulpit was free-standing, as I started to move to the other side, my weight pulled the pulpit over and I fell off the altar rail at the same time as the pulpit was falling. I tried to catch myself on the prayer bench as I fell between it and the platform, and my left hand caught the edge of the prayer bench but had almost completely slipped off before the pulpit fell on it. The end of my second finger on my left hand was all that remained on the bench and the pulpit smashed it flat as a pancake. No one had a car, so Mom and Grandma put me in my red wagon and hauled me the couple blocks to the doctor's office. To this day, the nail on that finger still grows below the end of the finger.

On December 23, 1945, Dad wrote, "Christmas is in the air. Gerald can hardly wait for Santa to open his presents. He keeps good track of the days that are left. He is a little too excited for his own good and is inclined to be a bit nervous. He needs more outdoor exercise and fresh air than is afforded in a large city. In fact, as 1946 closes, I hope we will be out of Chicago before much longer." We actually stayed another three and a half years.

In Dad's 1946 diary he reported we are still living on Barry Avenue and on January 8th he took Mother and me to the Field Museum. He wrote, "Gerald sure had a good time looking at all the animals." Mother wrote, "Gerald was delighted with all the animals and ran from one exhibit to another so excitedly, but we did not stay for long."

In my baby book mother wrote that in January 1946, the family was at the dinner table discussing the ongoing steel strike. They noted that if the strike continued some industries would have to shut down. Mother continued, "Gerald spoke up with, well you shouldn't steal anyway." A few days later she wrote that she reminded me that I had left my puzzles at the church, and I replied, "Isn't that just the limit."

On February 28, Dad wrote about driving to Zarephath for the funeral of the founder of the Pillar of Fire Church, Bishop Alma White. Her son, Arthur K. White became the bishop.

In Mother's 1946 diary, she wrote about my having my tonsils removed on July 15, which was very common for children in the 1940s. She said I had to stay overnight in the hospital. On July 20 she wrote, "Gerald says almost everything hurts his throat. Sunday calls him 'ice cream, popsicle Joe' because that is all he wants." Perhaps this was the beginning of my lifelong love of ice cream.

My parents went out for ice cream often as they reported in their diaries both before and after I was born. Mother wrote again on August 2 about my being sick. She said I had a fever and slept on the couch and did NOT want to get up. Mother called Dad, who was on a missionary trip, and they decided they should call their doctor. The doctor prescribed some medicine and said if I was not feeling better the next day to bring me in between 2:00 and 5:00 pm.

The next day, Saturday, August 3, Mother took me to the doctor, who diagnosed me with a mild case of Rheumatic Fever. He said I would need to stay in bed for four or five days and continue taking the medicine he prescribed.

My fifth birthday in 1946 was also Thanksgiving Day. Dad reported that all the Chicago workers came to the house for dinner and had a nice cake. He reported he had a new camera and "took lots of pictures."

In Dad's 1947 diary he reported we were still living at Barry Avenue. On January 2, he wrote that there was a large snowstorm and "Gerald had a great time playing in the snow." On February 22, we had professional pictures taken when Stanley was ten months old. On March 27 Dad reported taking me "out to the beach to fly his kite." This would be the first of several entries on kite flying over the next few years. On August 25, Dad stated that we are at the annual "Camp Meeting" at Zarephath where Mother and Dad anxiously awaited their assignment orders. Sure enough, they were directed to move, just a few blocks away to the church home on Oakdale Avenue.

The home on Oakdale Avenue was a two-story house with an unfinished attic. The attic had a floor, but I am told not much else. Mom told me about this house in describing this story. She said that I was forbidden from going into the attic, but the door had no lock.

One day I decided to go up to the attic, and as I walked up the stairs near the top, I saw a squirrel's tail sticking out into the stairwell. For whatever crazy reason I decided to grab hold of the squirrel's tail. As my mother told it, the bloody screams she heard from the attic brought her running. When she reached the attic, she found my arm scratched and bloody as apparently the squirrel had twisted around and scratched his way up my arm in an effort to escape. Mom says they took me to the doctor, where I received a tetanus shot. I was never again tempted to mess with squirrels.

In Dad's 1949 diary he writes that on January 22, he took Stanley and me sledding in the park. Dad always wanted us to "be quiet," but I believe we seldom were. On February 19 he wrote, "Gerald was home from school today and he and Stanley always have lots of fun on Saturdays, though they are pretty noisy." On March 31 Dad wrote a letter to the bishop suggesting that we were due a change from Chicago, but we stayed more than another year. On August 16 we visited Niagara Falls on the Canadian side on the way to the annual "Camp Meeting" at Zarephath where assignments of ministers and missionaries were made each year. We were directed to stay in Chicago!

In June of 1950, Mom and Dad were reassigned to the Pillar of Fire Church at 24 Beulah Street, San Francisco, CA. Two months later we were transferred to Oakland, CA where we lived in the missionary home at 2009 38th Avenue. Stanley and I went to the Pillar of Fire's Bethel School at 1957 Harrington Avenue in Oakland just one block down the hill from the missionary home, and later Charles did as well.

Even before we came to California, I had developed a love of reading books. Mother told me I could read aloud simple children's books before my 4th birthday. In Oakland we were within walking distance of the library, so Mother and I would check out six or eight books every week. Obviously at first, they were very simple books with lots of pictures, but as I got older, I began enjoying books like the Bobbsey Twins, the Hardy Boys, and Tom Swift, all series books that had many titles. Later in life I began collecting these books.

Dad did extensive work developing the steep hillside behind the missionary home in Oakland by building two large retaining walls across the property and down the sides, putting cement steps from the top to the bottom, and installing a playground on the lower level. We had three neighbors at the bottom of the hill, and at age ten I became well acquainted with two of them.

The first family raised chickens, and I would go over to their house and help feed and water them. They ate the chickens on occasion, sold a few when they needed to, and collected eggs on a daily basis. I would sometimes be given a few eggs to take home.

Apparently, it was common practice to use an ax to cut off the heads of chickens, but our neighbor had a much simpler solution. He would catch a chicken, grab it by its head, and then spin it around four or five times until the body separated from the head and went flying across the yard. Amazingly enough the chicken would jump up and run around for a few minutes until it collapsed on the ground. This phenomenon must be where the expression "running around like a chicken with its head cut off" comes from.

Our neighbor had two nails without heads in the wall of his garage and he would go pick up the chicken and push one leg over each nail with the neck hanging down to drain out the blood while he plucked out the feathers which he saved in a sack, I understood his wife used them to make pillows. He would put the innards in a metal bowl and put the chicken in a refrigerator in his garage. As I look back on this experience, it seems like cruel and unusual punishment, but for me it was simply the only way I ever saw a chicken killed.

The second family was an elderly couple. I am not sure of their age, but the husband had spent a lifetime collecting stamps from around the world. He awakened in me a lifelong desire to collect stamps. My first album was an international stamp album, and our neighbor would often give me large envelopes of used stamps from a multitude of countries. I would spend hours first trying to find the country and then where in the album the stamp belonged.

In the early days of my stamp collecting, we used what were called "hinges" on the back of the stamps. You would lick them and stick them in the album. By the time I joined the army, I had quit collecting international stamps and had only an American Stamp album. Collectors also progressed to using sleeves where the stamp was slipped inside and then the sleeve was pasted in the album. This allowed for collecting unused stamps called "mint" stamps. Using a hinge on a mint stamp ruined its value as the glue on the back of the stamp damaged the stamp and caused it to be valued the same as a used stamp. Today I have a wonderful collection of mint US stamps from the early 1900s up until 1987 when they started using self-adhesive stamps.

Mother's 1951 diary records that we were still living on 38th Avenue in Oakland. One of the souvenirs I still have is a 1940s booklet on Muir Woods. It was a favorite place to visit and picnic, and Mother wrote that on March 22 there were four carloads of folks from both San Francisco and Oakland who went for a picnic to Muir Woods and that Dad "took lots of pictures."

On May 8, I came down with my first of two cases of measles. Mother wrote, "Gerald really looks a sight; he is all broken out thick with measles head to toe." On May 18 Stanley came down with an equally bad case.

Eileen Gail was born on July 23, 1951, in Oakland. She was born at 11:03 pm (Just three minutes later than the time I was born). Mother wrote on August 23 that "after a month, Eileen still does NOT weigh as much as Charlie did at birth."

Mother's 1952 diary notes that I was fitted for braces on my teeth in April. This started a year-long saga of going to the orthodontist every few months. Mother faithfully saved the receipts — seven visits for $25 each. I remember this being a painful process as each time I went they tightened the braces to bring my teeth into alignment. For a week or two after, it simply hurt.

In June my mother wrote that she received a letter from our uncle, Ray Ritchie, who was enjoying shore leave in Rio de Janeiro, Brazil while serving on the famous battleship Missouri. The Missouri served in every US war since being commissioned in 1944, and is now a museum in Honolulu, Hawaii.

My first school presentation, saved by my mother, was about a trip we took to Lassen Volcanic National Park over the 4th of July weekend, 1952. I wrote and presented the following.

A TRIP TO LASSEN VOLCANIC NATIONAL PARK AND SHASTA DAM

After we ate breakfast, we packed the car and left our house at 8:45 am, Wednesday, July 2, 1952. We went toward the Oakland Bay Bridge but turned off into Berkley. It is about 30 miles to Crocket where we crossed the Crocket Bridge. On the edge of the water was a big C and H sugar refinery.

For about an hour we traveled through little towns and many fields of rice and hay. At 12:30 pm we stopped to eat lunch and get a little exercise, before traveling on. As we entered Lassen Volcanic National Park, we saw miles and miles of fence made out of volcanic rock. There were rocks all over the land from the eruption of 1915. We rode around trying to find a place to stay for the night. We finally found a cabin called Fire Mountain Lodge. This was on Wednesday night.

Thursday morning, we went up into the mountains. As we drove along, we saw a deer down over a 200-foot precipice. About this time, we were stopped by a ranger who had stopped about ten other cars ahead of us. The road had only been open for traffic for about four days and only one lane of cars could go through at a time. We had to wait nearly an hour. Finally, we got to move on. The ice in some places was 30 feet high on either side of the car. There were animal tracks going up the side of the mountain. The snow was high, but the sun was warm.

We ate our lunch on the other side of the mountain. We could see Cinder Cone, Lassen Peak, and Crater Peak. After we ate, we went back along the road and took some moving pictures. As we went on, we were stopped again and had a pleasant time eating snow and throwing snowballs. When we got back to the lodge, we were tired and glad to eat some supper and go to bed.

We got up at 6:00 o'clock on Friday morning, ate a bite of breakfast, and packed the car. We left Fire Mountain Lodge about 9:00 am. I liked the cool forest, but we soon came to hot rocky country. There were square hills with valleys on every side, also over-hanging cliffs.

After traveling over a hundred miles, we came to the little town of Lake Head where we were glad to get an air-conditioned cabin. In this neighborhood we found a wild-west park. There was a gold crusher, the first one ever used in that part of the country. There was a dummy barber who talked, also a jail that had a statue of a man in it that had "claim jumped." They had a fairy gold mine where a man had dreamed that he would get rich in a week. Water was running from a mystery faucet where there was no pipe or hose. We were entertained by singing cowboys.

The next day, Saturday, we got up about 6:00 o'clock and left for Shasta dam. It is 602 feet tall and is the highest overflow dam in the world. The base's thickness is 880 feet, and the crest is 37 feet wide. It has two elevators from the top down to the power plant. We visited a small harbor and watched some speed boats. We arrived home at two in the afternoon, tired but feeling we had an interesting trip.

Mother's diary records that on Sunday, July 6, 1952, Dad and I got poison ivy on the trip, and I got a case of intestinal flu. When we went to the doctor's office to get medicine to stop the itching, the doctor told us there was a series of pills we could take that would prevent us from getting poison oak or poison ivy ever again. Dad purchased the pills, and we took them over about a month-long period. The medicine worked and I never again got either kind.

Wednesday, July 9 was Mom and Dad's 12th anniversary, and July 23 was Eileen's first birthday. We took a family trip to Denver, CO in early September even though mother was nearly eight months pregnant. We visited grandma Ritchie, and our uncle Woodrow (Dad's brother) and Aunt Fern.

My brother, Robert, was born in Oakland a month later, on October 12, 1952, just seven days before Charlie's fourth birthday.

Our elementary school teacher in Oakland was deathly afraid of snakes. I recall finding a two-page spread in an old National Geographic magazine with photos of perhaps 50 different kinds of snakes. I tore out the pages and found an opportunity to place the pictures in the top drawer of the teacher's desk when no one was in the room. When the teacher opened the drawer, she jumped up out of her chair screaming and left the room. Someone came in and removed the pictures and we were all interrogated as to who had put the pictures there. I did not confess, and as I recall, I never told anyone I was the culprit.

One other memory I have is of a much larger boy who loved to bully all the smaller children, including me. I do not recall exactly how it happened, but I fell down right next to a chain link fence and the boy fell as well, landing face up on top of me. I wrapped my legs around his arm on the far side and grabbed his other arm holding it tight. He struggled to get loose, but I was able to hold on. I discovered that if I twisted my body under him just a bit his head would hit the metal post on the fence. After the second or third time he started screaming bloody murder and all the students gathered around. One of the teachers also came over, and I recall her telling the boy he deserved what he was getting as she knew he bullied the younger children. She made him promise to stop his bullying after I was able to hit his head on the post a time or two more. I was a hero to the younger kids. I do not remember having any more trouble from him after that.

1953 found the family still living in Oakland. Dad wrote in his diary that the "first day of 1953 is NOT very quiet with five children racing around the house—all but Robert who is only three months old." Mother wrote in her diary that they enjoyed watching the inauguration of President Eisenhower on January 20 on a borrowed TV.

On January 31, she notes that Dad took Stanley and me up into the hills to fly our kites. Mother reported in her diary that on February 20 Dad got her a new Wedgewood Stove with grill, broiler, and glass oven door. It was one of the most modern stoves of its time. She baked a cake while I mixed and baked some cookies all by myself. She notes that I always liked to help bake cookies, but at 11 years old, I was now old enough to do it by myself. We were all expected to work growing up.

On June 7, Mom wrote that I helped her paint a small closet, and the next morning I helped her paint the ceiling in the living room. That afternoon I helped Dad paint the woodwork in the living room. I not sure if this was my first time painting, but over the years, I have painted hundreds of times, inside and outside of houses. I suspect this started a lifelong habit of doing self-help projects around our homes.

Mother then wrote about a memorable trip to the orthodontist. I was fitted with a neck brace that was designed to move my lower jaw forward to give me a better bite. This to me was like a torture device, and any time no one was looking, I would either unhook the rubber bands or simply take it off. By mid-April my parents gave up trying to force me to wear this device and had my braces removed. As you will later read, I got to have a second round of torture a couple of years later. I found a picture of Ellen DeGeneres as a child wearing a brace almost exactly like mine and have included it in the photo section.

In March of 1953, Dad sold two old trailers. He then purchased a new frame and wheels. He spent the next two weeks getting it built up, assembled, and painted. It was 16 feet long and five feet wide. It had an angle iron frame, a plywood floor, and four plywood sides four feet tall. The plywood back had handles inside and out and could be lowered into place with a tarp to close the top three feet. There was a curved three-foot top and front covered with aluminum. There was a plywood bed constructed at the four-foot level that was eight foot long. It had a mattress and pads on the side and a short removable railing across the back. There was a ladder to climb up to the sleeping area. We used this trailer for many years until it was destroyed in the same Boulder, CO fire that killed Robert on December 29, 2021.

There was a family in Oakland by the name of the Glaziers. They became our good friends, and we visited their house and they visited ours often. Their two sons were the same age as Stanley and me. Apparently, I could be a colossal pain in the butt sometimes. Mother wrote on March 31, "Had a time with Gerald. Hope he gets under conviction soon."

I got my second case of measles on May 16, but it was a much milder case.

A number of my classmates in the seventh grade belonged to a local Boy's Scout Troop. They would tell me about camping trips, visits to various places, and about their weekly meetings where they worked on earning badges by learning certain skills like knot tying, first aid, chess, stamp collecting, or leather work. I asked my father a number of times to take me to a meeting and allow me to join, but he always had one excuse or another not to let me to do so. I do not remember exactly what happened, but as I recall my mother prevailed on Dad to allow me to join. On the evening of a scheduled meeting, we drove to the church where the scout troop met in a basement activity room. My father led the way down the steps and just as we reached the door of the activity room, Dad stopped, turned around, took my hand and led me back up the steps and out of the church. I was protesting and asking what was going on. All my father said was, "Get in the car, you will not belong to any organization where adults are smoking." The Pillar of Fire Church was adamantly opposed to not only smoking, but the evils of drinking, movie going, eating meat, and wearing of brightly colored clothing which will add flavor to my later life stories.

I purchased my first tennis racket on June 3, 1954, and would play off and on for most of my life. On Monday, July 26, the family took a trip to Yosemite National Park in western California. The occasion was a visit by Norman and Margaret Sillett, who had been good friends of Mom and Dad for years. Grandmother Ritchie was also visiting at the time.

In two cars, we made the six-plus-hour trip virtually straight west. Norman Sillett left early in the morning with Margaret and Grandmother. Dad had to attend an 8:00 am board meeting in San Francisco with the Bishop and Reverend Wolfgang, so we did not leave until 11:00 am, arriving in Yosemite about 6:15 pm.

The family camped in the trailer that Dad had outfitted with Coleman lanterns, a Coleman two burner stove, ice chests, and chairs and tables. We found a campground to stay in and figured out how the seven of us were going to sleep. As I recall, Mom and Dad plus Robert (age 1 ¾), Eileen (age 3), and Charlie (age 5 ½) were on the top. Stan (age 8) and I (age 12 ½) had sleeping bags on folding cots on the bottom.

On Tuesday morning we went sightseeing around Yosemite, and in the afternoon, we went swimming. On Tuesday evening there was an outdoor concert near the Upper and Lower Yosemite Falls that included community singing. The evening concluded with a large bonfire at the top of the upper falls. After all the logs were burning brightly, they were pushed over the edge of the falls, creating a shower of burning logs and fire down the mountainside.

On Wednesday morning, we had breakfast cooked on the Coleman stove and began to pack up the trailer. Dad never wanted to travel with fuel in the tank of the Coleman stove, so he disconnected the tank, released the pressure, and removed the lid. He walked away from the table down a small road we had used to drive into the campsite and began sprinkling fuel on the ground as he walked.

There was a box of matches on the table with the stove and I took the box, lit a match, and threw it on the ground. The gasoline lit and with a swooshing sound raced toward Dad. He happened to turn and see the fire racing toward him and threw the tank off the road. Thank goodness the tank got a few feet away before it burst into flames. As I recall I had a hard time sitting for the rest of the day and on the trip home and for a few days later as Dad whipped my butt until it was bright red, asking me a number of times if I was trying to kill him!

The Silletts and grandmother left early in the morning for Los Angeles, about a nine-hour drive. We stayed and went swimming and did some more sightseeing and left for Oakland about 6:30 pm arriving home at 2:30 am Thursday morning.

I have a box of souvenirs from my childhood. One item is from Knotts Berry Farm, an early amusement park in the Los Angeles area. On August 10 and 11, 1945, we visited our aunt Irene (Mother's sister) and uncle Fred who lived in Chula Vista. We also spent a day in Tijuana, Mexico where Mother reported we all purchased Mexican sombreros and wore them for a picture I have found for the photos at the end of the chapter.

On one memorable day, the family had gone to San Francisco and stopped at a small farmers market to purchase some fresh vegetables, including some brussels sprouts. As a child, I really liked brussels sprouts. I would eat them cooked or raw. As we drove across the Oakland Bay Bridge, I took the bag of brussels sprouts to eat some as a snack. After popping about the third one in my mouth and starting to chew it, I looked down at my hand and saw little black bugs starting to crawl out from inside the sprout.

I quickly rolled down the window and spit out the sprout in my mouth and threw out the one in my hand. I looked inside the bag and saw many little black bugs. I handed the bag to my mother in the front seat and told her I had eaten a few of the sprouts. I do not remember what she said, but she rolled down her window and dumped the sprouts and bugs out the window. I did not eat another brussels sprout for ten plus years after that experience.

Reading Mom and Dad's diaries now sheds light on things I never knew. As an example, Mother reported they received a letter from Bishop White on August 22, 1954, asking them to move back to Chicago. Mother reported they told the bishop that they did NOT want to return to Chicago but would rather be assigned to Colorado. She reported looking for an answer back, but never said what it was.

They must have convinced the bishop, because in September 1954, Mom and Dad were reassigned to the Pillar of Fire Church's Western Headquarters at Belleview College in Westminster, CO. We first lived in a small apartment in the store building, and about a year later moved to our home at 8300 Irving Street where Mom and Dad lived for the rest of their lives. Another thing I had not realized until recently reading Mother's diaries was that Mother did not know how to drive a car until November of 1954. On November 1, she reported taking her first driving lesson and studying for the written test. On November 3, she received her learners permit after scoring 80 on the written test.

During my eighth-grade school year, I was taught by Miss Margarete Stump. Mother wrote in her 1955 diary that Stanley and I participated in a school program on Friday, March 18. She also wrote that she spent two days creating an Indian costume for me to wear. On the day of the program, she wrote, "Gerald's and Stanley's program was tonight. It was about the most interesting one I have ever attended. Both our boys did fine." Just three days later she wrote, "Gerald made cookies after school." And then on the 23rd wrote, "Gerald was sick today—ate too many cookies!"

I should note that one of my strongest memories of the 8th grade was that Miss Stump was a strict disciplinarian. If she was displeased about something we did, she would tell us to hold out our hand, palm up, and she would smack our hand several times with a wooden ruler she carried. Another time she came up behind me, grabbed my ear, and pulled me up out of my seat. She then sent me home (which was just upstairs in the same building). I went upstairs crying and Mother asked me what was wrong. I told her Miss Stump had pulled my ear and it hurt! I am not sure I explained what I had been doing that displeased Miss Stump badly enough to have her pull my ear in the first place.

Later, when Dad came home, Mother explained to him what had happened, and Dad stormed out of the apartment. School was out, but apparently Miss Stump was still in her classroom. We could hear Dad shouting and a bit later he came back upstairs and told me I had better stop acting up in school or he would "tan my hide." I later overheard him tell mother that he did not think Miss Stump would ever touch my ears again, as he said he told her if she did, she would no longer be teaching! She never touched my ears again. My sister Eileen told me that she had Miss Stump as a teacher some years later and she was still the same very strict disciplinarian.

Dad reported at the beginning of his 1955 diary that we were still living in North Hall, sometimes call the Store Building above the church store and the 8th grade classroom. On January 15, Dad notes that Ray Sharpe, his wife Leta, and daughter Carol Rae (my classmate all through 8th grade and high school) moved out of North Hall to a small house on campus. Dad wrote on February 8 that we were told to move to Ray Sharpe's old apartment at the front of North Hall, even though it was a smaller apartment. Over the next six months Dad tells of several projects he carried out in the apartment including installing new kitchen cabinets and replacing the linoleum on the floors.

On February 16, Dad recalls an article in the Saturday Evening Post (SEP) entitled, "The South Will Like Integration" written by a professor at the University of Virginia. He immediately fired off a letter commenting on the article to the SEP and notes in the same entry, "I of course think otherwise." This is one of the earliest manifestations of my father's deep prejudice against Black people. On March 1, he reported in his diary, "I received a letter from the SEP saying they are to publish the letter I wrote to them in their March 26th Issue." He notes again that the letter is about integration. Unfortunately, they only published a short bit. We have been unable to find a copy of the full letter.

Dad also wrote that on March 18 he took me to see an Orthodontist, Dr. Klein, who thinks he can help me, although he notes my "condition is rather unusual." I believe he was referring to my large overbite and my teeth no longer being straight. Mother reported in her 1955 diary that in June I had once again gotten braces put on my teeth, as I had failed to wear my retainer two years earlier and my teeth had lost some of their "straightness."

Zarephath, New Jersey, Headquarters of the Pillar of Fire Church. Below, the home where I was born on Sunset Farm. My great grandfather, Wallace Lewis, cutting wood behind the home

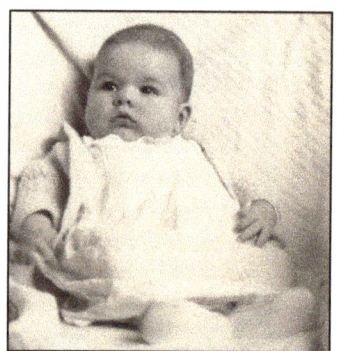 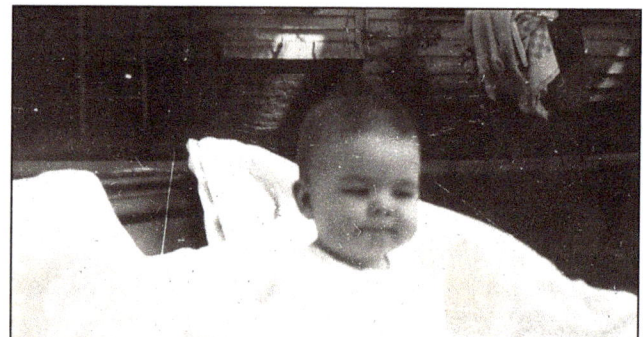 

Pictures taken of me at Zarephath in early 1942

Baby pictures include Mom & Dad holding me in front of the WAWZ Radio tower at ZA

Child Dedication led by Rev. Ray White in Assembly Hall at ZA in 1942: (L to R) Sunday, Thelma & Gerald; Nathaniel Wilsons; Henry Snellings; Donald, Mary Jane Cruver & Donald Richard; and Elmer Smiths

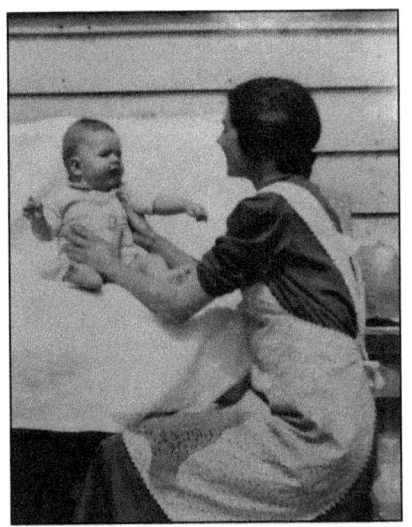

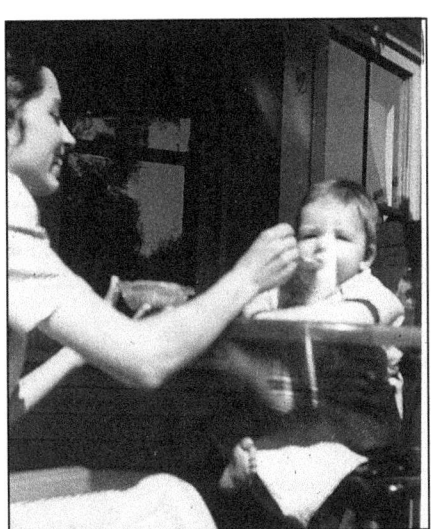

More photos from Zarephath in 1942

College graduation, June 1942: Left to Right: Donald Wolfram, Lois Stetson, Vernon Kutz, Phyllis Hoffman, and Dad

Dad holding me at Zarephath

Above pictures are all from Zarephath in 1942. Last picture is of me standing on Dad's trailer just before the move to Providence, RI.

Above photos from the Missionary Home in Providence, RI

Taken near Hazel's home in NC in December 1942 with her dog, Wimpy. My cousin Burton describes him as the smartest dog ever, who could hunt and fetch anything on command, including wood for the stove from the wood pile outside. I am wearing the same pants as below.

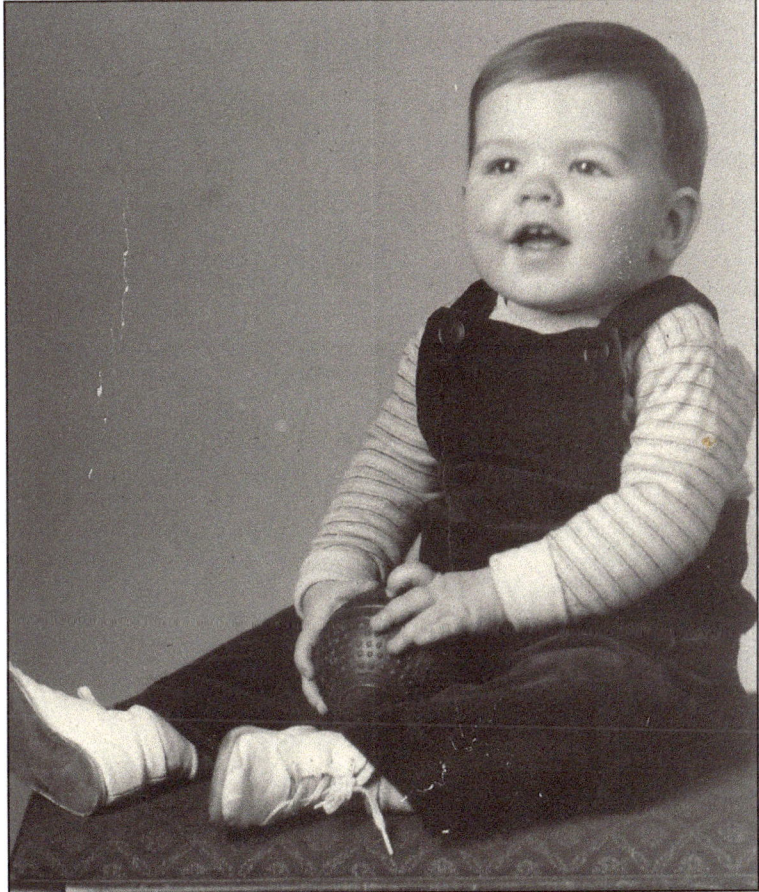

Taken on my 1st birthday in Reidsville, NC near where Mom and Dad were living

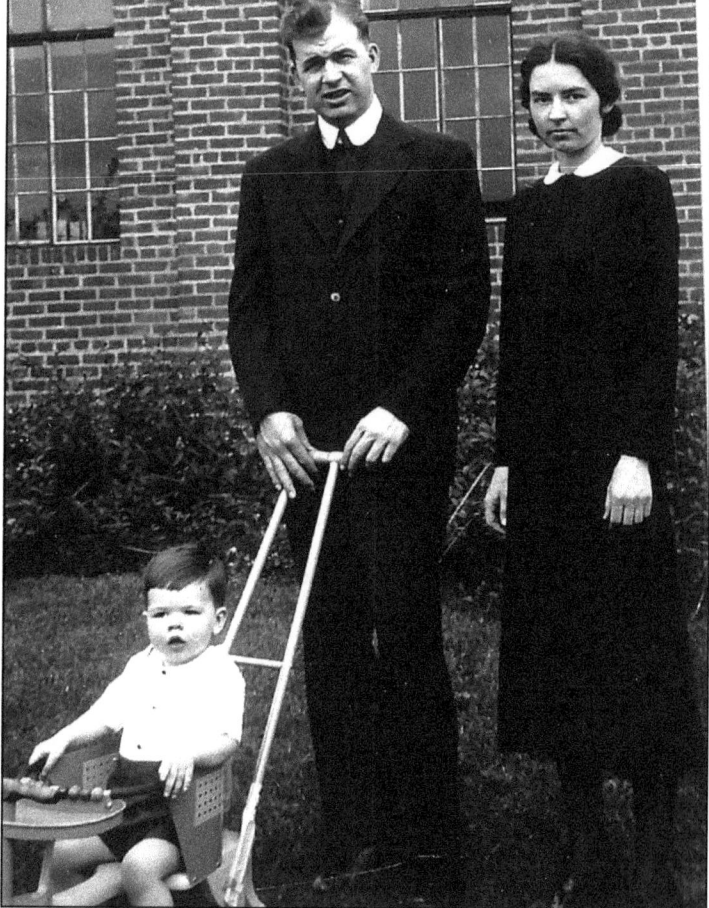

Pictures taken in Detroit where Mom & Dad were assigned after Dad recovered from his appendicitis. I am in my new "kiddy car" at the Detroit Church and left is on a trip to Windsor, Canada. Below, left, is Phyllis Hoffman, one of Mother's good friends, with me at about 18 months of age

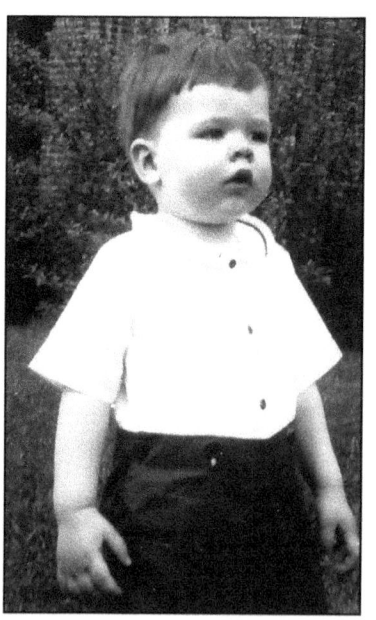

SUNDAY SCHOOL - 10:00AM
PREACHING - - - 7:30PM
CHILDRENS MEETING
WED. 7:00PM
BIBLE STUDY
THURS. 7:30PM
ALL WELCOME

Taken outside the Detroit Church. Appears to be a 1942 Dodge with 1942 Michigan license plate.

Start of photos from Cincinnati, OH. Same car behind them as above and on right Mom & Dad are standing in front of the building where we lived.

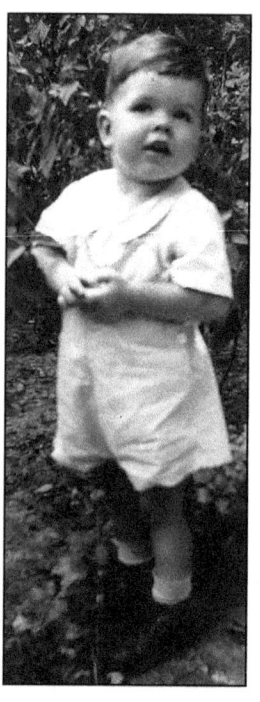 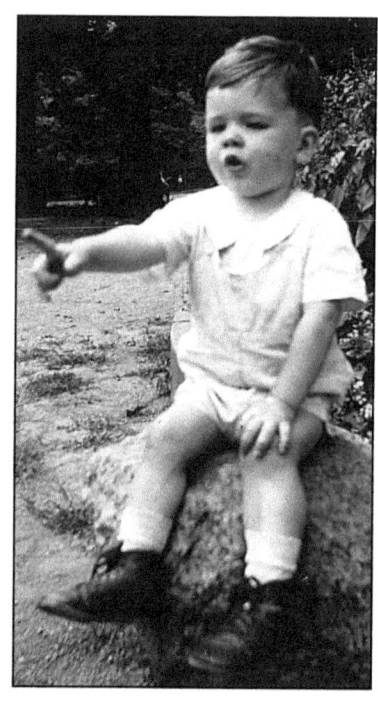  

My mother's sister, Irene with me.

*More photos
from Cincinnati,
Ohio—June 1943
to June 1945*

*Top right: one of
several photos in
which I am with a
group of girls*

Top right: Norman Sillett and his sister Margaret with me Center left: Hunter Ritchie and Grand-mother Ritchie with Mom and Dad Center right: Me in a wagon with my first "military" uniform Bottom left: Grand- mother Bessie Sharpe holding me Bottom right: Mother's good friend, Betty Summers with me holding my toy dog I received at birth.

May 1943

Pillars my father built

Uncle Fred Thornton

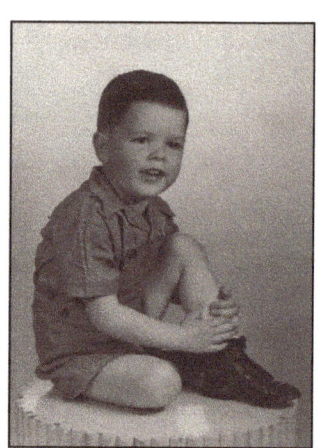

April 1945

Start of Chicago Photos, July 1945

Mom holding Stan, age 6 months: Dad holding Charlie, then Mom, with Stan and me at Lake Michigan

PILLAR OF FIRE
JUNIOR

Pillar of Fire, Publishers   Vol. 40. No. 10.   Zarephath, N. J.   May 15, 1946   Issued Semi-Monthly 50c a Year. 3c a Copy

Entered as second-class matter, November 1, 1922, at the post office at Zarephath, N. J. Acceptance for mailing at the special rate of postage provided for in the Act of February 28, 1925, authorized

EUGENE FIELD MONUMENT IN THE LINCOLN PARK ZOO, CHICAGO

It was erected in 1922 by the school children and citizens aided by the Benjamin F. Ferguson Fund. Picture by Rev. Arthur K. White. The little boy is Gerald Sharpe, son of the Pillar of Fire pastor in Chicago. *See page* 3.

"BUSHMAN," WEST AFRICAN GORILLA IN CHICAGO
LINCOLN PARK ZOO

Mr. White placed his camera on a rail, opened up the lens wide, for the big monkey house was rather dark, and exposed "Bushman" for a second. The gorilla moved in the first exposure, and he had to try again.

Postcard from 1946

Now children, look carefully at the picture of the big gorilla. Is he not a handsome looking animal? He seemed to be popular, for many people were standing by his cage watching him. His name is "Bushman." When he was one year old he was captured in Africa by Mr. J. L. Buck and brought to America. He weighed at that time thirty-eight pounds and was kind and gentle, but as he grew older and bigger he was hard to handle and for the last ten years has been kept in his cage.

This huge gorilla is six feet two inches tall and now weighs about 530 pounds. He eats twenty-two pounds of food each day, consisting of fresh fruit, whole wheat rasin bread, whole milk and vegetables. He cost $3,500. He has an enormous mouth, a big head and body. His great toes are dangerous looking and I am sure he would use them for weapons on anyone who dared to disturb him. In his cage he looks mild and submissive; but what do you think would happen if he were annoyed? I fear he would become angry, as girls and boys sometimes do when their wills are crossed, and their actions suddenly change.

Church and home at 1115
Barry Avenue in Chicago

According to my mother this was one of my favorite animals at the zoo. In the same magazine where I am on the cover, the Bishop's wife describes her visit to the zoo.

Left to Right: Charlie, Stanley and me standing

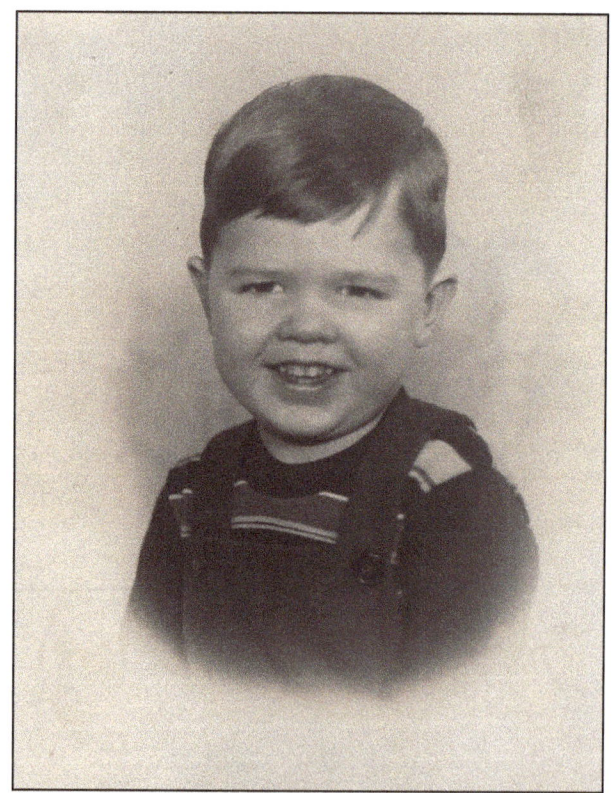

Top Left: Mother, Dad, Bertha Hollander, Me, Harry Ross, and Edna Clark at the 622 Oakdale Avenue home. Upper Right: Mother holding Stanley with me on the porch, Bottom Right: Bertha Hollander, Mom holding Stan, Me, Agnes Kubitz, and Edna Clark on the steps.

Start of California photos, June, 1950. Okland Church, photo following church service

1) School Program 2) Me 3) Me 4) Me

Oakland School Yard

The rear of 2009 W 38th is above left and 2011 W 38th is above right. My father built the retaining walls and steps with very little help. The garage behind 2001 and the back of 2009 can be seen below

Dad sitting on the steps with Stanley and Charlie behind him

Stanley & Charlie w/ Aunt Irene

There were two homes in Oakland, at 2009 38th Avenue (behind us in the right photo) and 2011 38th Avenue in the left photo (We are sitting on the front steps) A garage was behind 2011

Above: with the two Glazier boys

At the San Francisco Botanical Gardens

Snow at Lassen Volcanic National Park

Mother's New Wedgewood Stove

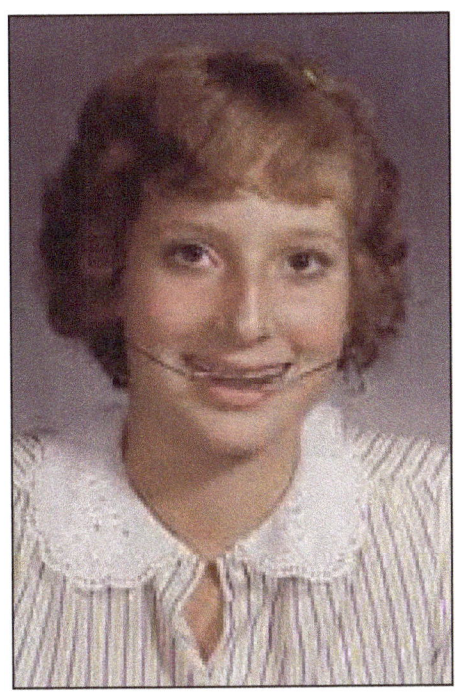

Ellen DeGeneres-Braces

Knots Berry Farm

Me, Charlie, and Stan

San Francisco zoo, elephant enclosure

Aunt Irene Thornton with Stanley, Charlie, and me

Yosemite Fire Falls

When the Church purchased this property in Pacifica, south of San Francisco in the late 1940s, there was only a handful of homes in the valley. Now it is a thriving community with more than 1,000 homes. Left: Construction of the first of many church and school buildings at Pacifica, started while I was in grade school. Below, dormitory building at Pacifica, photos were taken by Dad in December of 1955

MUIR WOODS

GEORGE MATHIS
AND DENIS MUIR

MUIR WOODS

Just over Golden Gate Bridge from San Francisco is a very Ancient Forest

Muir Woods is a venerable forest which grows in a canyon about fifteen miles north of San Francisco and seven or eight miles north of Golden Gate Bridge.

In 1908 it was given to the United States by Congressman William Kent and his wife and is now preserved as a national monument. It is unique in being the only considerable tract of ancient redwoods still growing close to a big city.

The spectacular trees, some of which are nearly as old as Christianity, are the chief attraction but there are many others.

During June and July fragrant azaleas, in magnificent profusion, bloom all along the floor of the valley.

During the winter when the salmon are running, one may, with luck, see the big fish, their backs out of water, struggling up the shallow creeks of the Woods to spawn.

Chipmunks are plentiful and so, for that matter, are deer, raccoons and skunks although these bigger creatures do not usually come into the popular places during the day.

However anybody who wants to see wild animals has only to walk for a few miles through the huge reserve which joins the Woods on the north. He is certain to see plenty of deer and squirrels by day. He may see raccoons and skunks in the early evening.

Muir Woods was a favorite place to picnic. It was a 45 minute drive and had historic redwood trees

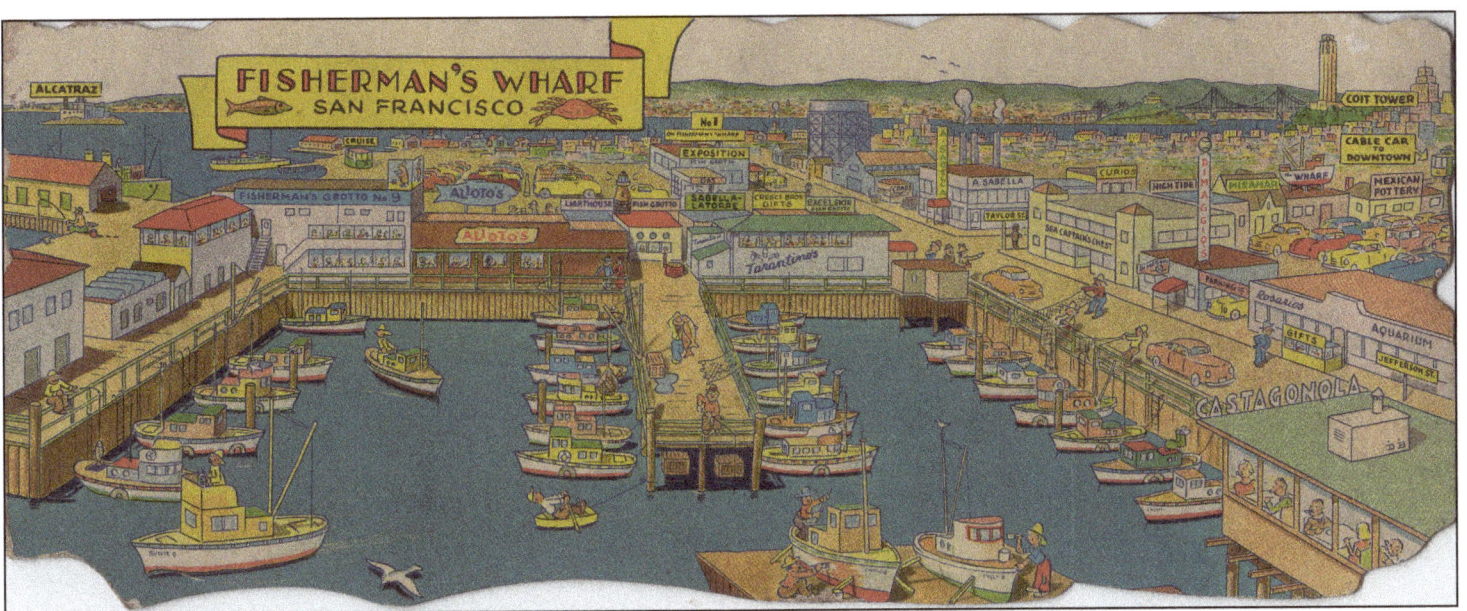

Fisherman's Wharf was another favorite place to visit, both to eat and see the seals that frequented the docks. I took piano lessons at least three times in my life and had little or no success in learning how to play. My parents insisted I practice, and I disliked doing it immensely!

F. F. GUARD'S
MUSIC PUPIL'S LESSON BOOK
AND PRACTICE RECORD

Name *Gerald Sharpe, 1954*

THEODORE PRESSER CO.   BRYN MAWR, PENNSYLVANIA

Copyright 1917, by THEODORE PRESSER CO.

418 – 40009   .15

COMMENCEMENT

Belleview College
Seminary
Preparatory School

ALMA TEMPLE
1340 SHERMAN STREET
DENVER, COLORADO
MONDAY EVENING, MAY 30, 7:30 P.M.

PROGRAM

Master of Ceremonies—Carl Jacobson

Director of Chorus—William M. Staats

Director of Orchestra—Ora Hardman

Processional: "War March of the Priests"—Mendelssohn

"The Star-Spangled Banner"   Audience

Invocation

Oration: "Hope in the Atomic Age"   Gary Rhae Cowan
                                        Belleview Preparatory

Chorus: "Done Found My Lost Sheep"

        (Negro Spiritual—arr. Rhea)

Sermon: "Inasmuch As—"   Mark Passmore Tomlin
                                        Bible Seminary

Processional: "Queen of Sheba"—Gounod   Orchestra

Oration: "Christianity vs. Communism"
                                Violet Joy Horner
                                        Belleview College

Cantata: "American Ode"—Purcell-Kountz   Chorus

Address to the Graduates: "Doors and the Door"
    Arthur Kent White, A.M., D.D.
                                President, Belleview College

PRESENTATION OF AWARDS

Faculty Award for High Scholarship
    Roger Sherwood Dunn
    Sharon Joy Mayhew
    Beatrice Viola Tomlin

Award of Merit—1954-1955
    Elaine Burr   Sharon Mayhew
    Ronnie Cartee   Charles Miller
    Janice Davis   Agnes Pinney
    Roger Dunn   Beatrice Tomlin
    Elaine Mayhew

Nelia Bray English Prize
    Sharon Joy Mayhew

Reader's Digest Certificate of Award to Valedictorian
    Jean Marie Bradford

The Bausch and Lomb Honorary Science Award
    Gary Rhae Cowan

Valedictorian
    Jean Marie Bradford

Salutatorian
    Gary Rhae Cowan

Spelling Awards
    Carol Allen   Michael McMeekin
    Ronnie Mason   Cheryl Lee Moreaux
    Esther Mayhew   Perry Ruby
    Eunice Mayhew   Eric Singer

Perfect Attendance
    Robert Bradford   Tina Lou Hall
    Harry Fitch   Esther Mayhew
    Nancy Gagne   Juliann Stiles
              Judy Wolfe

Orchestra: "Marche Slave"   Tschaikowsky

Benediction

CONFERRING OF DEGREES AND PRESENTATION OF DIPLOMAS

Candidate for Degree of Bachelor of Arts
    Violet Joy Horner

Candidate for Degree of Associate of Arts
    Violet Alice Kramer   Esther Mae Fetzer
    David H. Gross   Lynn Fred Schissler
    James Russell Cather

Candidates For Four-Year Diploma, Bible Seminary
    John Dean Cole   Margaret H. Sillett
    Ethel Florence Kramer   Mark Passmore Tomlin

Candidate for Two-Year Diploma, Bible Seminary
    James Russell Cather   Arlene Katherine Klein
    Eva Vivian Hancock

Candidates for Diploma, Belleview Preparatory School
    Jean Marie Bradford   Sharon Joy Mayhew
    Gary Rhae Cowan   Ray A. Oldenettel
    Richard James Derbyshire   Karl S. Trulin

Candidates for Diploma, Eighth Grade Graduates
    Naomi Bertha Akers   Richard Ervin Mason
    Robert Elsworth Bradford   Joseph J. Rogers
    Alva Lynn Cline   Lowell E. Schissler
    John Walter Dallas   Carol Rae Sharpe
    Donald W. Harvey   Gerald William Sharpe
    Arthur Myron Jacobson

BOULDER
    Ralph Boys Arnold   Shirley Jeanne Greenfield
    Gordon Ray Garrison   Marie Eleana Stewart

GREELEY
    Coena Joyce King   John Edward Stevens
    Nila Lorraine Rodman   Anna Louise Whittaker
              Thomas William Smith

LOVELAND   **CHEYENNE**
    Terry Dennis Cruise   Marilyn Y. Lee
    Neal Clifton McFarland   Robert Gregory Montoya

SALT LAKE CITY
    Sharon Bertha Hensley
    Wayne William Palculich
    James Richard Patterson

Bellview Elementary School, 8th Grade, Westminster, CO, 1954 - 1955

Store building on the right where we lived on the second floor for our first year on the Belleview campus.

My 8th grade classroom was on the 1st floor. Below is my 8th Grade Diploma

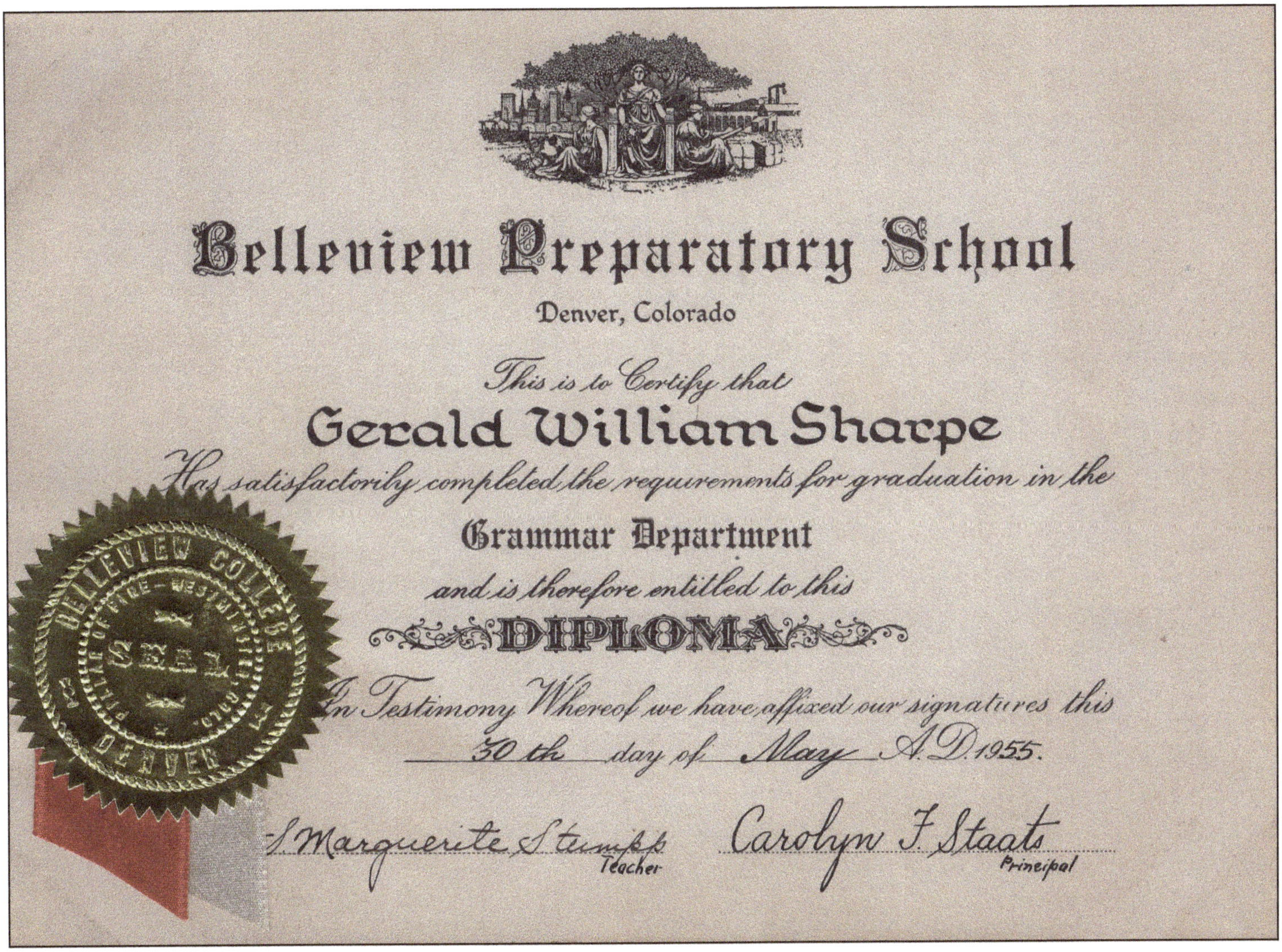

Belleview Preparatory School

Denver, Colorado

This is to Certify that

Gerald William Sharpe

Has satisfactorily completed the requirements for graduation in the

Grammar Department

and is therefore entitled to this

◆◇ DIPLOMA ◇◆

In Testimony Whereof we have affixed our signatures this

___30 th___ *day of* ___May___ *A.D.* 1955.

S Marguerite Stumpp
Teacher

Carolyn F Staats
Principal

2. BELLEVIEW HIGH SCHOOL

SEPTEMBER 1955 TO AUGUST 1959

WE MOVED OUT OF OUR APARTMENT IN NORTH HALL IN SEPTEMBER OF 1955. BILL AND BELLE Staats, the parents of my high school friend Willard, moved out of the home where we would live on September 10. It took about two weeks to move and then Reverend Robert Hartman moved into our old apartment.

We did not have a lot of sports in high school, but we did have a baseball field. We did not play other schools, but the students, faculty, and church members who lived at Belleview would participate in games. I was never very good at baseball. In my freshman and sophomore years, I was only about five feet tall and not very athletic so when the teams were chosen, I would usually be the last one picked. I had buck teeth, and my nickname became "Bucky beaver." I never liked the name but can remember the guys shouting out, "We'll take bucky beaver."

This might be a good time to point out that I was never called, "Jerry" in my high school years as my mother was adamant that my name was "Gerald." I can remember on more than one occasion my mother would hear someone call me "Jerry." She would step right up and say, "Excuse me, but his name is Gerald, if I had wanted that boy called Jerry, I would have named him Jerry. You will call him Gerald!" I do not think anyone ever made that mistake twice.

Mother reported in her 1955 diary that in October I had been assigned to play the snare drums in the church band. We practiced at least once a week and played in church services on Sundays at Alma Temple, the large Pillar of Fire Church just one block from the state capital building in Denver.

In the basement of the big red sandstone college building there were two ping pong tables. Willard Staats was one year ahead of me, and he and I became the best of friends throughout high school. He was close to six feet as a sophomore while I was barely five feet tall. We were called "Mutt and Jeff," two cartoon characters that were short and tall. We would sometimes play ping pong for five or six hours at a time in the evenings. When we got bored with the regular game, we would put a chair on the table over the net to make the game more challenging.

At Belleview we had a small pond that was used for irrigation and watering lawns. It had a small pier with a diving board on the end. I did not know how to swim, and for the first summer we lived at Belleview I watched one of our teachers/ministers, Mr. Reuben Truitt, teach young people how to swim. He scared me to death. His method was simply to scoop a young person up and say, "Today you are either going to learn to swim or you are going to drown. Now listen carefully, when you get in the water, kick your feet hard. That is all you must do."

He would then carry the child, kicking and screaming, to the end of the diving board and throw him or her as far out in the water as he could. I have to say I only saw him jump in the water once to save a child. Each of us, me included, would come to the surface sputtering, but kicking our feet, and we learned on the first try how to "dog paddle" back to shore.

About twice a year, we had various programs in school where we were expected to recite something, participate in skits, or sing. We were required to write stories we would later read. Even in grade school I had speaking parts that started to develop my natural ability for public speaking. My mother saved some documents from my childhood and later gave me a large envelope with birthday cards, letters, and a few typed speeches I gave at programs. The first of these was from 1955 as follows:

JOHNNY APPLESEED
HE LEFT A LEGACY OF APPLE TREES AND LEGENDS

Most of you have probably heard many stories about Johnny Appleseed and perhaps don't realize that he was a real person. Well, he was and I, John Chapman, was he. I was born and raised in New England and then wandered westward and reached Pittsburg. I lived there for twelve years in a log cabin I built all by myself. I planted an apple orchard and there I kept some bees. Since the bees didn't charge me anything for the honey, I gave most of it away.

At Pittsburg, I saw thousands of pioneer families come through on their way across the river to new settlements. And then I got a big idea — to plant apple orchards out on the frontier so that the children of pioneers could have apples to enjoy. It was as simple an idea as that. So, I left my home to spend the rest of my life planting orchards in the wilderness and to earn years and years of fame as "Johnny Appleseed."

I got my seeds from the apple-cider presses in Pittsburg, packed them in sacks and made my way westward, carrying the sacks on my back. I looked for good, rich loam in the forest clearings, planted seeds. I built fences of sticks to protect the young seedlings when they sprouted and moved on. When my seeds were all gone, I walked back to Pittsburgh for more.

I was welcomed in every pioneer cabin, and the settlers came to know me. I read the bible to them and even preached little sermons, sometimes, and occasionally I had little knick-knacks to give to the children. Since my death thousands of apple trees have kept my memory alive.

A portrait of Johnny Appleseed published in a Pittsburg newspaper. Born September 26, 1774—died March 18, 1845, Johnny Appleseed, was an American pioneer nurseryman who introduced apple trees to large parts of Pennsylvania. He became an American legend while still alive, due to his kind, generous ways, his leadership in conservation, and the symbolic importance he attributed to apples. He was also a missionary for The New Church (Swedenborgian) and the inspiration for many museums and historical sites such as the Johnny Appleseed Museum in Urbana, OH.

Another one from 1956 is as follows:

YOU AND SCIENCE

Have you ever stopped to think of what science really is? Although science is a word that is coming more and more into use, most of us cannot truly define it. Perhaps one of the best ways to express its meaning is in just one word, discovery—discovery of the way our world works. Whether we are students in the laboratory or scientists trying to solve complex problems, we are still, in our own way, trying to discover or get a better understanding of the world we live in.

Did you know that science influences your life every hour of the day? Does this idea sound strange? If it does, think of some of the common things you use every day: the telephone, your family car, a radio or television, some of the foods you eat, or medicines you buy when you are sick. Men, working as scientists, invented or developed these things for you.

Let us look at what science has done for us, say in the field of research against disease. One of our most recent and important victories has been the Salk vaccine, which has struck a decisive blow against infantile paralysis, better known as polio. Today because of its wide-spread use, polio's crippling effects, especially in young children, have been reduced to an all-time low.

Perhaps tomorrow, next week, or in the months to come, some scientist working in his laboratory may come upon another great discovery—a cure for cancer. The fight against this dread disease is becoming more important every day. Right now, one of our most distinguished statesmen, Secretary of State John Foster Dulles, has been overcome by cancer, because of the lack of a cure or even the knowledge of some way to arrest what has already developed.

Shortly after the start of my sophomore year, in August of 1956, my mother received a call from Miss Agnes Kubitz, the high school assistant principal, stating that I did not have to take Latin anymore. Mother wrote in her diary, "Gerald was so happy that he jumped up and down all over the house shouting—NO MORE LATIN!" I hated Latin as did most of us who had to take it.

In my freshman year, our teacher was Miss Caroline Staats, who later became the high school principal. We quickly discovered that she only marked up corrections to our translations on the first and last paragraphs. To test this theory, we first put in a few sentences in the middle of the translation that were NOT translated but were just fillers—nothing happened. We next only translated the first two and last two paragraphs and filled in the rest with paragraphs out of our Latin book. Finally, we only translated the first and last paragraphs and would have a page and a half of junk and would still get a grade with no marks anywhere else in the paper. I saved a paper for years that I turned in and got an "A!" It was nearly two pages long and everything between the first and last paragraph consisted of song titles from the top 100. I regret I have lost the paper.

We lived a life without a television, were not allowed to go to movies, and were closely monitored in everything we did, but we still managed to have fun. Our parents and most other ministers on campus had no objections to amusement parks. Denver had two, Lakeside and Elitch's. Both are still open today, but Elitch Gardens, as it is now known, was bought by Six Flags and has moved to a new location.

The parks had roller coasters and bumper cars. We would save our money to be able to go, and sometimes Dad would give me money for rides. Our favorite thing to do on the roller coasters was to wait to get in the front seat of the front car. As soon as the cars left the station, we would unbuckle our seat belts and stand up holding on only to the front of the car. It was a badge of courage to be able to ride the entire ride standing up. Of course, today, safety has become such a concern that bars come down in your lap and standing up would be impossible. We rode bumper cars as much as we could until our money ran out. I developed a life-long love of both roller coasters and bumper cars.

We were also allowed to attend Denver Bears baseball games. The Bears were the senior farm club for the New York Yankees. During my four years in high school, I probably attended more than 50 games. My father would take us on occasion, and many of the ministers on the Belleview campus attended when their children wanted to go. We were very fortunate to witness the development of many players and a manager who would become Yankee legends.

Ralph Houk took over as manager in 1955 during my freshman year and had players like shortstop Tony Kubeck, second baseman Bobby Richardson, first baseman Marv Throneberry (who won three home run titles), catcher John Blanchard, outfielder Whitey Herzog, and pitchers Don Larson and Tommy Lasorda (The now famous Dodger's coach). All did well and all were promoted to the Yankees and became stars. Admission was only 50 cents, and parking was 10 cents a car, although there were lots of complaints when the parking was raised to 25 cents.

In my freshman and sophomore years, at about five feet tall, I would get in with a 25-cent child's ticket for ages 14 and below. The team under Ralph Houk had winning seasons in 1955, 1956, and won the league title in 1957. Ralph's nickname was "The Major." In WWII, he fought in Bastogne and the Battle of the Bulge, and rose to the rank of major. Ralph became first base coach of the Yankees in 1958 and was hired as GM when the Yankees fired Casey Stengel after losing the 1960 World Series.

The major league accomplishments of these players and coach are too numerous to mention, but Don Larson, now known as Don "Perfect Game" Larson, pitched the only perfect game in World Series history during game five of the 1956 world series. I stayed home "sick" that day and listened to every minute of the game.

One other tidbit from Denver Bears baseball, I ate my first piece of meat in my life when I had a hot dog. My parents and the entire Pillar of Fire Church were, at the time, vegetarians. I remain a Yankees fan to this day!

While we were not allowed to go to movies, there was a drive-in movie theater called the North Drive-in on 72nd Avenue, a little more than a mile from Belleview. We lived on 84th Avenue on a high hill, and we could see the screen from our house. We would sometimes talk former students with cars into taking us to the movies by telling our parents we were going to someone's house.

A couple of times we "borrowed" one of the church's flatbed trucks, and six or eight of us would go. We would back the truck into the space and sit on the bed on lawn chairs. Amazingly, we only got caught once. That evening we left the theater around 11:00 pm and had not driven but a block or two when the truck stalled. After trying unsuccessfully to get the truck started, two of the older boys hiked back to Belleview and drove a second truck down to tow the first one in. Just as we were driving behind the barn where we intended to unhook the chain and put the trucks back in line where they belonged (by driving one and pushing the other), we saw someone with a flashlight running up the road. We all jumped off the truck with our lawn chairs and ran off toward our homes, including the driver who left the keys in the ignition.

The next morning, a Saturday, all of us who worked on the farm were called down to the barn to meet Albert Wolfe, who ran the farm. After we were assembled, he simply smiled and said something like, "Guys, I am not stupid. I suspect at least some of you took a truck to the drive-in and then had to tow it in. I figure that the best way to solve this problem is to simply collect all the keys and start locking them up. So, for those who have keys, give them to me now. Then (he named the two oldest boys) you two take the trucks up to the garage, unhook the chain, and bring the good truck back here. The rest of you go up and help push the truck into the garage so they can see why it is not running." Needless to say, we never again "borrowed" a truck to go to the movies.

I argued endlessly with Dad asking him to allow me to drive his car around the school campus so I could learn to drive before I got my driver's license. Even though I drove farm trucks to pick up hay out of the fields and drove tractors all over the campus, Dad was adamant that I should not touch his car until I got my license.

One day, about a year before I could get my license, he asked me to wash his car and I told him I would if he would allow me to drive it up to the school garage a block away as they had a vacuum and pressure hose. He finally agreed, and I very carefully drove the car up to the garage, washed it, and cleaned out the inside. A number of my friends were standing around and I decided I would see if I could get the wheels to spin. The car was pointed at the garage, perhaps eight or ten feet back away from the building. I had the bright idea that I could cut the wheel hard to the left, floor the accelerator, and peel out for home with Dad never being the wiser.

Unfortunately, the garage had columns that stuck out away from the wall about eight or ten inches, and as I turned, I misjudged the distance to the wall and the car caught the edge of the column and scraped the right fender almost back to the door. I never got to drive Dad's car again until after I got my license.

Every student at Belleview had to perform various jobs as part of our education and as the way things got done without hiring employees. Mother cooked at the college building twice a week, Dad occasionally helped at the dairy to milk cows at 5:30 in the morning, and Stan worked at the dairy on a regular schedule. Charlie was too young to have regular jobs, but I had several.

My main one was to work on the farm along with four or five other high school boys. We plowed and disked fields, planted seed, fertilized, and cut and baled hay, then loaded it on long trailers, and unloaded it into the barns. I also assisted Ray Sharpe, the minister who repaired things around the college and in church homes on campus. he had a warehouse of repair items on the second floor above the barn where we stored hay for the dairy after the usual storage areas were filled up. It was an amazing place. He went to army surplus often and in those days, he could take anything he wanted for free with his church/school card. He would come home with a truckload of all kinds of things such as rope, canvas, screws, nuts, bolts, plumbing supplies, electrical items, and tools. The barn was one hundred feet long, and he had long rows of tables where he sorted things into plumbing, electrical, carpentry items, and many other categories. On the walls on both sides hung ropes, chains, ladders, scaffolding, and anything he might need.

One of my favorite jobs, starting in my sophomore year, was announcing the two classical music broadcasts each week on radio station KPOF, owned and operated by the church. Each broadcast was two hours long, and I got to choose which 78 RPM records played on turntables in the control room. The church had a record library of more than a thousand albums. I still love to listen to classical music because of this early indoctrination.

Charlie told us the following story during a family reunion in 2014 that we had in Breckenridge, CO about a trip we took to Yellowstone sometime during my high school years. He was not sure when the trip took place but said that we stayed in the same travel trailer Dad had purchased in California.

Charlie said that I carried a small camera with me all the time and took tons of pictures. One afternoon, Stanley, Charlie, and I stayed in the camp and were in the trailer. He didn't say where our parents were, but we were hanging out when we heard a noise outside.

We looked out the back of the trailer and saw a bear with its head buried in a metal trash can.

Charlie said I rushed outside with my camera and was taking pictures of the bear. While standing in front of the bear, I was trying to get closer to get a better picture. At that point, Stan snuck out of the trailer and ran around it to come up directly behind the bear. Stan then hauled off and kicked him as hard as he could in the butt and then ran back to the trailer leaving the bear to come roaring up out of the trash can with me standing directly in front of it. I then ran quickly to the trailer and climbed over the tailgate to get away from the bear. Charlie said that Stan had to run from me the rest of the day as I was very upset that he could have gotten me mauled by the bear!

Dad was a master at getting things donated. I remember Bishop Arthur K. White (often called simply "AK") coming to the house in September 1957 and telling Dad he needed to leave for Jacksonville, FL right away because Hurricane Irma (the worst hurricane to ever hit Jacksonville) had damaged the roof of the Jacksonville church on Sunday the 10th.

He said part of the roof had been completely torn off and tarps were needed to keep the rain out, plus lumber and roofing materials to rebuild it. I remember Dad asking the bishop why he needed to go to Jacksonville. As I recall the bishop said he did not know any other way to get the job done. Dad told the bishop he would take care of it and would let him know later in the day whether or not he would have to go to Florida.

For the next two hours I watched in fascination as my father made numerous telephone calls to businesses and companies which he found by simply calling "411 Information" which in those days was a free service provided by the phone company. He called companies both in Jacksonville and in other places in the south. As a result of his many calls, a hardware store owner agreed to donate the tarps and have his employees put them on the church roof, a lumber company agreed to donate lumber and plywood as well as nails, a construction company agreed to rebuild the roof in preparation for roofing, a roofing company in Georgia that Dad had gotten shingles donated from previously agreed to donate needed shingles and rolls of tarpaper, a trucking company agreed to haul the shingles and tarpaper to Jacksonville the next day, and finally, after calling a number of roofing companies, one agreed to put the shingles on at no cost.

I remember Dad sitting back with a big smile on his face as he called the bishop. It was a short conversation as I recall, something like, "The church will have a new roof before the end of the week, and I will not need to go there to have it completed." This was only one of many times I listened to Dad talk to businessmen about the good work the church did and explain how everyone dedicated their lives to the church with no pay. Then he would simply say that any donation they could make would be greatly appreciated.

I started early in life, probably as a sophomore in high school, practicing the skills I learned from observing my father. I would volunteer to get refreshments for class events and visit grocery stores and get the manager to give the school ice cream, candy, cookies, and/or soda. I got the manager of a Goodwill store to agree to donate any baseball equipment he got, and we received a number of gloves, bats, and balls from him. Later in life I perfected this skill while serving in several Kiwanis Clubs in the USA, which I joined while in the army, and even later while serving in and leading a Rotary Club here in the Philippines.

In Dad's 1957 diary he noted that we were still living at Hopewell on Irving Street. In mother's diary from January 4, she stated that she hoped her mother would come to visit and wanted to fix up the basement. She wrote, "Gerald is painting tonight on the basement walls. He already did the back stairs and bathroom." The next day she again wrote, "Gerald painted a good part of the day and this evening in the basement." Then on the 8th she added, "Gerald is painting in the basement again today." Finally, on the 9th, mother indicated that Dad did some patching in the basement so I could finish the painting. Apparently, I did such a good job on the basement that Dad got me to paint the gym in the basement of the college building on the 10th of January and later, in early March, I re-painted the metal and plywood on the family trailer.

In March of 1957 my father bought us a Shetland pony. We did not get a saddle, but we learned to ride him bareback. Stan, while still in grade school, was very good at riding him. I would normally only walk or trot him up to the college building. On the way home across the ball field, he would sometimes sense we were heading home and suddenly bolt and begin galloping for home. I am not sure where he learned the trick, but he would jump in the air at full gallop, come down with all four feet planted, and I would go sailing off him into the soft dirt on the ball field. He would then calmly wander on home like he was telling me, "I do not like it that you are riding me." He did this a few times, catching me unawares until one day Mother was in the front yard and saw him throw me off. I think the next day Dad sold the pony. Later Dad was to buy two 80-acre farms and on one he boarded horses. I never much cared for horseback riding after my experiences with that pony.

At about the same time, Dad decided to put a two-bedroom addition on the back of our home and on June 6th had 500 concrete blocks delivered to start the job. For the next six months the addition was steadily constructed, with our father doing much of the work.

That summer the family took a trip to the Grand Canyon. Two events stand out in my mind besides Stan trying to scare everyone to death on the edges of deep ravines by hanging out of small trees. First, on the trip there, Mother was asking Dad a question, but he seemed zoned out not paying attention to what Mother was asking. She asked a second time, still with no response. As Mother would later tell the story, I leaned forward and tapped her on the shoulder. When she turned around, I said, "Mother, you should not bother Dad, he is asleep." She started laughing so hard that Dad turned and asked what was going on. Mother described what had happened and everyone had a good laugh.

Our car did not have air conditioning, so we often rode with all the windows down. On the trip home across the Arizona and New Mexico desert, it was very hot, and Dad was driving fast to get more wind coming in the car to make it cooler. At one point, a wasp flew in Dad's window and he started swatting at it to try to get it to go back out the window. He had no success but about a minute or two later Dad slammed on the brakes, pulled off the road, jumped out of the car, and proceeded to take his pants off on the highway. Turned out the wasp had flown up his pants leg and was stinging him. While Dad was in pain from the bee stings, we were all laughing at seeing him standing on the road without pants.

Every summer the young men from the high school and several of the younger instructors and ministers took a camping trip to the mountains where we climbed one of the 58 14,000+ foot mountains in the state. This summer it was decided we would climb one of the Maroon Bells, a group of three peaks, the highest of which was 14,153 feet. In mid-June we took a flatbed truck loaded with camping equipment and three or four cars to a site near the base of the mountain. These trips were designed to be walking climbs with no ropes required so that we could make it to the top and back down in one day.

We had perhaps 25 men and boys, all high school age or older. We arrived mid-afternoon and set up camp, then sat around a campfire listening to the older men tell stories. We rose at first light, had a good breakfast, and started up the mountain. It was an easy trail for a while and then we came to a rockslide perhaps several thousand feet long. It was not very steep, but the rock was loose and every once in a while you would climb up one step and slide down in cascading rock for a step or two.

We had one young man with us who was a boarding student from somewhere back east, New Jersey I think. He was a sophomore and we considered him to be somewhat of a "sissy" but he was allowed to make the trip. As we were going up the rockslide he would stop and say he was not going any further as he was scared. We convinced him to get to the top of the rock side which was perhaps two thirds of the way up the mountain, but he did not want to go any further. One of the older ministers said he would stay with him and wait for us to come back down the mountain.

There were a number of different trails leading to the top of the tallest mountain. We separated into three groups. I believe there were six or seven of us in my group, and we agreed to meet back at the rockslide in two or three hours. As I recall it was a steep rise but an easy climb until we reached a place where there was a very steep rise in the mountain and the only way around it was on a ledge that was perhaps a foot or two wide, with a very long drop off into a valley below us.

We discussed going back and finding another trail up the mountain, but no one had seen one for some time. The ledge looked wide enough to move on with our backs to the mountain, and it looked like it opened up into a large meadow in about 100 feet or so.

Not sure why, but I was the first person on the ledge. About three quarters of the way along it a small piece of the ledge had collapsed leaving a gap about a foot wide. I told the others that I was going to turn around to face the wall, find a hand hold, and simply step across the gap.

I found what I thought was a solid rock to hold onto and started to lift my leg to step across the opening when the hand hold broke away and I came within a hair's breadth of plunging into the valley below. Fortunately, I still had my weight on solid ground and my handhold on the near side was solid enough to keep me in place. I think I screamed and everyone behind me wanted to know what happened.

It took me a minute or two to regain my composure as my heart was racing and I was hugging the wall for all I was worth. After a few minutes we discussed going back and simply giving up getting to the top. I recall saying that as soon as I could breathe again, I was going to try a second time.

I made it the second time as did all the rest, and we were the second group to reach the top. We waited a few minutes for the third group and asked the other groups about the difficulty of the route they took after telling them about our experience. No one on our team wanted to go back the same way so we followed one of the other groups back to the rockslide.

We had lunch, and the leader of the group, I think it was Giles Cather, the first husband of my stepmother, who told us that everyone was on their own getting back to camp. We all started down the rockslide, jumping and sliding in the loose rocks, and having fun doing so.

Unfortunately, we noted that the young man who did not want to go up the rockslide was still sitting at the top and making no move to come down. We shouted at him and told him it was going to be dark in a couple hours and he really did not want to be on the mountain after dark. He made no move to come down and I remember Giles telling me to go up and get him. He said something to the effect that he did not care how, just get him down the rockslide. I climbed back up and tried to talk him into coming down. He was a much bigger boy than I was, but I was much quicker on my feet.

Having no luck convincing him with reason, I started calling him names like "sissy" and "cry baby" and would run up and slap him on the head. He got madder and madder and for the next hour I got him to chase me down the rock slide telling me that he was going to "kick my ass." Every once in a while, he would realize he was running down the rockslide and would sit back down. I would have to climb back up and torment him some more until he would again be mad enough to chase me.

After three or four of these episodes, we were at the bottom of the rockslide, and I took off for camp with him in hot pursuit. I beat him to camp and briefly explained to Giles what I had done. When the boy raced into camp, Giles tripped him and sent him sprawling on the ground. When he got up, several folks grabbed him. I recall Giles telling him that if he even thought about touching me, he was going to "tan his hide." I think the boy went home for the summer and was not allowed to return.

During my freshman and sophomore years I was the shortest boy in my class, at about five feet tall. In the summer after my sophomore year, in late June 1957, I was asked to go to Pacifica, south of San Francisco where the church was adding to the school campus. It was on a hill overlooking the Pacific Ocean, south of the city, in a place called San Pedro Valley. Ministers in the church from all over the country were asked to send their sons for the summer if they were old enough to work.

I rode a train called the Zephyr by myself to San Francisco. I picked a seat and sat down. Within minutes, a group of sailors in their white uniforms and white hats boarded and took seats all around me. I was sitting in a window seat and one of the sailors sat in my row on the aisle. Before long the sailor sitting in my row pulled out a comic book from his bag and started reading it. When he finished it, he asked me if I would like to read it. I said something like, "Sure, but I have never read a comic book before as we are not allowed to have them."

Almost immediately the sailor stood up and announced to his friends, "This young man has never read a comic book in his whole life." Long story short, I had a pile of comic books on the seat next to me within minutes, and I spent hours reading them. I finished the last one as we were pulling into the Salt Lake train station, and I gave them back to the sailor who had given me the first one. I stayed on the train, but all the sailors got off as we were told we would be in the station for more than an hour. When they returned, they handed me a bag with about 20 new comic books in it. I put them in my carry-on bag, and they provided great entertainment over the summer. At home, I hid them under my mattress.

On the leg from Salt Lake to San Francisco, I went to the restroom and there was a group of four or five older men sitting on the floor playing cards for money. This was also something new for me, as we were not allowed to have playing cards at our house. I stood and watched for a while. One of the men, who was heavily bearded and appeared to have been drinking, turned and looked up at me and said something like, "Son, go get me a drink."

He handed me a twenty dollar bill. I asked him what he wanted, and he said he did not care; he was just thirsty. I went to the dining car, got him a large drink, and returned and handed him the change and the drink. He put the drink on the floor next to him and gave me back a five dollar bill. He said he had been winning while I was gone

A minute or two later he picked up the drink and took a large swig. He immediately spat it all out and turned to me and with a string of expletives like I had never heard before, wanted to know what the _____ I had gotten him. I said, "grapefruit juice" and I turned and got out of there as fast as I could. I was happy to be among a group of sailors. Fortunately, I did not see him again on the trip.

In San Francisco, I was picked up by Reverend Reuben Truitt, the same person who taught me to swim at Belleview. We drove to San Pedro Valley, an area that later became the town of Pacifica.

I quickly learned that all of the cement was being mixed in one-bag cement mixers and then taken by wheelbarrows to where it was needed. There were about 15 young and older men assembled to work. Several of the workers were about my age. There were four or five older church ministers in the group. The younger ones of us got to fill buckets with sand and gravel and carry them to the cement mixer and dump them in. The older men operated the mixers, put in the water and bags of cement, and dumped the cement into wheelbarrows. The older boys did the wheeling of wheelbarrows full of cement.

We worked from sunup to sundown with breaks for breakfast, lunch, and supper. Dad came to visit in mid-July for a few weeks as he needed to get things donated for the construction. During the next month and a half of six-day work weeks, I grew nearly four inches. When I stepped off the train back in Denver on August 20, my pants were up to my calves and half my forearms stuck out of my sleeves. I had to have all new clothes to start my junior year in high school.

Dad wrote in his diary on August 24, 1957, "The boys are taking turns selling corn." We always had a garden at our house and Dad planted one on one of the two farms he purchased north of Denver. Mom and Dad raised corn, tomatoes, beans, lettuce, strawberries, and other vegetables at our home. Dad raised enough corn so that we could sell it along the highway two blocks from our house.

Stan, Charlie, and I would take turns managing the roadside stand, and for some reason Stan always brought home more money than either Charlie or I did. That went on for a while and one day I decided to go down and watch what was going on. We had signs that said, "$4.00 A DOZEN" for the corn, but Stan had removed the signs and was standing alongside the road with corn in his hand. If a potential customer slowed way down and yelled out the window while moving, "How much?" he would say, "$3.00 a dozen" and they would often pull over and come back and buy corn. If they actually stopped in front of the stand and put down the window and asked how much without getting out of the car he would say, "$4.00 a dozen" and they would often pull over and come back to buy corn. If they pulled off the road, walked back to the stand, and asked how much he would say, "$5.00 a dozen."

Sometimes they would buy but sometimes they would say, "That is too much" and he would say, "Well then, how much are you willing to pay?" They almost always bought corn for at least $4.00 a dozen, but often for more. I think these early sales lessons helped all three of us to become better salesmen later in life.

Dad reported in his 1957 diary on the progress of building the addition to our house and in early October of that year the roof was finished, and the new addition was "watertight."

We pulled a few pranks while in high school. Once, when there was a grade school program being held in the third floor auditorium of the college building, we put all the cars in the parking lot behind the building into a huge half circle. We began by pushing the cars together, first one against the building on the far side, then each one after that bumper to bumper in a half circle. Since no one locked their cars in those days, and virtually all were standard transmissions, we were able to put each car in neutral, thus allowing us to maneuver the cars until there was only room for a Volkswagen beetle that a number of us managed to slide in place with the front touching the building and the back touching the bumper of the car behind.

We hid inside the garage that had windows looking out on the parking lot and watched the parents and teachers trying to figure out how to get at least one car moved. After about a half hour they apparently came to the conclusion that they would have to work together to slide the VW out the same way we got it in.

All of the high school students (about 25 of us) spent the next day in the chapel being lectured and made to pray. No one confessed and by nightfall, we were all sent home.

At a second program someone drove a tiny car, even smaller than a VW. We got the bright idea to put it up on the flat roof that covered the left entrance to the college building by the parking lot. About 20 of us were able to lift it up and put it on the roof. Once again, we hid in the garage and watched people mill about.

Fortunately for them, Albert Wolfe had attended the program, and he apparently told the owner he could get the car down safely. Albert walked down to the barn and brought up the forklift we had. He lifted the car off the roof and set it on the ground. One again we spent all day in the chapel instead of attending school but also, once again, no one confessed and about 10:00 pm we were sent home with dire warnings that nothing like this had better ever happen again.

One of the things the older boys did when a new male student arrived from back east was to organize a snipe hunt. This was a type of practical joke from as early as the 1840s, in which an unsuspecting newcomer is duped into trying to catch an elusive, nonexistent animal called a "Snipe." We would set the stage for the hunt over some weeks with imaginative descriptions of the snipe. We would tell the newcomer that the snipe resembled a cross between a jackrabbit and a squirrel and that it only comes out very late at night. We would say that the snipe would only come out on the darkest of nights and it reacts to noise by hiding. We would describe how to catch the snipe by saying that it will run from people so we would form a large circle and slowly but quietly walk toward the hunter. We would all carry long sticks and say we would use them to swish in the grass in front of us while closing the circle, forcing the snipe toward the person holding the bag.

Boys who had been the hunter made up stories of how they caught a snipe and how much they enjoyed the experience. On a night with no moon, we would lead the hunter with his large pillow-case-like bag to a spot out in one of the hay fields that was down in a draw so no lights could be seen. The hunter would be told he must crouch down, remain very quiet, and as his eyes became accustomed to the dark, he would be able see the snipe hopping toward him. We would tell the hunter we were going to walk away from him, form the large circle, and drive the snipe right to him.

The last instruction was there could be no noise or the snipe would simply hide. We would then leave the newcomer alone in the field to at some point figure out the joke. The next morning, we would all tell the person he was a good sport as many of us had carried out a snipe hunt at some point upon our arrival at school.

For a period of time, Stan grew like a weed and I did not. We used to wrestle all the time until my mother would break us up with a broom or a belt. These good-natured wrestling matches sometimes turned ugly, and I would chase him and he would chase me. One time we were wrestling on the grass on the lawn outside the back door. I do not remember what I did but it made Stan mad. I jumped up and ran around the back of the house and out across the field behind our house. He was chasing me, but I could run faster.

One of the things we did for fun was find long sticks, sharpen the tips, and throw them like spears at targets we hung on two hay bales in the field behind our house. In this instance Stan picked up one of these spears and threw it at me. It was a lucky throw and it hit me in the back of the knee and the pointed stick stuck in my leg. I went down to the ground screaming as it hurt a lot. Stan came running and pulled the spear out of my leg and he helped me hobble back to the house. I had a scar and a knot in my muscle on the back of my leg for 20+ years.

Another time we got into one of our wrestling matches just before supper and Dad came out to break us up. Dad always sat at the end of the dining room table closest to the living room and Mother sat on the end closest to the kitchen. The five of us children sat on two sides of the table (Milton was not born until just before I went off to college). This particular evening, I was sitting next to Dad, and Stan was on the opposite side of the table next to Mother.

I do not remember exactly what happened, but our argument spilled over into words at the dinner table. I said something that Stan did not like, and he picked up a fork and threw it at me. In the blink of an eye, it imbedded itself in the skin right above my right eyebrow. Stan got a whipping from our father, and I had a scar on my forehead again for more than 20 years.

During my junior year in high school, Willard Staats and I would draw mazes on 8 ½ by 11-inch bond paper in study hall as neither of us needed to study more than we already did (see a partially completed one in the photos at the end of the chapter). We considered having to sit in a study hall for 50 minutes a waste of our time.

We used a pencil, and we would spend several days drawing the maze and making sure there was only one way to get from the beginning at one corner to the exit at the opposite corner. We drew very small channels and we had to pay close attention to where we blocked or unblocked paths by erasing and then redrawing lines to create another blockage. We would exchange puzzles and use a sharpened stick to follow the paths. After we exchanged mazes, it sometimes took a day or two to figure out the only path to the exit. In mid-December 1957 I was bent over a maze in a study hall erasing blocks and putting new ones in to change the solution to a maze when suddenly an arm came over my shoulder and snatched the maze off my desk. I jumped up and snatched it back. It turned out it was the principal of the school, Miss Carolyn Staats.

She immediately tried to take it back but missed the maze, grabbed my shirt, and jerked on it. She tore my shirt and in amazement I said to her, "Keep your dirty rotten hands off of me." She did not take kindly to my outburst and told me to "Go home now" as I was kicked out of school.

After explaining to my parents what happened I was told I would have to apologize to the principal and as I recall I said I would do so *only* if she apologized to me for grabbing my maze and tearing my shirt. Over the next week before Christmas vacation, my father met with the bishop several times to discuss what would happen to me. At one point it was suggested that perhaps I should go to North Carolina and live with my uncle Woody who had a son a little younger than I was.

The local high school there was directly across the street from Uncle Woody's house. This was the same high school Dad attended years previously. At some point I was told I needed to go talk to the bishop myself. He lived in a house on the Belleview campus called Rose Hill on the next street over from our house on Irving Street. I walked over and was ushered into his office and the bishop told me to take a seat in front of his desk. He immediately began telling me I reminded him of my grandfather, Hugh Ritchie, and began describing him in very unfavorable terms. After a few minutes of him saying nothing but negative things, I stood up and told the bishop, "I did not come over here to listen to you speak ill of my relatives." I turned on my heels and walked out.

By the time I got home my father had received a call from the bishop saying I was "extremely disrespectful." I told Dad what had happened, and he seemed unhappy but went out, got in his car, drove over to Rose Hill, and about 30 minutes later he returned and said, "You will write a letter of apology to Miss Carolyn and you will give this note to Miss Carolyn when you return to school, on probation, after Christmas vacation."

I still have the letter of apology I wrote and the note from the bishop. The note said, "Please accept Gerald back in school." It was signed by A. K. White. I found out years later that he had apparently told the bishop he was NOT sending me to North Carolina and if they would not let me back in school, he would pack up the family and we would all leave the church and move to North Carolina. I guess the threat worked because I returned in January and graduated a year later.

At the beginning of 1958 Dad wrote in his diary, "Still at Hopewell, now three plus years in Denver. Gerald has been out of school for a while because of some difficulties with Miss Carolyn Staats but will return to school next week." He also noted that on January 2nd he began laying hardwood floors in the new addition. On the 13th he wrote, "Finished putting down the hardwood floors by myself and picked up materials to sand and finish the floors. Sanded all day." He noted that after sanding for three days he finally completed sanding and on the 16th of January began applying varnish to the floors. On the 21st he reported that he was painting in the new master bedroom.

On February 4 Bishop A. K. White called everyone in the Denver area together and directed that all work on projects in church homes cease until all the church projects under way were finished. Dad noted that the bishop specifically was unhappy with the addition to our home and said work must cease. While Dad does not say so in his diary, I believe he had a meeting with the bishop to try to get him to allow the addition to be finished as it was very close. I always suspected that part of the problem was that the bishop did not appreciate Dad using threats to get me back in school.

Apparently, that meeting did not go well and on February 12, Dad reported that he was told he would no longer be broadcasting on the church radio station. Dad noted, "This is like the last straw in a series of events which has brought about a virtual estrangement between me and the bishop." Finally, on March 26, Dad wrote, "Had a one-on-one meeting with Bishop A. K. White and now seem to have a little better understanding about several things. So, we shall see how things work out."

In March of 1958 I was tasked in my English class to write a short report on the life of a Confederate general and I chose Robert E. Lee. I was then tasked to recite the story at a high school program as follows,

MILITARY GENIUS AND HERO OF THE SOUTHERN CONFEDERACY

In the summer of 1825 a tall, handsome boy of 18 traveled by stagecoach and river steamer to West Point, the Military Academy of the United States. There he took his entrance examinations along with some two-hundred others who wished to enter the academy. Several days later he received word that he and some eighty others had passed the examination.

As a little boy he lived in Alexandria, Virgina, and attended the Alexandria Academy, where he studied Latin, Greek, and mathematics. When he was eleven his father died of poor health and worry over money matters. This tragic event left his mother very sorrowful, and soon she became an invalid. Because of this young Robert had to take over some of the tasks which his mother normally performed.

When he was eighteen, he received an appointment to West Point. At first life at the Academy seemed hard, but soon he became used to the daily routine. Even under the hard life at the Academy he graduated second in his class of forty-six. During his four years at the Academy, he had not received a single demerit, a record that stands to this day.

He graduated on July fourth, 1829, and immediately joined the Army Corps of Engineers. But his happiness and success were marred by the death of his beloved mother. Soon after, in June 1831 he married Miss Mary Ann Custis, whom he had known most of his life.

Some years later our country declared war on Mexico, and he was ordered to join the forces in San Antonio, where he built roads and did other engineering jobs. More than ten years later he was appointed Superintendent of West Point, a position he held for three years. Lee did not like the work of superintendent; what he wanted was to see active duty, so he was delighted when he was given command of a cavalry regiment stationed in Texas.

While he was there the Southern States withdrew from the Union. He was then recalled from Texas and asked by President Lincoln to accept command of all Union forces, but he felt that it was necessary to decline on behalf of his native state of Virginia. Soon he assumed command of all Confederate forces and throughout the Civil War his daring, courage, and experience won him many battles and an equal amount of fame.

Near the end of the war, while trying to defend Richmond, he had to retreat for the first time because of the overwhelming numbers of Grant's forces, but it was not until the Battle of Gettysburg that he was truly defeated. This decisive battle won the war for the North and forced Lee to later surrender to Grant.

But Lee's story does not end here, for he accepted a position as President of Washington University and during the years he was there he greatly enlarged the campus and built-up attendance from 50 to nearly a thousand.

Near the end of his term as president he became very ill and died on October 11, 1870. The then famous and now immortal Robert E. Lee will be remembered always by Americans as being as glorious in defeat as most men are in victory.

At the end of my junior year, on June 4, 1958, the day after the last day of our classes, Dad had asked me to clean out the gutters on our house. I got the ladder out, leaned it against the house, and started cleaning near the front on the drive-way side, moving toward the back. I had moved the ladder about five or six times and was working a short distance before the back door. I would start on the left side of the ladder, then remove the leaves and debris under the rungs of the ladder, and then clean out the gutter on the right side before moving. I was holding on to the left side of the ladder with my left hand and as I reached out to the right, I apparently reached a little too far and suddenly the ladder simply flipped over causing me to fall face up toward the ground.

Unfortunately, there was a propane tank that I hit first before tumbling to the ground. I am told I was screaming bloody murder and Dad was away at the time. Mother called Miss Stump, my 8th grade teacher, and she drove mother and me to St Luke's hospital in Denver. I broke two ribs and punctured my left lung. A tube was inserted in my chest to drain the air. Dad wrote in his diary on June 6, "Gerald has to remain in the hospital for at least a week or ten days. They are using a pump to relieve his chest cavity of air and he must breathe four times each day with a pressure apparatus over his mouth."

When I was discharged the hospital staff told me I would never climb any mountains or run any marathons as my lung was permanently damaged. Over the years I have done both, climbing three 14,000-foot mountains and I ran a marathon suffering no more ill consequences than making every muscle in my body sore for days.

After healing from my ladder fall, Dad took Charlie and me, along with Mark Tomlin, one of the young church ministers, on a two-week driving trip to North Carolina to visit his mother who was not doing well health wise. We left on August 6 and drove first to Omaha, NE; the next day on to Chicago; and on day three on to Cincinnati. We left Cincinnati about 4:30 a.m. on the 9th for North Carolina and drove all day. We arrived in Wentworth about 4:00 p.m., which was very good driving time. On August 16 Dad wrote that, "Gerald painted some in Mother's cottage." We drove home by way of Chicago and were back in Denver on August 20. It was fortuitous that we were able to visit because Dad's mother died on November 20, 1958, just three months later.

During both my junior and senior years our students participated in a nationwide *Reader's Digest* reading comprehension contest. *Reader's Digest* sent contest forms to the schools, and each student taking the test had a number of minutes to read a long paragraph, then the paragraph was turned in to the instructor, and two pages of multiple-choice questions were handed out. We answered the questions in the allotted time, and the answer sheets were mailed back to *Reader's Digest*. I think we did this four times during the year and schools were issued certificates for those students who nationally scored in the top five percent. I was one of four students who received a certificate during my junior year as well a free year's subscription to the *Reader's Digest*.

In the Pillar of Fire Church, something else we boys were not supposed to do was date girls. We still found ways to associate with girls by playing tennis at night, using the many dark places in the basement of the college to make out, occasionally going to a drive-in movie or to one of the two amusement parks, or just hanging out around the campus. After I got my driver's license, as a senior in high school, Dad would occasionally let me borrow his or Mother's car to go somewhere in the evenings. One time early in my senior year, Dad agreed to let me drive a car to attend a concert at the Red Rocks Amphitheater in Morrison, CO, not far from Belleview. It was a date with my high school classmate, Carol Rae Sharpe. We attended the concert, and after it was over, we sat in our seats until most of the crowd left.

There was a huge parking lot above the amphitheater that looked out over the lights of Denver, and it was a favorite place for young people to go have a place to be alone for at least kissing and petting. After leaving our seats, we got into the car and drove to a spot where there were no other cars around us and started making out. We had only dated a few times and had never done much more than kiss and touch each other on the outside of our clothes.

This particular evening, I had high hopes of doing more. We engaged in some kissing and heavy petting, and I was able to unzip her dress and pull it down, exposing her underwear. So, much to my shock and disappointment, Carol Rae was wearing a corset that started above her knees and went almost up to her neck that her mother had apparently laced up tight in the back with knots in the strings. I had no success in even beginning to get the thing loose. After a few minutes of trying, we simply gave up and went home. We quit dating shortly after that, but we remained good friends throughout the rest of her life. I regret that she passed away a few years ago.

I religiously kept a diary from January through April in 1959 and I listed my expenses in the first few pages through May. Reading it now reminded me that I had made an agreement with Albert Wolf that since I was not getting paid to do farm work, he would let me use an empty chicken coop to raise chickens to make a little bit of money.

According to the diary I bought 100 three-day-old baby chickens for $27.25. For the first four weeks, feed was $4.50 a week and as they got older it became $9.50 a week for the last four weeks before they were sold as broilers at eight to nine weeks old. I seem to recall that I sold them for $1.40 each which meant I got about $55.00 in return. I was able to raise two groups before graduation in early June. That was more money than I had ever had at one time, so I was pleased with it. If I factored in my time and work, I probably lost money as I had to feed and water the chickens twice a day, seven days a week in all kinds of winter weather and I had to clean out the saw dust and straw at least twice during the eight weeks.

One spring day I was cleaning out the coop and was wearing cowboy boots with the pants tucked into my boots so that they would not get covered with chicken poop. Just as I was emptying out a wheelbarrow load of poop, I felt something on my leg and looked down just in time to see a small mouse disappear down my pant leg into my boot. I immediately lay down, put my feet in the air, and tried my best to shake him out. He was not letting go of my pants leg for anything. Next, I sat down on an old pile of lumber and tried to get my boot off without squashing the mouse inside and ruining my boot. Long story short it took about 15 minutes to maneuver my foot to where I felt the mouse get past my ankle and drop down under my instep. Fortunately, with a little shaking to keep him under my foot, I was able to remove my boot and allow the mouse to return home!

There is an entry in my diary from Tuesday, February 10, 1959, that reads, "Went to school, created quite a riot, have a gob of girls liking me, my favorite is Cheryl. Mary is trouble and she is mad." Cheryl Treinen was a junior and I do not remember Mary. On Wednesday, the 11th, I wrote, "I'm in good with Cheryl, going to a party, Mary might go home."

Not sure what I did, but it may have had something to do with girls as I wrote on Thursday, the 12th of February, "Out of school. Fined $5.00. Talked with the Bishop three times, he told me about Mert." This was Mert Weaver, and he indicated he had had lots of trouble with him back at ZA much like me. Friday the 13th, "Attended a party, loads of fun." On Saturday, the 14th of February I reported, "C grounded for six weeks. We are going steady!" (Assume this is Cheryl). All through the rest of February and virtually all of March I report on, "calling Cheryl" most evenings. On Easter Sunday, March 29th, I reported being invited to attend Cheryl's church and have dinner at her house. (Grounding must have ended.)

Must not have been a very successful visit as on Monday, March 30th I wrote, "Cheryl and I aren't going steady as of this morning." I do not mention Cheryl again in my diary for 1959, but I dated her off and on for three more years. My high school classmate, Art Jacobson, has provided me a picture from a birthday party at his house that was across the street from the Belleview campus. I believe it is the only picture I have of Cheryl.

My youngest brother, Milton David, who would be known as "Milt," was born in Denver, CO on April 19, 1959, two months before I graduated from high school. All my brothers and my sister attended school at Belleview and all, but Milt graduated from high school at the Belleview Preparatory School (Milt graduated from Westminster High School).

After graduating from high school in June of 1959 I took a trip back to the Pillar of Fire Church's Eastern Headquarters at Zarephath, NJ, the last week of June. Edwin Staats, the older brother of my good friend Willard, drove a car back there from Denver, and I went with him in return for sharing expenses for gas, which cost me $11.00.

I had a good time visiting our many relatives who lived in New Jersey, including uncle Ed and aunt Gertrude and their son Burton; aunt Mary Jane and uncle Donald Cruver and their two children, Donald Richard "Hoppy" and Beverly "Bev," and aunt Ivy and uncle Ted Rystedt.

I was asked to join a paint crew at Zarephath painting the outside of a building, but when I found out they were only willing to pay $2.00 a day, I declined.

Near the end of the second week of July I found out one of the ministers was driving back to Belleview and I was able to catch a ride with him. I think Dad gave him some money for gas.

Dad kept several bee hives at our home at 8300 Irving and also on the farms he purchased north of Denver. One of my jobs during the summer was to assemble bee hives. They came in a box of about 100 pieces of wood, a bag of nails, and a roll of wire. They were fairly easy to assemble as the sides were tongue and grooved and the frames came with holes drilled in the sides that wire was passed through back and forth about a half inch apart which was where the bees built their honeycomb. Today kits come with a bee's wax sheet instead of wire. We always had honey in the house and Dad would give honey for Christmas presents to the various families living at Belleview.

When I returned from Zarephath in mid-July I began looking for a job. Someone told me that they were building a large Catholic church and school down Federal Boulevard about a mile — just across the four-lane road to Boulder. I was told they were hiring workers and so I went and asked about a job. I was hired the same day for what was called a "carpenter's helper," someone who carried lumber, held wood to be cut, and simply did whatever the carpenters didn't want to do themselves. I do not remember what it paid but since I had never had a paying job or much of an allowance, I was thrilled with what they were offering. We worked five and a half days a week, getting off at noon on Saturday.

One of my memorable experiences during this summer was being able to buy two tickets to a Ricky Nelson concert at the Red Rocks Amphitheater on August 14, 1959. My girlfriend, Cheryl Treinen, was a huge Ricky fan and I had told her I was reluctant to attend as she and most of the females attending would likely spend the entire concert screaming. She assured me she would NOT do that. Elvis Presley was the other large heart throb at the time, but during 1958 and 1959, Rickey had twelve hits on the charts in comparison with Elvis Presley's eleven. In the summer of 1958, Nelson conducted his first full-scale tour, and his 1959 tour sold out across the nation. Ricky was the first rock and roll performer to ever perform at the Red Rocks. At the time, the Ricky Nelson International Fan Club had 9,000 chapters around the world. We had seats on the far right of the amphitheater, and I made the mistake of sitting between Cheryl and the stage. Contrary to her protestations earlier, she, and most of the females in attendance, did scream throughout the event. This was my first large concert and years later I had a similar experience at a John Denver concert with my 1st wife.

My classmates in high school have remained friends for life. Willard, (a year ahead of us) Lowell, and I shared many experiences in high school and hung out together often. Art Jacobson and Carol Rae were also a big part of my life while the others, from outside the church, communicated often.

We are now down to just five of us: Willard, Lowell, Art, Wayne, and me. I lost track of Wayne Palculich for several years and then found out he had changed his last name to Lunquist. In early 2023, I found a telephone number for Wayne in Pennsylvania and called him. His wife Linda answered the phone. I asked if I could speak to Wayne and she said yes, but he will likely not know who you are.

She related that two years previously Wayne had had a dental appointment one morning and he did not come home. By evening she called the police to report Wayne missing and they told her they were sorry, but someone had to be missing for 24 hours before they would process a missing person's report. The next day she filed the report, and an all-points bulletin was sent out with his description and the make, model, color, and license plate number of his car. Linda next called their bank and requested info on her husband's credit card usage, hoping that would help locate him. She said they told her they could not release that information.

Linda said she went to the bank with the missing persons' report and demanded to speak with the manager. The manager finally agreed to pull a report and it showed Wayne had charged in four different states. The last charge was at a gas station in New Jersey. This information was passed to the police and information about the location of additional charges was also sent to the police over the next two days with no success in locating him.

On day four the story was picked up by television news organizations and Wayne's picture and pictures of his car and license plate were shown on the evening news up and down the east coast.

On the morning of day five, Linda got a call from the police in Philadelphia, PA, about an hour south of where they live in Landsdale, saying her husband was at the station, but he did not know who he was or where he lived. They said a lady had seen the missing person's report on TV and when she went out of her home, she observed the car parked in front of her house with a man asleep inside. She called the police, and they came and picked him up. His credit card had his name on it and they had a copy of the missing person's report.

Linda went and picked up her husband, had a friend drive his car back, and she took Wayne to a hospital where he was diagnosed with the sudden onset of dementia with almost total loss of memory. I had a short conversation with Wayne and sure enough he apologized for not knowing who I was.

1956

November 1957

With Stanley, 1958

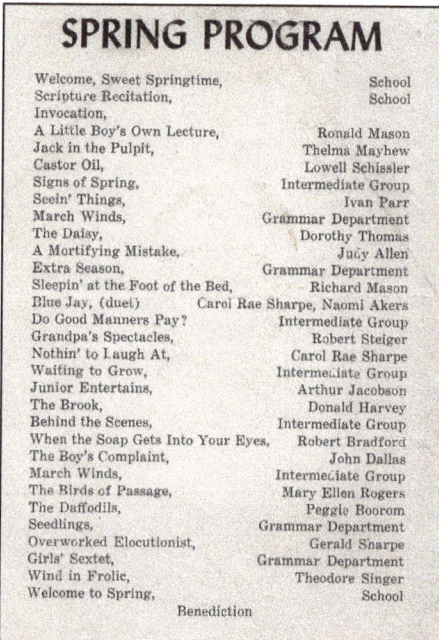

SPRING PROGRAM

Welcome, Sweet Springtime,	School
Scripture Recitation,	School
Invocation,	
A Little Boy's Own Lecture,	Ronald Mason
Jack in the Pulpit,	Thelma Mayhew
Castor Oil,	Lowell Schissler
Signs of Spring,	Intermediate Group
Seein' Things,	Ivan Parr
March Winds,	Grammar Department
The Daisy,	Dorothy Thomas
A Mortifying Mistake,	Judy Allen
Extra Season,	Grammar Department
Sleepin' at the Foot of the Bed,	Richard Mason
Blue Jay, (duet)	Carol Rae Sharpe, Naomi Akers
Do Good Manners Pay?	Intermediate Group
Grandpa's Spectacles,	Robert Steiger
Nothin' to Laugh At,	Carol Rae Sharpe
Waiting to Grow,	Intermediate Group
Junior Entertains,	Arthur Jacobson
The Brook,	Donald Harvey
Behind the Scenes,	Intermediate Group
When the Soap Gets Into Your Eyes,	Robert Bradford
The Boy's Complaint,	John Dallas
March Winds,	Intermediate Group
The Birds of Passage,	Mary Ellen Rogers
The Daffodils,	Peggie Boorom
Seedlings,	Grammar Department
Overworked Elocutionist,	Gerald Sharpe
Girls' Sextet,	Grammar Department
Wind in Frolic,	Theodore Singer
Welcome to Spring,	School
Benediction	

Spring 1956

Bellview High School and College campus seen from the women's dormitory (winter)

Many photos from this chapter have been generously provided of my high school classmate, Art Jacobson.

Left: We lived on the Belleview campus beginning in 1954. The Belleview High School and College shared the large red building seen here from our front porch (summer). The garage where I wrecked my father's car is on the right. The ball field where we played baseball and where I rode our pony is between our home and the garage. Right: Kirkwood, the women's dormitory.

Belleview Campus, Left view from college tower inlucding the Pond where we learned to swim on right.

Store Bldg, where we lived 54/55.

1) Art Jacbonson's home 2) Willard's home 3) Boy's dormitory 4) Pond where we learned to swim on right 5)Store building where we lived 6) Garage 7) One-story roof

Where we put cars in a half circle and put one on a roof (7).

The large red limestone "castle-like" building that dominates the northern skyline of Denver was built in 1891/2 as a Presbyterian University to be the "Princeton of the West." Classes began in 1908, but in 1917 virtually all the students went to serve in WW I so the school closed. It was sold to the Pillar of Fire Church in 1920 and the name changed to Belleview College in 1925. It is now on the National Register of Historic Places.

The home at 8300 Irving St. where we lived starting in 1955, My parents remained there for the rest of their lives. My father built the addition on right side of house and the three-car garage.

Above: Train ticket to return home from working in California, August 1957.

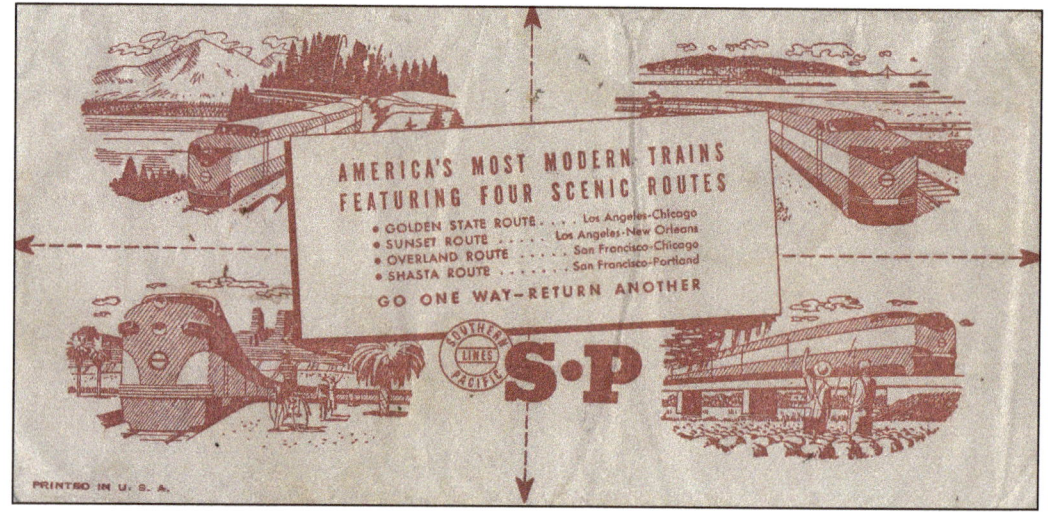

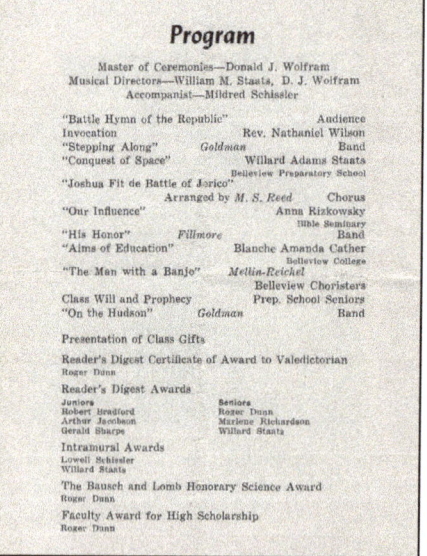

Left: Graduation program from 1958: I received a Readers Digest Award as a junior.

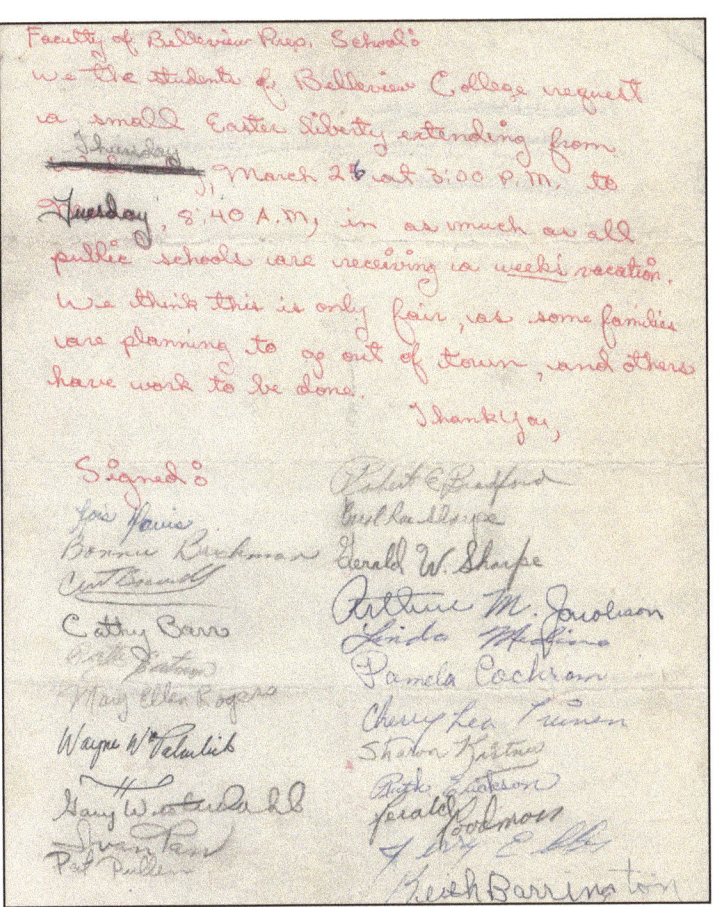

A student petition requesting a 4-day Easter break fell on deaf ears.

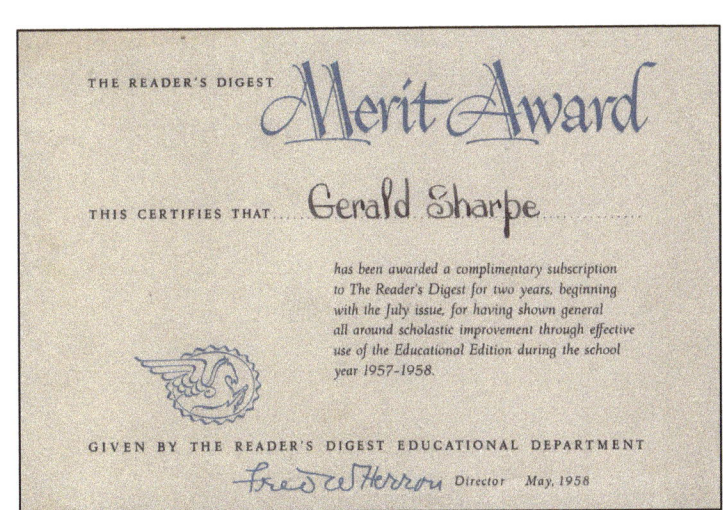

Reader's Digest Award

Programs in the Belleview auditorium on the 3rd floor of the College Building.

Birthday party at Art Jacobson's house. From left, Art, Bobby Bradford, Me, Carol Rae Sharpe, Cheryl Treinen, Linda Medina, and Wayne Palculich, who later changed his name to Wayne Lundquist. I think Cheryl was my date and a junior in school. This picture, sent by Art Jacobson, is the only one I have of her.

Belleview High Students Attend H.S. Conference

Five seniors from Belleview High School attended the discussion on "The United States and Africa" at the University of Denver, Dec. 5th, as a part of the 25th annual International Relations conference of Colorado high schools.

Belleview students participating were Bobby Bradford, Wayne Palculich, Gerald Sharpe, Lowell Schissler and Arthur Jacobson.

Left to right: Renee Thornton, Me, Eileen, Milton, Charlie, Jeff Thornton, Robert, & Stanley.

Bear Stadium, Home of the Denver Bears, a New York Yankees farm club

57/58 autographed balls for sale on ebay

An unfinished maze. A completed one was snatched off my desk by the principal and my reaction got me kicked out of school

Stars and Stripes Forever

ENGLISH PROGRAM

Invocation	Rev. Robert B. Dallenbach
STARS AND STRIPES FOREVER Sousa	Chorus
OUR AMERICAN HERITAGE	Carol Rae Sharpe
MAN'S OLDEST DREAM	Robert Bradford
ATLANTIC ZEPHYRS Simons Trombonist, Ivan Parr	
SCIENCE AND YOU	Gerald Sharpe
GIVE ME LIBERTY	James Clark
STOUT-HEARTED MEN Hammerstein-Romberg	Chorus
"UNCONDITIONAL SURRENDER"	Curt Boswell
YOUNG ABE LINCOLN IN ANECDOTE	Lowell Schissler
CLAIRE DE LUNE Debussy	Pianist, Cathy Barr
THIS I BELIEVE Helen Virden	Ruth Erickson

SKIT: RUSH TO THE ROCKIES

Cast: Arthur—Arthur Jacobson
Gerald—Gerald Sharpe
Cowboy—Curt Boswell
Skier—Lowell Schissler

VICTORY	Paul Yoder	Chorus
Benediction		Rev. Ellsworth Bradford

Master of Ceremonies: L. Ray Sharpe

Thursday Evening, 7:45, February 26, 1959, at Belleview

I gave a presentation and participated in a skit

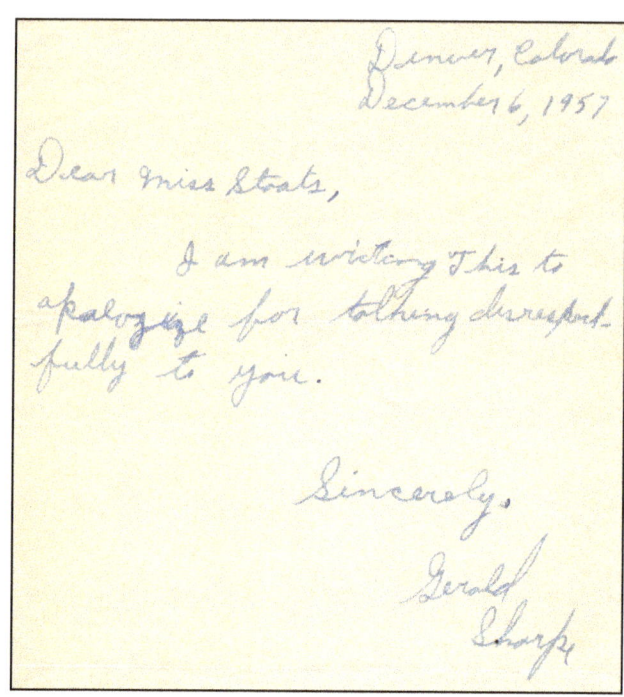

Above: My apology letter to the principal. Below: Note from the Bishop giving his permission to let me back into school starting in January 1958

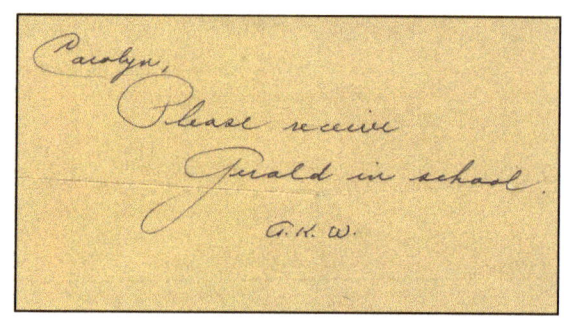

Graduation Invitation

Belleview
Commencement
1959

You are cordially invited
to attend the
Commencement Exercises
of the
Graduating Classes
of
Belleview College, Bible Seminary,
and Preparatory School
Monday Evening
June 1, at Eight O'clock
Alma Temple, 13th and Sherman
as well as
other programs listed on
the accompanying
calendar

CALENDAR

Belleview Kindergarten Commencement Alma Temple, 1340 Sherman Street, Denver, Monday, May 18, at seven-thirty o'clock.

Class Night Program, Belleview College Auditorium, West 83rd Avenue and Federal Boulevard, Westminster, Friday, May 29, at eight o'clock.

Baccalaureate Sermon by Bishop Arthur Kent White, Alma Temple, Sunday, May 31, at eleven o'clock.

Commencement of Belleview College and Associated Schools, Alma Temple, Monday, June 1, at eight o'clock.

Katie Gast, bottom right above, had to attend summer school before graduating. She often babysat for my younger brothers and sister

CURTIS BOSWELL
TREASURER

WAYNE WM. PALCULICH
VICE- PRESIDENT

LOWELL SCHISSLER
PRESIDENT

DR. DONALD J. WOLFRAM
SPONSOR

BARBARA J. PALUSKY
SECRETARY

Belleview Preparatory School

Class of 1959

GERALD SHARPE

CATHY A. BARR

ARTHUR JACOBSON

CAROL RAE SHARPE

ROBERT ELSWORTH BRADFORD

UNIVERSAL STUDIOS

Me

Bobby Bradford

Carol Rae Sharpe

Our graduation dinner in the college building library

Wayne Palculich

Katie Gast

Barbara Palusky

Carol Rae Sharpe

Me, Bobby Bradford, & Lowell Schissler

Willard Staats

Photos taken during graduation week.

My graduation portraits.

Class Night Exercises

Belleview College
Seminary
Preparatory School

May 28, 1959, at 8:00 p.m.
COLLEGE AUDITORIUM

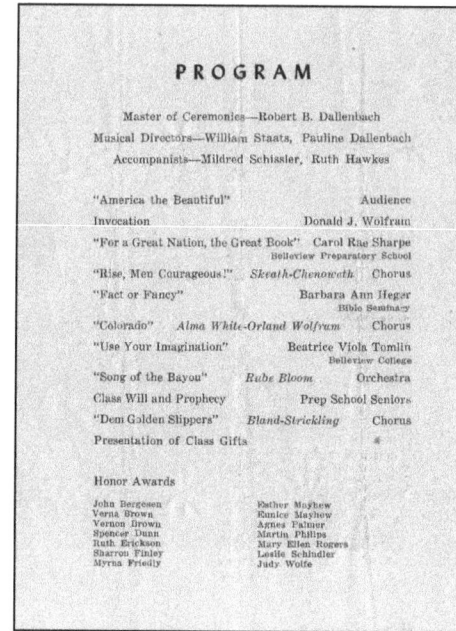

PROGRAM

Master of Ceremonies—Robert B. Dallenbach

Musical Directors—William Staats, Pauline Dallenbach

Accompanists—Mildred Schissler, Ruth Hawkes

"America the Beautiful" .... Audience

Invocation .... Donald J. Wolfram

"For a Great Nation, the Great Book" Carol Rae Sharpe
Belleview Preparatory School

"Rise, Men Courageous!" *Skrath-Chenoweth* Chorus

"Fact or Fancy" .... Barbara Ann Hegar
Bible Seminary

"Colorado" *Alma White-Orland Wolfram* Chorus

"Use Your Imagination" .... Beatrice Viola Tomlin
Belleview College

"Song of the Bayou" *Rube Bloom* Orchestra

Class Will and Prophecy .... Prep School Seniors

"Dem Golden Slippers" *Bland-Strickling* Chorus

Presentation of Class Gifts

Honor Awards

John Bergesen
Verna Brown
Vernon Brown
Spencer Dunn
Ruth Erickson
Sharron Finley
Myrna Friedly

Esther Mayhew
Eunice Mayhew
Agnes Palmer
Martin Philips
Mary Ellen Rogers
Leslie Schindler
Judy Wolfe

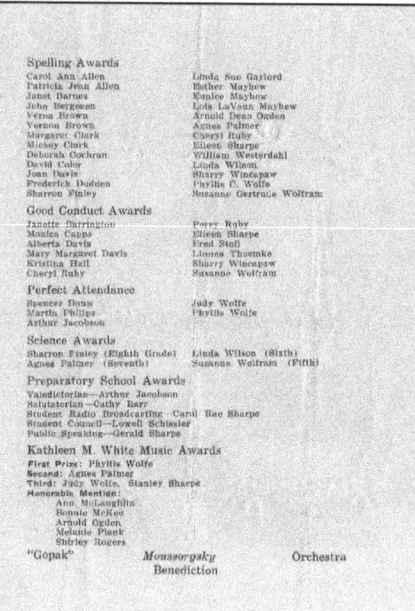

Spelling Awards

Carol Ann Allen
Patricia Jean Allen
Janet Barnes
John Bergesen
Verna Brown
Vernon Brown
Margaret Clark
Mickey Clark
Deborah Cochran
David Color
Joan Davis
Frederick Dodden
Sharron Finley

Linda Sue Gaylord
Esther Mayhew
Eunice Mayhew
Lois LaVaun Mayhew
Arnold Dean Ogden
Agnes Palmer
Cheryl Ruby
Eileen Sharpe
William Westerdahl
Linda Wilson
Sherry Wincapaw
Phyllis C. Wolfe
Suzanne Gertrude Wolfram

Good Conduct Awards

Janette Barrington
Monica Capps
Alberta Davis
Mary Margaret Davis
Kristina Hall
Cheryl Ruby

Perry Ruby
Eileen Sharpe
Fred Stoll
Linnea Thoemke
Sherry Wincapaw
Suzanne Wolfram

Perfect Attendance

Spencer Dunn
Martin Philips
Arthur Jacobson

Judy Wolfe
Phyllis Wolfe

Science Awards

Sharron Finley (Eighth Grade)   Linda Wilson (Sixth)
Agnes Palmer (Seventh)   Suzanne Wolfram (Fifth)

Preparatory School Awards

Valedictorian—Arthur Jacobson
Salutatorian—Cathy Barr
Student Radio Broadcasting—Carol Rae Sharpe
Student Council—Lowell Schissler
Public Speaking—Gerald Sharpe

Kathleen M. White Music Awards

First Prize: Phyllis Wolfe
Second: Agnes Palmer
Third: Judy Wolfe, Stanley Sharpe
Honorable Mention:
Ann McLaughlin
Bonnie McKee
Arnold Ogden
Melanie Plank
Shirley Rogers

"Gopak" *Moussorgsky* Orchestra
Benediction

Our Class of 1959 graduation program.

The graduates of the Classes of 1959 at Alma Temple in Denver

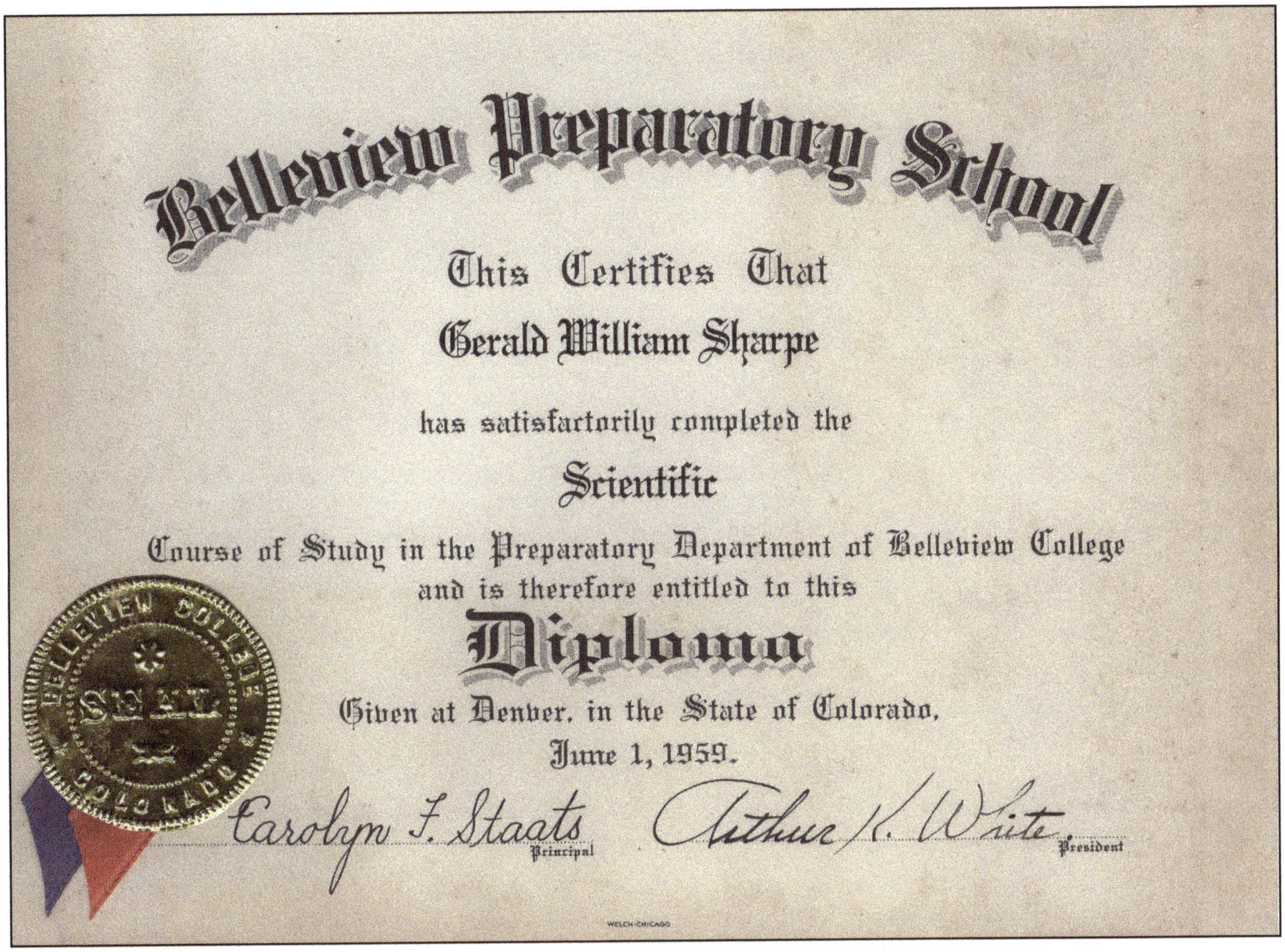

Belleview Preparatory School

This Certifies That

Gerald William Sharpe

has satisfactorily completed the

Scientific

Course of Study in the Preparatory Department of Belleview College and is therefore entitled to this

Diploma

Given at Denver, in the State of Colorado,
June 1, 1959.

BELLEVIEW COLLEGE SEAL COLORADO

Carolyn F. Staats Principal   *Arthur K. White* President

WELCH-CHICAGO

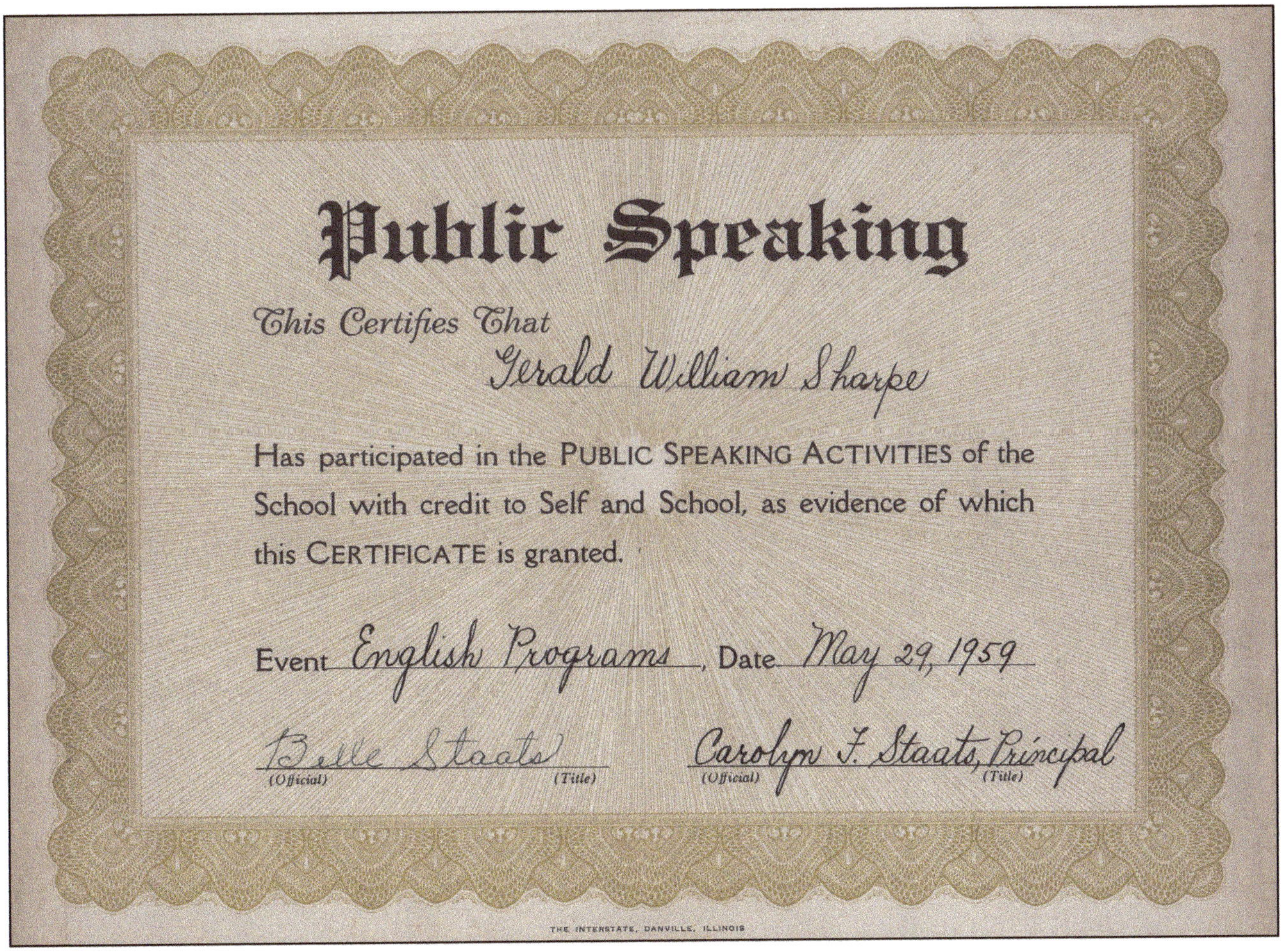

Public Speaking

This Certifies That

Gerald William Sharpe

Has participated in the PUBLIC SPEAKING ACTIVITIES of the School with credit to Self and School, as evidence of which this CERTIFICATE is granted.

Event *English Programs*, Date *May 29, 1959*

Belle Staats (Official) (Title)   *Carolyn F. Staats, Principal* (Official) (Title)

THE INTERSTATE, DANVILLE, ILLINOIS

We climbed the center peak of the Maroon Bells by going up the rockslides and then around the back of the peak on the right where the trails did not require ropes.

RICK NELSON

Rick is a member of America's best known and best loved family. His father, Ozzie, one of the few multi-faceted talents in show business, is producer, director, chief writer, and star of the Nelson's TV series.

Rick's mother, Harriet, a lovely and beautiful woman, enjoyed a highly successful career as a singer, recording artist, and movie star before becoming a member of "The Adventures of Ozzie and Harriet" almost 15 years ago.

Talented and handsome brother Dave has co-starred in the family comedy series in radio and TV for almost a decade and recently distinguished himself in a featured role in the 20th Century-Fox smash hit film, "Peyton Place."

I attended the Rick Nelson Concert at the Red Rocks Amphitheater on August 14, 1959, with my girlfriend, Cheryl Treinen, who was suffering from cancer. I have the original program we purchased at the concert. This program is currently listed on eBay for $110.00.

3. WESTERN STATE COLLEGE

SEPTEMBER 1959 TO AUGUST 1961

AS THE SUMMER NEARED ITS END, I STARTED THINKING ABOUT COLLEGE, SO I LOOKED AT A few options, but did NOT turn in any kind of application. A friend I was working with had applied to Western State College in Gunnison, CO, about a 200-mile drive southwest across three mountain passes. I talked to Dad about college, and he indicated he might be willing to pay for my education.

A few days before classes were to begin, I told Dad I had decided to attend Western State and wanted to know if he would drive me to school. He said he would if I would agree to his conditions for attending. I asked what those conditions might be, and he said, "No smoking, no drinking, and not having any 'nigger' friends." The first two I could agree to easily, but the third caused a rather lengthy discussion that had my father trying to convince me that Negroes were inferior, and that he had lived in the south where he said he observed that they simply were not equal to white people. He quoted the writing of Thomas Jefferson to justify his views. I finally agreed to "take a wait and see attitude on the subject," and this seemed to satisfy him.

The day of registration arrived, and we departed home a little before 4:00 am for the drive to Gunnison. We arrived a little after 8:30 am and found a parking place. We asked someone where registration was being held and we were pointed in the proper direction. In just a few minutes, we came upon a line of several hundred students and parents winding its way up to the large administration building. We were told the doors did not open until 9:00 am. Dad was wearing his black suit and clerical collar and looked every bit the role model of a minister or priest.

As we got in line a very distinguished gentleman came walking up the sidewalk behind us carrying a briefcase and Dad stepped out of line and said something like, "Excuse me, my son has not registered in advance, is this the proper line we should be standing in to get him enrolled?" The gentleman simply said, "Why don't you come with me?" We proceeded to walk past all the people in line and the gentleman began carrying on a conversation with Dad as I trailed behind. We went up the steps into the administration building and turned left down a long hallway. At the end was an office that had a sign that said, "Peter Mickelson, President." We went into the office and President Mickelson told us to take a seat.

After hanging up his coat and putting his briefcase behind his desk, he stepped out of the office for a minute and returned with a gentleman he introduced as the Dean of the College. He said to me, "Young man, Dean Kjosness will get you registered while I talk to your father." Sure enough, I was taken to the Registrar's office a few minutes before the doors were opened to begin registration and in about 45 minutes I was registered and signed up for five freshman classes, assigned a room in Ute Hall (one of several dormitories on campus), issued a student ID card, and signed up for a part-time job working in the cafeteria. The Dean escorted me back to the President's office and told him I was finished. I thanked the Dean, and Dad thanked the President for his assistance and for the donation he had given Dad for the church's good works. We went and found my dormitory room on the second floor of Ute Hall, and I brought my suitcases to the room and emptied them so Dad could take them home.

Thus began my two years at Western State College. I later learned that Ute Hall was the second building built on campus in 1931 and was the first residence hall built. It had about 70 rooms with two students to a room. I met my roommate, Sam Rhodes, who was two quarters ahead of me, and he encouraged me to join several campus organizations. In a letter to Mom and Dad I wrote,

September 2, 1959

"School here is getting into full swing, and I really have a load. On Mondays, Wednesdays, and Fridays (mornings) I have chemistry, algebra, history of civilization, physical education, and at 1:15 pm English. From 2:15 pm on I have time off. On Tuesdays and Thursdays, I have new student orientation at 8:00 am, from 9 to 11 I have trigonometry, and from 2:15 to 5:05 pm I have chemistry lab."

Later in the same letter I wrote,

"Initiation started last Wednesday and since then it has been worse. We must wear our pants inside out and backwards, we must wear little green and white bennies, we have to wear signs on the front with our name and hometown on it, and one on the back with "Beat Greeley" on it. We play Greeley in football this Saturday and it is homecoming week. Football is a big thing around here and I sure was glad we beat Nevada 14 to 13 last week. We also cannot walk on any grass, we must call all Lettermen (they are the football players in charge of initiation) sir, and we have to know all the football cheers. As if to top it all off we must earn 500 points by doing odd jobs for the Lettermen. I have 250 points that I got by letting three Letterman into line in front of me to supper."

Finally, in the same letter I wrote,

"I got a letter from Mrs. Treinen and was glad to hear she has gotten Cheryl into Children's Hospital. I sure hope something will be discovered before it is too late. But I guess the best thing to do is just pray and hope things will work out for the best."

I dated Cheryl in high school, and we became good friends. She was diagnosed with Leukemia a short time later and she battled it for nearly four years before passing away.

Twice a week, a student refilled the two vending machines (one with candy and one with soda) on each floor of our dorm. I quickly learned that they would be empty within hours of being refilled.

One day I asked the student who he worked for, and he said there was a company that had the contract, and they kept a small supply room in the basement of the building next door. He said the manager restocked the supply room every Saturday morning. He told me he was simply too busy to fill the machines more than twice a week and was planning to quit. The next Saturday I loitered outside the dorm next door until I saw a van pull up and a man started loading sodas on a dolly. I went up to him and introduced myself and asked him if he would like to double or triple the amount of money he was getting from the vending machines. He asked how I could do that, and I said by filling up the vending machines at least once a day if not more often. I asked how much he paid his workers, and he described a small commission on everything that was sold. He got an employment application out of his van, and I filled it out while he was unloading.

I did not hear from him all week, so the following Saturday I waited for him a second time. As soon as he saw me, he said the student who was doing the two dorms quit and he would like to hire me on a trial basis. He showed me the inventory forms that had to be filled out, showed me how to account for the money removed from the vending machines, gave me the keys to the storeroom and vending machines, and wrote down the combination to the small safe where I was to put the money. I asked if he could restock the room twice a week if I ran short of inventory and he said he could. Long story short, I sometimes filled the machines two and three times a day — particularly on the weekends during my freshman year — and he actually raised the amount he delivered each Saturday about three times until I was never short of anything. I had considerably more spending money than ever before. Unfortunately, when I returned for my sophomore year, the school had signed a new contract, and the person filling the machines was not a student but rather a company employee.

Working in the school cafeteria was a blessing, as I earned enough money to pay for my meal plan which provided three meals a day. As a result, Dad's cost to send me to school was greatly reduced. We rotated tasks: serving food, cleaning tables, mopping the floors, operating the dishwashers, and unloading trucks when they brought food. The cafeteria often catered events for student organizations, and the cafeteria manager, Mrs. Mickkelson, would hire the best people on the staff to work at these events. I made extra money by doing so.

In October I wrote home telling Mom and Dad I had eaten shrimp on the 12th and had become violently ill. This was to turn into a life-long allergy.

Two of my friends played intramural softball to earn their required one credit hour for physical education. I signed up and asked to be put on their team, the "Ute Hall Beatnicks." I tried fielding and hitting and was not very good at either, but I was asked to learn how to pitch underhanded. I practiced a lot and could do OK for a few innings. In my freshman year I think we were in the middle ranking of the league by year end. More on softball later.

I previously mentioned my college roommate, Sam Rhodes, from Denver. We each had half of the dorm room with a bed, dresser, and desk. I quickly learned that studying in my room was nearly impossible for me as every evening in the dorm was one big party with students going in and out of each other's rooms, drinking, playing loud music, and generally bothering anyone trying to study.

Sam and I were both members of the Young Democrats Club and the International Relations Club. He was very popular on campus. He moved off campus for the Spring Quarter. He also had a car, and I rode with him on occasional trips to Denver.

In January 1960, Sam went home for a week after he was notified that his father had a serious heart attack. Fortunately, his father recovered.

Also in January, I was selected as one of three speakers at an International Relations Club meeting. We all three spoke about Cuba and how the USA could better interact with the country's leadership. A newspaper article at the time reported there was lively discussion following our three talks. The following is a transcript of my talk,

FIDEL CASTRO

Just a little over a year ago Fidel Castro made his triumphant entry into Havana, Cuba after having succeeded in wresting the government away from the none too popular Dictator, Fulgencio Batista.

It was in 1947, while he was a law student at the University of Havana, that Fidel first got involved in politics and efforts to overthrow a government by force. At that time, he joined an expedition to help depose Dictator Rafael Trujillo of the Dominican Republic. After this attempt failed, he returned to Havana and completed his studies, receiving his law degree in 1950.

Three years later he was fighting again. This time with his brother Raul and a small group of men he attacked a Cuban army barracks. During the fight both Fidel and his brother Raul were captured and sentenced to 15 years in prison, but only served two because of an amnesty.

After their release they came to the USA for a short time but then moved to Mexico where they planned and organized an invasion of Cuba.

To say the least their revolution started badly. Out of 81 men, only 12 survived the surprise attack during their landing on December 2, 1958.

Today, only three of these men remain: Fidel, Raul, and the powerful Ernesto "Che" Guevara, an Argentine by birth.

Raul Castro is a very dedicated communist and also is violently anti-U.S. His wife, even though she attended an American college, is also a first-class hater of the United States.

Raul is also known to be a far more capable organizer and administrator than Fidel—thus making him potentially more dangerous. He acts only after carefully thinking over what is to be done, not in the quick-tempered and emotional way Fidel does. He is also the military boss of Cuba's 35,000-man army. It seems Raul is well qualified to take the place of his brother. This is not too unlikely either since Fidel has named Raul as next in line if something should happen to him.

As for Ernesto "Che" Guevara, he has control of everything not held by the Castro's. He heads the National Bank, the Cuban Police, the new Industrialization Program, and the National Agrarian Reform Plan, the agency directing the massive redistribution of land to the country's peasants. He has been called many times the "Economic Czar of Cuba."

"Che" Guevara is a highly trained medical doctor but in his own stated opinion: "It is better to save countries than lives." He has participated in many Latin American revolts, but this was by far his most successful. These three people now hold Cuba in their hands. What they will do and accomplish will be of the greatest interest in the years to come.

As a member of the Young Democrats Club, I had the opportunity to attend a two-day conference in Denver on February 26 and 27, 1960. It was a seminar sponsored by The Citizenship Clearing House and it focused on current problems in government. There were two of us from the Young Democrats and two from the Young Republicans Club as well as a faculty advisor and his wife. Also in February, a test was given in all the English classes on campus called The National Listening Test. It was like the Readers Digest Tests in high school. When the results came back, I was informed by my advisor that I had achieved the highest score in the school and was in the 98th percentile in the USA.

One day I saw a sign on the bulletin board that said, "Organization meeting and election of officers for a chess club." It gave a date, time, and place for the meeting. I had grown up playing chess and had brought a chess board and an inexpensive chess set with me to school. I attended the meeting with only four or five others and we each talked a little bit about our experience playing. I got elected President of the club and my room and the room next to me became places where we would play on our study desks.

James Sasaki, from Taiwan, lived in the room next to mine and his roommate played chess. James spoke perfect English and he could study even when mayhem was taking place in his room. He had turned his desk so that his chair was in the corner and the desk was across the corner facing out into the room. He would sit in his chair and study regardless of what was going on in the room. Even when someone tried to talk to him, he would ignore them, and he appeared to be able to simply block out the noise around him.

James would, however, often watch us play and would occasionally ask a question or two. He told me he had never played before but would like to learn. A few weeks went by and he had watched perhaps 40 or 50 games. One evening, only one person showed up to play and after one game he left. James's roommate said he had to go somewhere, and James asked me if he could play a game with me. I said he could, but I thought I was going to have to tutor him on the finer points of the game. It quickly became apparent that his observation of games had paid off, as he did not think about his moves and took only a few seconds to make his next move. He beat me that evening. In fact, I never won a game against him our entire freshman year. The next year, I was the vice president of the club and James transferred to some school in California.

We tried to register the Chess Club as an official organization of the college but found that the requirements included the club's being in existence for one full academic year. I still have the constitution and by-laws we prepared as one of the requirements. We gave up on the idea and determined we would do it the following year. As I mentioned, I joined the Young Democrats Club, which was a registered organization, and I still have my membership card. I also belonged to the International Relations Club, another registered organization..

In early March 1960, I received a letter notifying me that I had not yet registered for the draft and needed to do so without delay. So on March 9, I filled out the form I had been sent and mailed it in. A couple of weeks later my father called and said I had another letter from the selective service. I asked him to open it for me and it turned out to be my selective service registration card which I still have today.

Since I went to college without a car it was necessary to hitch rides home with other students. There was a bulletin board where students would post where they were going, how many passengers they were willing to take, when they were leaving and returning (usually to depart Friday afternoon and return Sunday afternoon), how much they wanted to be paid, and how to get in touch with them (usually a dorm room and building). I tried to go home every few weeks and if Sam was not going, I would look at the board and try to find someone who was going somewhere north of Denver so I could be dropped off on the way.

A couple of these trips turned out to be rather harrowing experiences. One time I went with two seniors. They stopped in town and purchased two six packs of bottled beer. The road to Denver went over three mountain passes. The first was Monarch Pass about 40 miles outside Gunnison, then another 80 miles to Red Hill Pass, and then about 20 more miles to Kenosha Pass where the road down the far side was winding with sharp hairpin curves for about 30 miles. I was sitting in the back seat, and we had no problem getting over Monarch Pass but by the time we reached Red Hill they were on their second six pack of beer. I asked them several times to slow down, and they just laughed at me. They were throwing their empty bottles in the back seat and seemed to take pleasure in driving on the wrong side of the road where you could look down into the valleys. By the time we reached the summit of Kenosha Pass and started down the other side both six packs were gone, and the driver was all over the road. I finally laid down on the floor of the back seat, and the driver seemed to take pleasure in twisting the wheel back and forth trying to throw me from one side of the car to the other. I was sure I was going to die that afternoon, but we finally got to our home in Westminster. I told the two boys NOT to bother to come pick me up on Sunday. After describing the drive home to Dad, he agreed I should take a bus back to Gunnison on Sunday, which I did.

There were a couple of other memorable trips home. One of the young men who was in the chess club was a black freshman from back east somewhere, may have been Pittsburg or Philadelphia. Western State was a virtually all white college with only three or four non-white students in the whole school and not a single black faculty member. The minority member of the chess club and I became friends, and he played chess in my room often.

One evening I told him about my conversation with my Dad about Negros, and he said he was not at all surprised as in high school he had friends whose fathers felt the same way. Over the following few weeks, we hatched a plan for him to visit home with me. After a month or two, we found a ride and left together on a Friday afternoon. I told Mom and Dad that I was bringing a friend home with me for the weekend, but I neglected to tell them he was black. When we arrived, Dad was not home but after a look of real surprise, Mother welcomed him into our house and showed him the bedroom where he would be staying. I asked Mother if I could borrow her car to show him around the Belleview campus and she agreed.

We came home about an hour later and I could see Dad's car was in the driveway. I told my friend I was not sure how Dad was going to react, but if it got ugly, Willard Staats, to whom we had just talked, still lived in a big house on campus and he had offered to let us stay in his home if needed. Apparently, Mother had briefed Dad on the situation, and I suspect warned him he had better not make a scene. We went into the house, I introduced my friend to Dad, and low and behold we sat in the living room and had a perfectly civil conversation. Mother cooked a nice supper and later Willard picked us up and we went to a movie. We went with Dad out to our farms and did a little work for him on Saturday. We did not go to church with the family on Sunday as our ride was going to pick us up about noon. As I recall I never remember Dad even mentioning the visit again although this was the only time that I pushed my luck.

One other memorable ride was in January or February of 1960. There wasn't much snow on Monarch Pass on the trip going to Westminster but apparently it had begun snowing around the pass by mid-morning on Sunday. There were four of us in the car as two of us had paid to go. There are

COL(RET) GERALD W. "JERRY" SHARPE

certain times of the year that tire chains are required to drive over Monarch Pass. We arrived at the pass about 3:00 pm and the road already had a coating of snow. We were stopped at the checkpoint, and as I recall, were told that chains were not yet required but were recommended. The driver told the policeman that he was used to driving in snow and would at least try it without chains. We went off the road once, but with a little pushing from the three passengers we got going again. We finally got to the top of the pass with a lot of slipping and sliding.

We started down the other side and as slow as the driver tried to go, we slid off the road twice. We finally convinced the driver to stop and put on the chains on one of the flat areas going down. We ran out of snow near the bottom and pulled over and removed the chains, but we should have put them on at the checkpoint before starting up the mountain. I did my best after that not to look for rides if snow was forecast for the weekend.

Western State College has what is claimed to be America's largest college emblem. On the side of what is called Tenderfoot Mountain above the campus is a large "W" built by students in 1923 with rocks carried up the mountain. It stands 450 feet from top to bottom. Students whitewash the W annually in early May on what is called "W Day" when they carry 50-pound sacks of lime and water in 5-gallon cans up the hill to be mixed on the site. I recall "W Day" on May 10 as one big picnic and a day off from school. It was also a day of a few lime burns, as lime is very caustic.

In May of each year elections are held for the student council for the following year. I filled out a nomination form and got the required 50 student signatures to co-sponsor my nomination. I was later pleased to learn that I had been comfortably elected and would assume my position at the start of my sophomore year.

After my freshman year I wanted to find another summer job. I checked with the construction company I had worked for previously and was told they were not hiring. Dad had a friend named Jim Arthur who used to be in the Pillar of Fire Church but had left with his family of five children and moved to Wyoming. He was working for the Argo Oil Corporation outside of Thermopolis.

He came back to Colorado for a short visit, and Dad asked him if he thought there might be a job for me in the oilfield. He said he did not know but the company had its offices on the top floor of a tall office building in Denver, and he would get Dad an appointment with the president. A day or two later we went downtown and took the elevator up to the 20 something floor. We were escorted to a very large corner office that overlooked the city of Denver. Once again Dad was wearing his black suit and collar, and the president asked us to sit down.

Dad explained a little about the church, and as I recall the president said he enjoyed listing to classical music on KPOF, the church radio station at Belleview. Dad told him that since ministers received no salary, he was struggling to pay for college and anything he could do to help by providing me a job for the summer in the oilfield in Wyoming would be greatly appreciated. The president, after a time, said give me a minute and he called someone and asked about adding another employee to a special crew. After he hung up, he told us that yes, they could give me a job and I could start the following Monday. He said they had four sons of corporate vice presidents working on a special project and I would be the fifth person on the crew. He also said it paid $2.00 an hour but I would have to find a place to live. That was more than what I had received as a carpenter's helper, and we left ecstatic that the visit had been successful.

We drove to Thermopolis on Sunday June 19, 1960, and Jim Arthur agreed to let me stay at his home which was on a small farm about ten miles from the oilfield and about 15 miles from Thermopolis. He agreed I could go to work with him every morning and his wife would make me lunch. He showed me the bedroom where I would sleep. It had two bunk beds, and I was to have a top bunk and share the room with three of his younger children. Dad worked out an agreement for me to pay a monthly rent for food and lodging in the small three-bedroom house. I spent Monday in-processing, and on Tuesday I joined the crew of the four other young men, all of whom were sons of senior officers in the corporation.

This was in the days before automatic controls, so the oil flow had to be manually stopped when the 10,000-gallon tanks were full. Unfortunately, the people operating the controls were either lazy or did not know their jobs very well as it appeared that every tank had oil running down the side below the lid that was reached by metal steps going up the side of the tank. It was clear the flow was not being turned off in time. We were told we would spend the summer cleaning this oil off the tanks with a sprayer contraption that had a large tank of an oil softener they called "gunk" and a second large tank filled with solvent. We were given long handle brooms and were shown how to operate the engine, turn it on and off, and switch tanks.

We learned the process was to start at the top spraying gunk on the oil to soften it, then using the long handle brooms we brushed the oil loose, then using the solvent we would wash the tank clean. As I recall it took about a day to clean one tank and there were about 30 tanks in the field. By the time we reached the last one, the previously cleaned tanks looked like they had never been cleaned.

The boys on the crew were treated with great respect and were called Mr. XXXX. I was a mystery, as no one could find a Sharpe on the Argo Oil Corporation organization chart and I was perfectly happy to leave it a mystery for both the boys on the crew and the workers. It took about a month for the story to come out as to who I really was and after that I was simply referred to as "Sharpe" rather than Mr. Sharpe.

I am not sure why, but I decided to return to Denver for the weekend of July 1 and 2. In a letter home on June 29 I said, no I did not get fired but had decided to come home for the weekend. I wrote that I would arrive by train at 8:00 am Saturday morning and asked if Dad could pick me up at the main Denver train station. I also asked mother if she could arrange a dinner for Cheryl Treinen and her parents on Saturday evening at our house, and if they could not attend a dinner, to ask Cheryl's mother if she could arrange for Cheryl to be at home on Saturday night, without telling Cheryl I was going to visit. I have no recollection of what happened on the visit. I have been told recently that I was taken back to work by Willard Staats's older brother Edwin and his new wife, Marilyn Schissler, who has recently passed away. Edwin reported that I volunteered to pay them about what my train ticket was going to cost and that it was enough money to almost pay for their honeymoon to Yellowstone National Park.

I saw a notice on the bulletin board on the 19th of July that there was a room for rent right on the oilfield property. There were about 20 small houses for the staff. The house address was listed as well as the name and position of the lady advertising the room for rent. She was the secretary of the field manager, a divorced lady with a four-year-old daughter. I went to see her on our morning break, and she said she could show me the room at lunchtime. I asked about meals, and she said she would be happy to add meals to the agreement as she already had to prepare three meals a day.

It was a nice three-bedroom, two-bathroom house which meant I would have my own bedroom and bathroom. I immediately agreed to pay $50.00 a month, and we agreed I could move in the next morning. I told Jim Arthur on the way home and he had no problem with it as he understood sleeping on a bunk bed for more than three weeks and sharing the one bathroom in the house left a lot to be desired.

After moving in I quickly learned that the daughter had her mother very well trained to do what she wanted her to do. On my first night, the mother put her daughter to bed about 10:00 pm, about an hour later than she said was her bedtime. The mother had no sooner left the room than the little girl stood up in her crib and started screaming at the top of her lungs. The mother just ignored her for a while but finally went in and got her out of bed, held her for a while until she calmed down, and then she put her back in bed. This sometimes happened two or three times before the little girl finally went to sleep. In addition, whenever the mother went out the door to hang up clothes or talk to someone in the front yard, the little girl would also stand at the door and just scream at the top of her lungs. When the mother came back in, she would pick her up, sit down with her in a rocking chair, and get her calmed down after which she was perfectly happy to play with her toys.

About a week after moving in, the mother asked me if I would mind babysitting her daughter while she went on a date with one of the assistant managers. I assured her I would love to babysit her daughter. The mother got ready to go, kissed her daughter, and went out the door. The daughter immediately went to the door and started screaming. I stood behind the daughter waiting for her mother to get in her date's car and watched them drive away. I picked the little girl up, put her in her highchair still screaming at the top of her lungs, and went to the sink and got a glass of water.

I walked back over to the little girl and told her to stop screaming. It had no effect on her, so I threw the entire glass of water in her face. She immediately stopped screaming and got this shocked look on her face. I got her out of the highchair and took her into the living room where her toys were and played with her for a couple of hours. Her mother had suggested I put her to bed at 9:00 pm, so I told her it was time to go to bed.

We put her toys away, and I carried her into her bedroom and laid her down in her crib. Before I even got to the door she was up screaming as she did every night. Once again, I went to the sink, got a glass of water, and walked into her bedroom. I said, "Stop screaming." It had no effect on her whatsoever, so I brought the glass of water out from behind my back and once again threw the whole glass full in her face.

She immediately stopped screaming and again had this same shocked look on her face. I again laid her down and told her it was time to go to sleep. I did not hear another sound from her. When her mother came home a little after 11:00 pm she asked if I had any trouble getting her to go to sleep. I said I had none and did not elaborate. From that day forward if I was home, all I had to do was hold up a glass and she would stop screaming.

The next two nights I suggested to her mother that she let me put her daughter to bed. When I put her in her crib she would lie down quietly. Her mother asked me what I had done to her daughter, and I said, "Nothing." I only told her that I had convinced her daughter I was not interested in hearing her scream. She was very pleased with her daughter's change in behavior as once or twice when her mother put her to bed and she started screaming I got an empty glass, walked to her door, and held it up. She would immediately stop screaming and lie down. The rest of the summer was very pleasant, and I enjoyed playing with the little girl most evenings and I babysat for her about once a week.

The entire oil field was covered in chunks of rock. At every break and at lunch after eating I would practice throwing rocks underhanded. By summer's end, I could pretty much hit whatever I was throwing at and I could throw farther underhanded than a couple of the boys could throw overhand. This practice served me in good stead when I returned to pitching intramural softball in my sophomore year at Western State. I worked until the first weekend in September and Dad drove up to bring me home.

When I returned to Western State in September of 1960 for my sophomore year, I was assigned a dorm room on the third floor of Ute Hall. Because of the large influx of new students, the dorms were very crowded, and they had replaced one of the single beds in some larger rooms with bunk beds to allow three students to live in the rooms. After much negotiation with the other two students in the room, I was given a place to hang up my clothes and I put my things away.

When I went to work at the dining facility, one of my coworkers told me about several places off campus to rent. After the evening meal, I looked at one that was a large one-bedroom studio apartment with two beds, two closets, a stove, refrigerator, kitchen table, two study desks, two bookcases, and a couch. As it turned out, my roommate would be one of my good friends from my freshman year. I asked the lady who owned it how much the rent was, and she said $22.50 a month. This turned out to be a few dollars less than my dorm room. I told her I had to go to the student housing office in the morning to get permission to move. She agreed to hold the apartment until lunch time. I got permission to move and went back and paid for the first month. Fortunately, my new roommate had a car and that evening he helped me move out of Ute Hall and into the apartment.

Girl watching was one of our favorite pastimes in the early weeks of my sophomore year. If we were not serving, we would stand behind the serving line and quietly grade the new freshman girls on looks as they passed down the line getting their food. There were only a few "9s" or "10s" but one particular freshman young lady got a unanimous "11" from us. Over the next couple of days there was a lot of discussion as to whether anyone had the nerve to ask her out on a date. No one had seen her with anyone but girlfriends around campus or in the cafeteria, so we were fairly certain she did NOT have a steady boyfriend. I did not participate in any of these discussions as my dating history left a lot to be desired and I quickly concluded that there were guys on the crew with a much better chance of getting a date with her than I did.

About day three or four of the discussions, no one had gotten up the nerve to ask her out. I am not sure what prompted me to do it, but I went over to her table where she was sitting alone, introduced myself and asked her name and where she was from. She told me she was Rosalie Lynnscott from Englewood, CO, very near Westminster. After a bit of small talk, I asked her if she would be interested in going swimming with me on Saturday evening. As I recall she smiled and said she did not know how to swim, but yes, she would be happy to go. I agreed to pick her up at her dorm. I was quite the hero when I went back with a big smile on my face telling the crew she had said yes.

On Saturday evening I walked over to her dorm, met her in the lounge, and we caught a ride with one of her girlfriends to a pool just outside of town. After changing into our swimsuits, we went to the pool and waded into the shallow end. I said let me show you how to swim first. I took two or three strokes kicking my feet in the shallow end and then stood up. I suggested she try it while I supported her by keeping a hand under her on her stomach. She tried this and swam a few feet and stood back up. She told me she had actually been swimming some in high school but was not a very good swimmer.

We had an enjoyable evening, and I was again quite a hero back at the dining facility. She agreed to go to the movies with me the following Saturday, but on the way there she told me 14 different boys asked her out during the week. Again, we had a good time, but she started dating one of the freshman wrestlers.

Soon after, I met a second girl, Sandi Jones from Denver, in my advisor's office (we had the same advisor) and I asked her out that evening to a student council get acquainted dance. Neither of us knew how to dance very well so we simply sat, had a few sodas, and got acquainted. I did not do any more dating the rest of the year that I can recall, either because I was too busy or too bashful to ask.

When I returned to Gunnison in the fall, I no longer had a job stocking vending machines in my dorm and the one next door, but I was fortunate to still have my job in the cafeteria. The college had a small six-lane bowling alley in the Student Union, and the pins were set by hand rather than by a machine. I had enjoyed bowling on occasion during my freshman year, and the first time I went back, there was a small sign on the desk where we paid to rent shoes and bowling balls that said, "Pinsetters Wanted." I asked the young lady behind the desk about the job, and she gave me an application form. I sat down and filled it out and had to take it to the administrative office. I was told they were looking for two or three more pinsetters so they could set up a schedule. I was hired that same day.

I forget what we were paid, but one pinsetter had to set the pins in two alleys. When the ball was thrown, you had to jump down in the pit, pick up the pins that had been knocked down and put them in the proper spot in the rack, and then pick up the ball and put it in an elevated chute that allowed the ball to roll back to the rack where the bowlers once again picked them up for their next throw. I can honestly say it was one of the most physically exhausting jobs I ever had, as I never stopped jumping back and forth between the two alleys for which I was responsible.

One night my two lanes were taken over by four students from Iran. After bowling their first ball they were supposed to wait for me to pick up the pins they had knocked over and then return their ball to them. When the first bowler threw his ball, I jumped down in the pit and started picking up the pins. I heard the sound of a ball being thrown and glanced up the alley to see a second ball coming down the lane. I dropped the pins and jumped up on the wall just as the second ball hit. The bowler on the second lane threw his first ball and once again I jumped down to pick up the pins, but looked up the alley to see a second ball headed my way once again. After jumping out of the way a second time, I then picked up all the pins in both alleys and put them back on the rack. Instead of putting the balls in the chute to return them, I threw all four of them back up the two alleys at the bowlers.

After a lot of yelling the manager came back and asked me what was going on. I explained what had happened and he went back out, got the four bowlers together, and explained that if they ever threw a second ball while the pinsetter was picking up the pins they would be banned for the rest of the year. I became the hero of the pin setters and from then on if someone threw a second ball while we were down in the pit, the two balls were thrown back up the alley. The managers already knew they did not need to come back and ask, they would immediately solve the problem. I quit the job in the bowling alley at the end of the winter quarter.

As I previously mentioned, near the end of my freshman year, I had been elected to the Student Council for my sophomore year. Sam Rhodes, my roommate for the first two quarters of my freshman year, was also on the council and later in the year he was one of 20 students from Western State to be listed in the Who's Who of College and University Students in America.

The council was broken down into four sub-committees. The first was the Student Affairs Council that was made up of ten professors and three students. I was appointed to this subcommittee at our first meeting. Second was the subcommittee on Campus Services. This had three professors and three students. I was appointed to this subcommittee by the chairman of the Student Affairs Council and was told to find two additional students that were NOT council members to serve. At the next meeting I proposed two names, and the council accepted them.

The third subcommittee was on Campus Life. This also had three professors and three students that included the Student Council President and Vice President. Fourth was the subcommittee on Scholarship and Admissions. This had three professors, three individuals from the admissions office, and three members of the council. In rereading the minutes of council meetings published in *The Top of the World,* the campus newspaper, it is clear that I was very active on the council. My name is recorded often in making or seconding motions and submitting items for the council's consideration. I was appointed to the planning committee for the annual Christmas party, I was appointed as a judge to the Student Traffic Court, and in March I was one of six council members appointed to revise the Council Bylaws. In May I was reelected to the council to serve in my junior year, and I was appointed as one of four council members to plan the October 6, 1961, homecoming activities (a dance, a parade, and the election of a king and queen).

During the first week of classes in my sophomore year, I was asked by one of my classmates, Don Gerardi, who also worked with me in the cafeteria if I would consider helping out as a sports reporter. He said he was the only one doing it and there were often events going on at the same time. I told him I had never done any reporting but would be willing to give it a try. In the second issue of *The Top of the World,* dated September 30, Don and I are listed as "Sports Reporters."

We had to turn our proposed stories into the editor for review, and he would mark them up in red ink to make them more readable. I quickly learned what he was looking for, and by the third week he asked me to wait while he read what I had written. As I recall I turned in two or three stories and he had almost no corrections. He asked if I would be willing to be the sports editor of the paper and have Don and anyone else I could recruit work for me. I agreed, and by the October 19th edition of *The Top of the World,* I was listed as the sports editor and Don as a sports reporter. I kept the position for the rest of the school year through three editors and considerable turmoil in staffing the paper.

The first sport of the year was football, and the team had recruited a 300-pound lineman named Vernon Singleton who had come highly regarded as a sophomore transfer from a junior college. The first game of the season was against Chadron State College from Chadron, NE, which had a 200-pound running back that had been all conference the year before. On the first three or four offensive plays, this running back ran straight at Vernon and each time seemed to knock him down and run right over the top of him. Around the fourth play of the game, Vernon was carried off the field on a stretcher and did not return. Chadron State won 13 to 7. To be fair, Vernon came back and played well the rest of the season and was selected as player of the game once or twice. As I read now what I wrote each week in the papers, I am surprised that I felt at liberty to comment in editorials about the school's sports programs and about the lack of student attendance at sports events. Both serving on the student council and working on the student newspaper prepared me for later tasks in the Army and beyond.

During my freshman year I became friends with two members of the college wrestling team. When

I signed up for my sophomore courses, I reviewed the classes that were available for the required physical education course, and saw that wrestling was being offered in the fall semester, so I signed up and really enjoyed learning about the sport. The freshman wrestlers, who were not competing on the varsity level, were tasked to attend our classes and help the instructor.

In the spring semester, I felt comfortable enough to sign up for intramural wrestling and competed in the 130-pound class. I do not remember my win/loss record, but I did OK. I rejoined the same intramural softball team, and we finished second behind a team made up of military veterans. My pitching had improved immensely over the summer, and I believe we lost only two games of the ten or so I pitched.

I also played on an intramural flag football team. This was supposed to be a no contact sport, as you were to simply grab one of the flags that were attached on both sides of a belt by snaps. Unfortunately, some boys thought that tackling you to get the flag was easier than trying to snatch it off your belt.

As a member of the Young Democrats Club, the 1960 presidential election was a big deal to me. John F. Kennedy was the first Catholic ever to run for president, and if he could be elected, he would become the youngest president ever. We handed out literature all over town and on campus for weeks before the election. Kennedy was opposed by Vice President Richard Nixon. Election day and night on November 8, 1960 found me working in the Gunnison County Democratic Headquarters. I had a car to use and delivered voters to polls and returned them home, delivered ballots to voting places, picked up copies of lists of people who had voted, and then checked the names of those registered. We would call those that had not voted and offer to take them to the polls and back home. After the voting places closed, we would call and try to get tallies from each precinct in the county and post them to a big board.

We, of course, watched television to try to keep up to date on the national race. We worked until 4:00 am and then closed up the shop and went home. We were thrilled to learn Wednesday morning that Kennedy had won in an extremely close election with only 117,000 votes separating the candidates. Kennedy won 303 electoral votes to Nixon's 219. In an unusual twist, 15 electors from Alabama and Mississippi voted for Senator Harry Byrd, Jr. It was the closest election since 1884 when Democrat Grover Cleveland won by only 24,000 votes. We were disappointed to learn in the morning that we had lost Gunnison County 1,296 votes for Nixon to 1,044 for Kennedy. Nixon won the state 402,242 to 330,629. I was invited to attend the Colorado Presidential Inauguration Ball in Denver but was not able to go. I have kept the invitation as a cherished souvenir.

On Saturday, December 17, 1960, I was riding back to Westminster with my high school classmate, Art Jacobson. He had enrolled at Western State a year after I did and lived in Ute Hall, just as I did as a freshman. In a recent message Art wrote, "I recall vividly, you and I were traveling to Denver that fall from Gunnison in my '50 Chevy, one Friday afternoon, to go to our homes in Westminster, when, a few miles west of the bottom of Monarch Pass, we heard on the radio that a large plane had crashed into a church called The Pillar of Fire. I looked over at you, as you said, 'That's where my grandfather is,' or words to that effect. I remember the look of surprise and despair on your face. What a coincidence."

The news reported that a crash between two airplanes had taken place over New York City the previous morning. A United Airlines DC-8 jet enroute to Chicago from Idlewild (now JFK) airport, collided with a TWA Super Constellation propeller plane flying from Columbus, OH to LaGuardia Airport. The TWA plane exploded on impact, likely instantly killing all 44 on board. The plane broke into pieces and fell onto Miller Field, a military airfield on Staten Island. The crippled United Airlines plane managed to remain in the air for another eight and a half miles before crashing onto Sterling Place and Seventh Avenue, setting fire to over a dozen buildings and killing five individuals on the ground, including my 90-year-old great grandfather, Wallace E. Lewis, who was the caretaker in the Pillar of Fire Church.

It was later reported that the plane was going at an estimated speed of 200 mph when its wing struck the roof of a brownstone at 126 Sterling Place, causing the fuselage of the plane to break into two pieces with one veering to the left and crashing directly, with tragic irony, into the Pillar of Fire Church across the street. The plane exploded in flames destroying the church and killing the passengers on board. The left wing, now on fire, sheared into an apartment building next door to the church, while the other section of the cabin, filled with screaming passengers, crashed into the McFadden's funeral home on the corner of Seventh Avenue and Sterling Place.

Also later reported, it was called a miracle that the two buildings destroyed held only one person, Wallace Lewis. If the plane had crashed just a few feet in any other direction, hundreds could have been killed in the crowded apartment buildings next to the church and funeral home. One additional miracle took place in that crash. It is believed that all the occupants were killed almost instantly upon striking the buildings; however, an 11-year-old redhead boy, Stephen Baltz, from Wilmette, IL, who had been put on board by his mother and father, was thrown from the plane and landed on a snowbank where residents rolled him in the snow to extinguish his burning clothes. He was conscious but sustained severe burns and broken bones. A lady by the name of Dorothy Fletcher shouted at her neighbors to throw down some blankets and a coat which she placed on the boy. She then took the boy to the nearby Methodist Hospital. The boy was conscious and described the crash to doctors and later to his parents and grandmother. He was quoted as saying, "I remember looking out the plane window at the snow below covering the city. It looked like a picture out of a fairy tale book. Then, suddenly, there was an explosion. The plane started to fall, and people started to scream. I held onto my seat until the plane crashed, and that is all I remember."

Doctors did all they could to save him, but unfortunately, he died the following afternoon. A picture of Dorothy, standing over the boy with an umbrella shielding him from the snow, appeared on the front pages of New York newspapers the next morning. A small bronze memorial to the boy, that includes the burned and blackened 45 cents that was in his pocket at the time of the crash, is displayed in the lobby of Methodist Hospital.

Each year two famous conductors were invited to lead the college orchestra in concerts. One was held in February and the other was held in March of 1961. During the February concert, Arthur Fiedler of the Boston Symphony was invited. The concert was a two-day affair with the college orchestra playing the first evening and the Colorado High School All-Star Orchestra playing on the second evening. This orchestra was made up of the students who won a statewide contest for their instrument or in some cases were in the top three or four.

I was tasked to write about the concert even though I was the sports editor of the college newspaper. I attended all the rehearsals I could when I was not in class. During one of the rehearsals for the high school orchestra, Mr. Fiedler banged his baton on his music stand during one of the loudest portions of the piece being practiced and everything went quiet. He came down off his podium and walked through the orchestra to the back row. He said to a very frightened young lady in a very loud voice that I could hear where I was sitting in the back of the auditorium, "You played a b flat and it was supposed to be a b sharp, please do it right!"

He went straight back to the podium, and they started over. I was mightily impressed that he could not only hear an instrument in the back row but could detect a minute difference in the note being played. I was very fortunate to be able to ask him a few questions after the concert for the newspaper article I would write, and he was kind enough to autograph my program, which I still have today. During March the guest conductor was Mr. Ferde Grofe. He conducted some of his compositions that included his *San Francisco Suite*. Perhaps his most famous piece is the *Grand Canyon Suite* that had been played in a concert he conducted during the summer band camp at Western State.

On one of my trips home during my sophomore year, I had a long discussion with Dad over his statement to me that the only true religion in the world was the Pillar of Fire Church. My reaction was one of amazement. I asked him if he thought God was a fair and just God? Dad said he thought he was. I then asked if he thought a fair and just God would pick one small religion of at the most about a thousand members to be the one true religion on earth and let the other hundreds of millions of people of other religions go to Hell because they did not even know there was such a religion as the Pillar of Fire. His answer was that people of all other religions could go to heaven as long as they believed in God and lived exemplary lives and had never rejected the one true religion.

I said OK just for argument's sake, let's say the one true religion in the world was really the Watusi religion in Africa, but you did not know this. One day your doorbell rings and a black man, a member of the Watusi tribe, appears at your door and asks if he could have a few minutes of your time to let him tell you about the one true religion in the world. I asked Dad what he would do. He said he would tell him no thank you and shut the door. I said, well, by your previous statement, you would now be going to Hell because you rejected the one true religion in the world. He had no answer so I also told him that it seemed to me that if he truly believed that concept, then sending Pillar of Fire Missionaries to Africa to tell them about the one true religion in the world was not very fair as it was likely 95% of the people would reject the idea that their religion was not the one just like he had. Again, I told him it seemed to me that what the missionaries were doing was simply sending people who rejected the message straight to hell. We did not get any resolution to this subject, but I honestly believe that Dad was convinced he was right.

After my sophomore year at Western State, I returned home and told Dad I would be looking for a job because one of the campus ministers, Reverend Ellsworth Bradford, who was also the father of Bobby Bradford, my high school classmates, had a shiny black 1941 Chevrolet coupe for sale that he had completely restored. I had talked to Rev. Bradford about buying it. Rev Bradford told me how much money he wanted, and I asked if he would hold it for me while I earned the money during the summer. He agreed to do so and I returned home to inform Dad I had to find a job to earn at least enough to buy the car as I was NOT going back to school without one. Dad said, "How about you work for me during the summer, and I will buy the car for you." I agreed and proceeded to work every day doing whatever my father wished. I cleaned out horse stalls, repaired railings on fences, raked the yard at Hilltop (one of Dad's two 80-acre farms), and dealt with boarders who had questions or needed help loading or unloading horses. I worked around our home as well, mowing the grass, watering the garden, and helping remove weeds from the vegetable garden.

As the summer neared its end and school would soon start at Western State College, I began questioning my father as to when he was going to get me the car he had promised to buy in return for my labor. I would remind him of his promise, and at first, he would say something like, "Let's talk about it later." After a few days of that answer, I brought it up at the dinner table and asked our mother to help. Dad then asked me if I remembered early in the summer asking him to get insurance on his car so he could be comfortable letting me drive his car. I assured him I did remember that conversation and then asked if that meant that he was planning on letting me take his car to college rather than buying me the car he had promised. He said no, that driving it during the summer was in place of his buying me the car. I got really mad and asked Dad if he thought I was stupid. I asked him if he really thought that I would ever agree to a deal where driving his car a few times during the summer would replace the car I had worked the entire summer to buy so I could take it back to college.

He basically said that he was not buying the car and that was the end of the discussion. I stormed out of the house and went up to Willard Staats' house where I found him walking up the road to the college building. It turned out he was having an argument with his parents about something as well. We talked until late in the evening and decided we would simply join the Army.

The next morning, we went to an Army recruiter's office in Westminster, and he told us that we needed to take an entrance exam, have a physical, and choose a job. He said if we were serious, he would drive us down to Denver to the main recruiting station where we could take our entrance exam and choose a job that same day. He said we would come back the next day and take the physical and if we passed, we would be sworn in after lunch and we would be put on a bus to Fort Ord, CA where we would take our basic training.

We agreed to be tested and pick a job. He drove us downtown and we took a two-hour test. We waited while the test was scored and then we were each called back in to see the recruiter one at a time. I was told I had scored very high on the test, and I could pick any job the army was offering at the time. The recruiter gave me a book that had one-page descriptions of each job being offered.

I was told to take the book and study it while he talked to Willard. A while later Willard came out with the same book and said he had been told the same thing. After about 30 minutes, the recruiter called me back in and asked what job I would like to choose. I asked him what job would get me as far away from Denver as possible. He told me they had an option called "Europe Unassigned." I would be assigned to Germany after I completed my basic training and advanced individual training (AIT). I would be given a job upon arriving and there was no guarantee what that job would be. I said that would be fine, sign me up. I forget what job Willard chose, but I think it was something in communications. The recruiter drove us back to Belleview, gave us his phone number, and said he would pick us up at 6:00 am at our houses where he dropped us off. He told us we were allowed to take one small bag with personal items.

Dad was not home when I returned, but after supper I said I wanted to talk to Mom and Dad privately. We went to their bedroom, and I proceeded to tell Dad that I was only going to ask him one more time about the car. I asked if he was or was not going to buy Rev. Bradford's car as he had agreed to do at the beginning of the summer. He said he was not. I said fine. I have decided to join the Army instead of returning to college. He said that threat was not going to change his mind, as he did not believe I would do such a stupid thing. I said, "Well, when I am gone in the morning maybe you will believe me. And by the way, Willard is going to join as well." I left the house and went up to talk to Willard. He said his parents had agreed to everything he wanted, and he was NOT going to join. He said he had called the recruiter to let him know. Willard did join a few years later and we served together at Fort Sill, OK.

The next morning the recruiter arrived in our driveway at exactly 6:00 am and I was waiting with my bag. Everyone in the house was still asleep. He said he was disappointed Willard had decided not to join but he was glad I had not changed my mind. After arriving at the main station, I joined about 30 other young men, and we proceeded to go to five or six stations to complete our physicals. A little before lunch time the recruiter told me I had passed but four young men had not. He said that until I was sworn in and signed my enlistment contract, I could still change my mind.

Right after lunch we were all sitting in a lounge waiting for the swearing in ceremony when an announcement came over the loudspeaker system. It said, "Mr. Sharpe please pick up one of the white telephones along the wall, you have a telephone call." I picked up one of the phones and an operator told me to please hold on for a minute while she connected me. A minute or so later my mother came on the line. She was crying and sobbing and said something like please come home, Gerald; Dad will buy you the car. I thought for a minute and then said to Mother, "Please tell Dad to shove that car up his ass." and hung up. Two hours later I was sworn in, had signed my enlisted contract, and had boarded a charter bus for Fort Ord, CA.

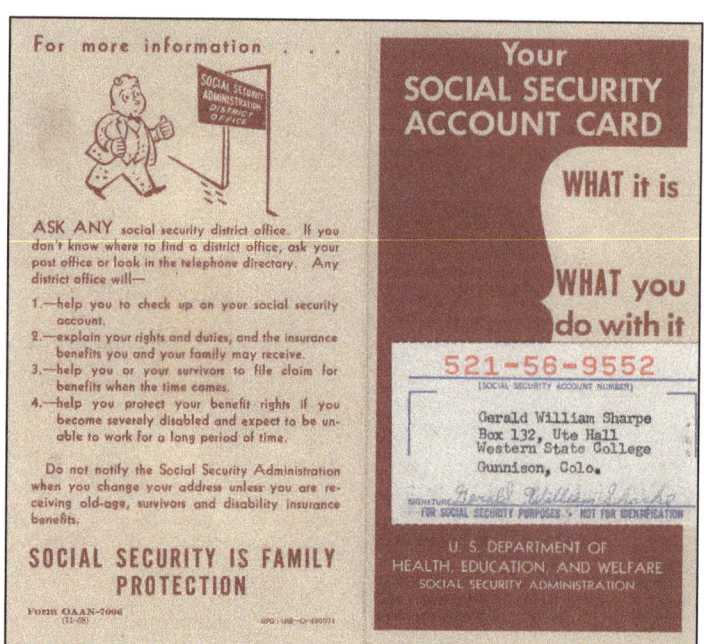

My only date to a football game.

First dance of the freshman year.

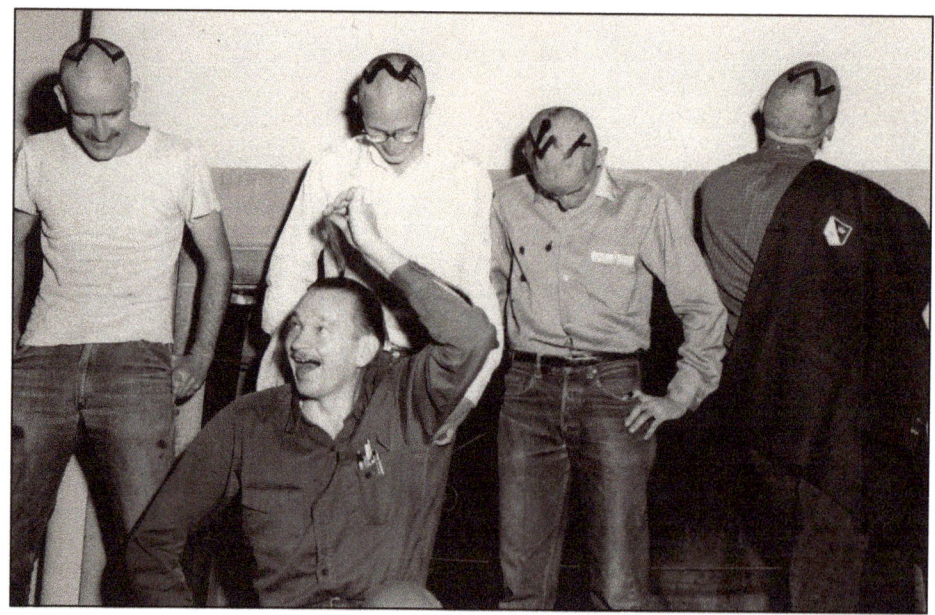

Freshman initiation, some decided to paint "W"s on their newly shaved heads.

International Relations Club

ROW ONE: Meredyth Hall, Derean Charlesworth, Sec.-Treas.; Elwin Powell, V. Pres. ROW
TWO: Gerald Sharpe, Rolly Gardner, Jim Heath, Louis Dwire, Graham Grassfield.

International Relations Club is primarily for History-Political Science students. It is concerned with stimulating consideration of International problems.

Phyllis Collins
President

Cafeteria manager, Mrs. Mickkelson

Taylor Hall, college offices.

Ute Hall, where I lived as a freshman.

Keating Hall — the cafeteria where I worked.

College Union Building, Bowling Alley.

Nancy Routt
Charla Ruland
Jim Schieb

Gordon Rowe
James Sasaki
Gerald Sharpe

Phillip Shirley
Velma Smith
Wayne Sprouse

Carol Shreeves
Emily Snell
Marna Stofford

Sharon Skoog
Jan Snyder
Dottie Steere

James Smith
Don Spickelmier
Martha Stenman

Class of 1963

197

1960 and 1961 yearbook pages with my photos.

| Mary Nay | Heskel Nazarian | Tim Neil | Leah Nelson | Bob Nichols | Tom Nomura | Joe Niutschell |

| Clyde Pearson | Lou Perkins | Judy Phillips | Jim Plank | Ronne Preston | Brian Rader | Dieter Rain |

| Byron Randall | Eileen Randolph | Jean Reed | Ray Reece | Earl Rettig | Mary Reynolds | Evie Rhodes |

| Jerry Rice | Judy Richardson | Dean Richerson | Carolyn Rook | Charlotte Ruland | Anthony Sabus | Herb Schlanger |

| Karl Schmidli | Tom Schoenke | De Lace Schwarz | Bonnie Scott | Gerald Sharpe | Penni Shepard | Sandi Shipman |

43

Top O' the World

Stan Tyler—*Editor*

NOT PICTURED: *Spring Editors*
Mary Ann Bunner and Jerry Lane.
Dr. James Kinneavy—*Sponsor*

Verdis Finch
Business Manager

Jan Bartlett

Gerald Sharpe

Don Girardi

Mary Wright

Al Silberman

My high school classmate, Arthur "Art" Jacobson also attended during my sophomore year.

Student Council

Darrel Barcus
President

JON STOUFFER
Vice President

KAREN CONKLIN
Secretary

EVELYN RHODES
GERALD SHARPE
MARYETTA SIMMONS

SUE ULLRICH
LORA LEE VAN HOUTEN

NANCY VERZUH
JO WEMLENGER

PEGGY WILSON
GENE YELLICO

On March 30, 1923, the Colorado State Normal School was renamed to Western State College. The school had a smaller "N" on a hill behind the football stadium. On May 2, 1923, the world's largest college emblem was constructed on the side of Mount Tenderfoot. In 1930 the college was allowed to purchase the entire side of the mountain from the state, 1,149 acres. Since then, the student body celebrates "W Day" the first week of each May and whitewashes the emblem.

A History of the World's
Largest College Emblem

Western State College's

W

Gunnison, Colorado

Edited and Compiled by
SAM RHOADS – VET'S CLUB

May 6, 1961

An example of the kind of pin-setting machines we had in the bowling alley in the College Union Building.

History of the W, written by my college roommate.

W Day

We Climb a Mountain . . .

Bubble, bubble, toil and trouble!

One for the money, Two for the . . .

to Whitewash the World's
Largest College Emblem . . .

Yo-o Heave Ho

College Young Democrats

Row One: Sally Walkers, Sandra Mowbery, Mary Stenman, Louise Perkins. *Row Two:* Allen Morris, Sam Rhoades, Gerald Sharpe.

Young Democrats of Western State College functions on this campus as well as in the community. They work with the senior party on their campaigns, drives and political functions.

The members attend state conventions, the legislative seminar, and the board meetings. Western State College hosts the seminar for high school students and junior colleges to teach them the two party policies.

The organization sponsors some key speakers during the year. High State and National officers are honored.

GUNNISON VOLUNTEER
FIRE DEPARTMENT
SPONSORED

BAND CONCERTS

FEATURING...

Arthur Fiedler

Renowned Conductor of the Famous
BOSTON POPS ORCHESTRA

FRIDAY March 25

EIGHTH ANNUAL:

Western State College Concert Band

ROBERT VAN NUYS, Director

ARTHUR FIEDLER, Guest Conductor

STUDENT NIGHT – I D Cards Honored   Adms: Adults 50c; Students 25c

All Tickets Sold at Box Office

SATURDAY March 26

FIFTH ANNUAL:

Colorado High School Honor Band

Conductors:

ARTHUR FIEDLER – ROBERT HAWKINS

GUNNISON NIGHT   Adms: Adults $1.00; Students 50c

BOTH CONCERTS – COLLEGE AUDITORIUM – 8:00 P. M.

WESTERN STATE COLLEGE
OF COLORADO . . . GUNNISON

P. P. MICKELSON, President

8TH ANNUAL

Firemens Concerts

WARREN MERGELMAN, Chief, Gunnison Fire Department

and

5TH ANNUAL
COLORADO HIGH SCHOOL

Honor Band

ROBERT VAN NUYS and PAUL TODD, Co-Chairmen

March 25, 1960
Western State College Concert Band

ROBERT VAN NUYS, Conductor

Guest Conductors:
ARTHUR FIEDLER — PAUL TODD

Soloist: RICHARD ABATO

March 26, 1960
Colorado High School Honor Band

Conductors:
ARTHUR FIEDLER — ROBERT HAWKINS

EACH CONCERT — 8:00 P. M. — COLLEGE AUDITORIUM

Western State College Concert Band Arthur Fiedler, Guest Conductor

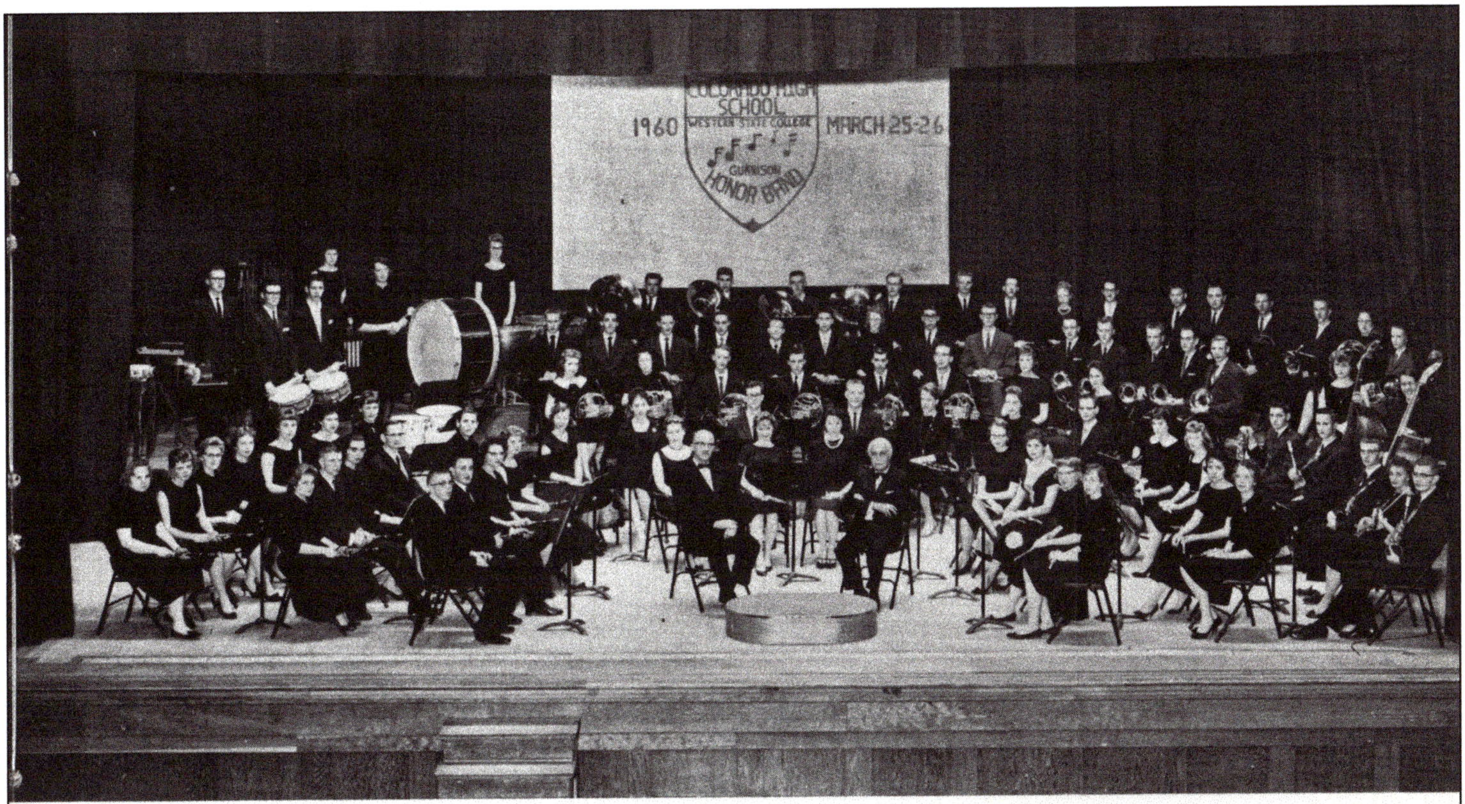

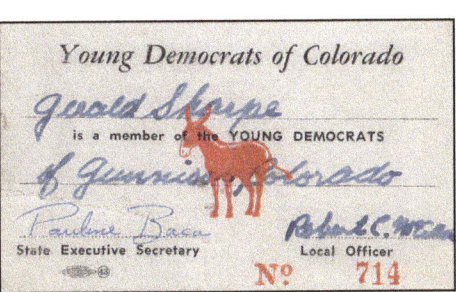

High School Honor Band

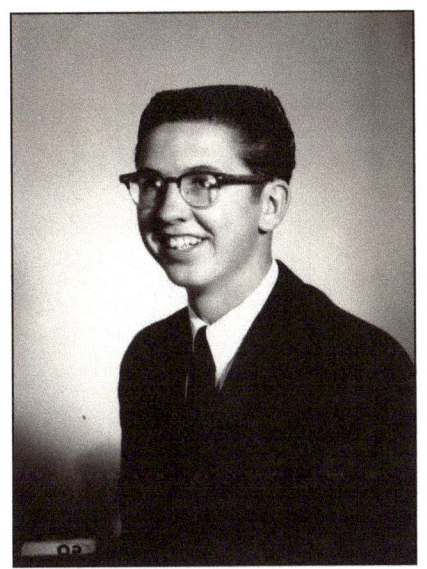

Yearbook Proof Photos

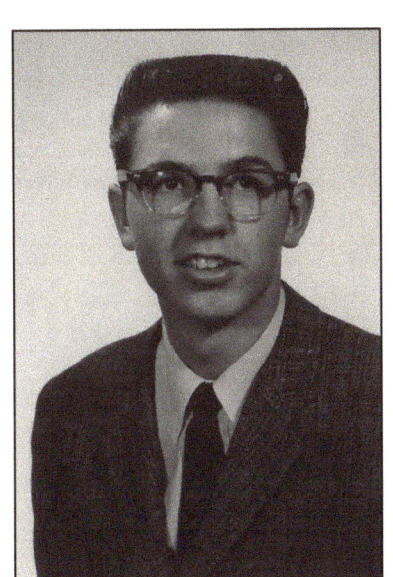

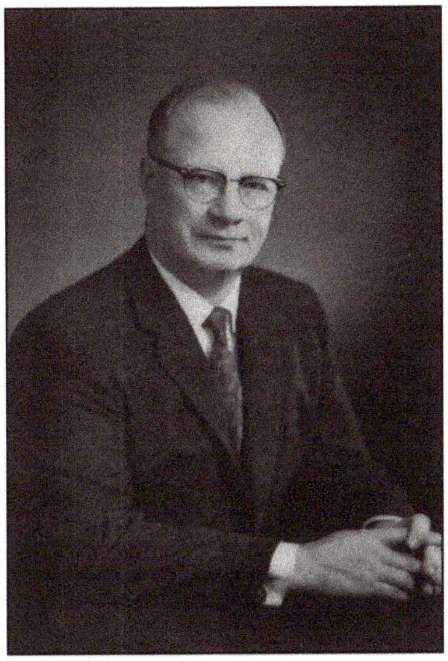

President Peter P. Mikelson

Young Democrats of Colorado

Gerald Sharpe

is a member of the YOUNG DEMOCRATS

of Gunnison, Colorado

Pauline Baca

State Executive Secretary

Local Officer

Nº 714

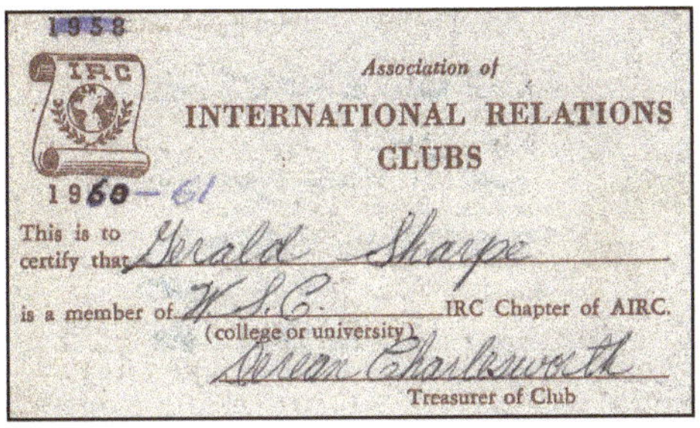

1958

IRC

Association of

INTERNATIONAL RELATIONS CLUBS

1960-61

This is to certify that Gerald Sharpe

is a member of W.S.C. IRC Chapter of AIRC.
(college or university)

Treasurer of Club

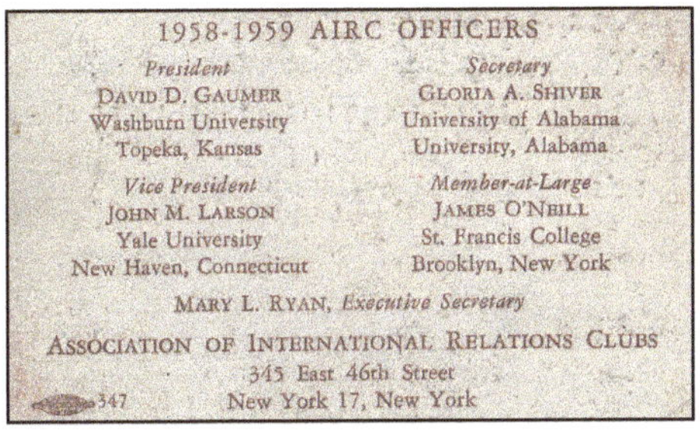

1958-1959 AIRC OFFICERS

President	*Secretary*
David D. Gaumer	Gloria A. Shiver
Washburn University	University of Alabama
Topeka, Kansas	University, Alabama
Vice President	*Member-at-Large*
John M. Larson	James O'Neill
Yale University	St. Francis College
New Haven, Connecticut	Brooklyn, New York

Mary L. Ryan, *Executive Secretary*

Association of International Relations Clubs
345 East 46th Street
New York 17, New York

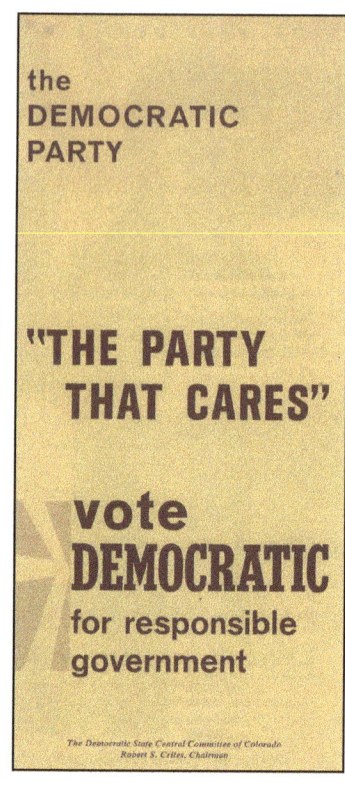

Campaign Literature for John Kennedy and Lyndon Johnson for the November 8, 1960, election.

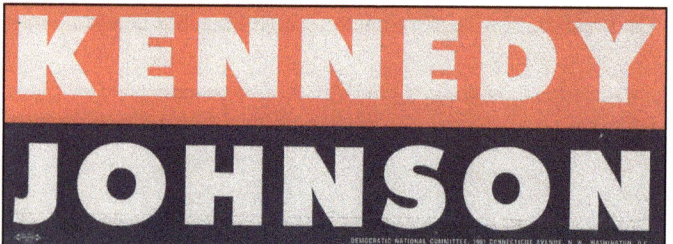

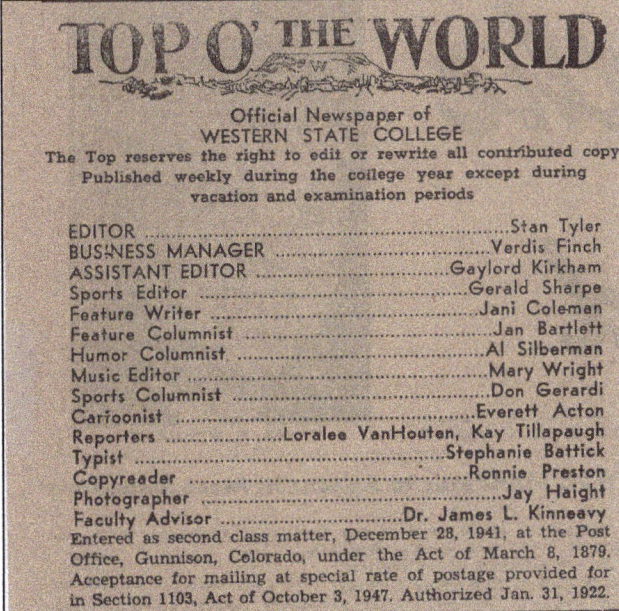

TOP O' THE WORLD

Official Newspaper of
WESTERN STATE COLLEGE
The Top reserves the right to edit or rewrite all contributed copy
Published weekly during the college year except during
vacation and examination periods

EDITOR	Stan Tyler
BUSINESS MANAGER	Verdis Finch
ASSISTANT EDITOR	Gaylord Kirkham
Sports Editor	Gerald Sharpe
Feature Writer	Jani Coleman
Feature Columnist	Jan Bartlett
Humor Columnist	Al Silberman
Music Editor	Mary Wright
Sports Columnist	Don Gerardi
Cartoonist	Everett Acton
Reporters	Loralee VanHouten, Kay Tillapaugh
Typist	Stephanie Battick
Copyreader	Ronnie Preston
Photographer	Jay Haight
Faculty Advisor	Dr. James L. Kinneavy

Entered as second class matter, December 28, 1941, at the Post
Office, Gunnison, Colorado, under the Act of March 8, 1879.
Acceptance for mailing at special rate of postage provided for
in Section 1103, Act of October 3, 1947. Authorized Jan. 31, 1922.

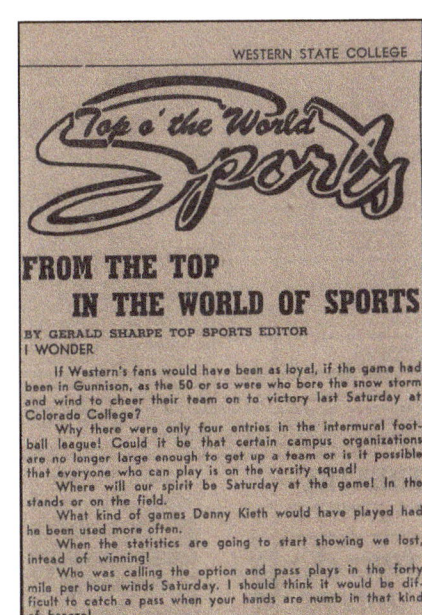

WESTERN STATE COLLEGE

Top o' the World Sports

FROM THE TOP
IN THE WORLD OF SPORTS

BY GERALD SHARPE TOP SPORTS EDITOR

I WONDER

If Western's fans would have been as loyal, if the game had
been in Gunnison, as the 50 or so who bore the snow storm
and wind to cheer their team on to victory last Saturday at
Colorado College?

Why there were only four entries in the intermural football league! Could it be that certain campus organizations
are no longer large enough to get up a team or is it possible
that everyone who can play is on the varsity squad!

Where will our spirit be Saturday at the game! In the
stands or on the field.

What kind of games Denny Kieth would have played had
he been used more often.

When the statistics are going to start showing we lost,
instead of winning!

Who was calling the option and pass plays in the forty
mile per hour winds Saturday. I should think it would be difficult to catch a pass when your hands are numb in that kind
of breeze!

How a football team feels after it has lost a game it
should have won?

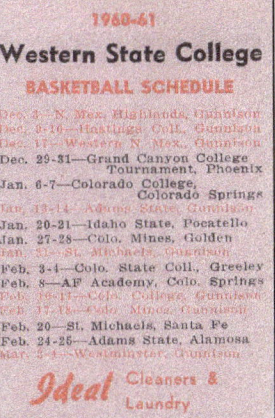

1960-61
Western State College
BASKETBALL SCHEDULE

Dec. 3—N. Mex. Highlands, Gunnison
Dec. 9-10—Hastings Coll., Gunnison
Dec. 17—Western N. Mex., Gunnison
Dec. 29-31—Grand Canyon College
Tournament, Phoenix
Jan. 6-7—Colorado College,
Colorado Springs
Jan. 13-14—Adams State, Gunnison
Jan. 20-21—Idaho State, Pocatello
Jan. 27-28—Colo. Mines, Golden
Jan. 31—St. Michaels, Gunnison
Feb. 3-4—Colo. State Coll., Greeley
Feb. 8—AF Academy, Colo. Springs
Feb. 16-17—Colo. College, Gunnison
Feb. 17-18—Colo. Mines, Gunnison
Feb. 20—St. Michaels, Santa Fe
Feb. 24-25—Adams State, Alamosa
Mar. 3-4—Westminster, Gunnison

Ideal Cleaners & Laundry
Phone 555   Gunnison, Colo.

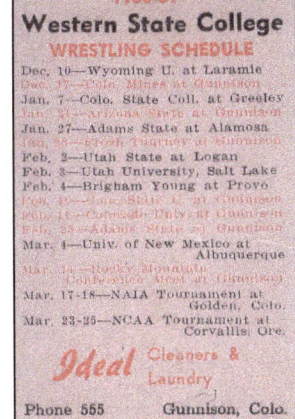

1960-61
Western State College
WRESTLING SCHEDULE

Dec. 10—Wyoming U. at Laramie
Dec. 17—Colo. Mines at Gunnison
Jan. 7—Colo. State Coll. at Greeley
Jan. 21—Arizona State at Gunnison
Jan. 27—Adams State at Alamosa
Jan. 28—Frosh Tourney at Gunnison
Feb. 2—Utah State at Logan
Feb. 3—Utah University, Salt Lake
Feb. 4—Brigham Young at Provo
Feb. 11—Adams State at Gunnison
Feb. 17—Colorado University at Gunnison
Feb. 25—Adams State at Gunnison
Mar. 4—Univ. of New Mexico at
Albuquerque
Mar. 11—Rocky Mountain
Conference Meet at Gunnison
Mar. 17-18—NAIA Tournament at
Golden, Colo.
Mar. 23-25—NCAA Tournament at
Corvallis, Ore.

Ideal Cleaners & Laundry
Phone 555   Gunnison, Colo.

Western State College of Colorado
Associated Students
Identification Card

No. 2780

SPRING 1961

SIGNATURE

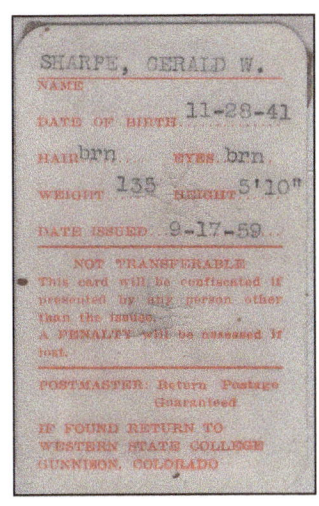

SHARPE, GERALD W.
NAME
DATE OF BIRTH 11-28-41
HAIR brn   EYES brn
WEIGHT 135   HEIGHT 5'10"
DATE ISSUED 9-17-59

NOT TRANSFERABLE
This card will be confiscated if
presented by any person other
than the issuee.
A PENALTY will be assessed if
lost.

POSTMASTER: Return Postage
Guaranteed

IF FOUND RETURN TO
WESTERN STATE COLLEGE
GUNNISON, COLORADO

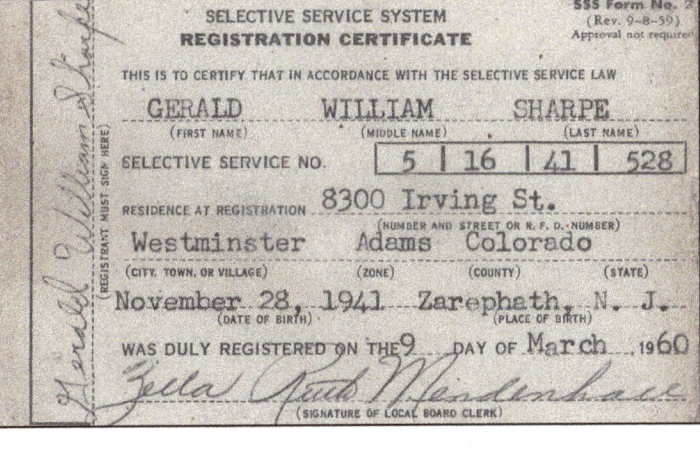

SELECTIVE SERVICE SYSTEM
REGISTRATION CERTIFICATE

SSS Form No. 2
(Rev. 9-8-59)
Approval not required

THIS IS TO CERTIFY THAT IN ACCORDANCE WITH THE SELECTIVE SERVICE LAW

GERALD (FIRST NAME)   WILLIAM (MIDDLE NAME)   SHARPE (LAST NAME)

SELECTIVE SERVICE NO. | 5 | 16 | 41 | 528 |

RESIDENCE AT REGISTRATION
8300 Irving St.
(NUMBER AND STREET OR R.F.D. NUMBER)
Westminster   Adams   Colorado
(CITY, TOWN, OR VILLAGE)   (ZONE)   (COUNTY)   (STATE)

November 28, 1941   Zarephath, N. J.
(DATE OF BIRTH)   (PLACE OF BIRTH)

WAS DULY REGISTERED ON THE 9 DAY OF March 1960

(SIGNATURE OF LOCAL BOARD CLERK)

(REGISTRANT MUST SIGN HERE)

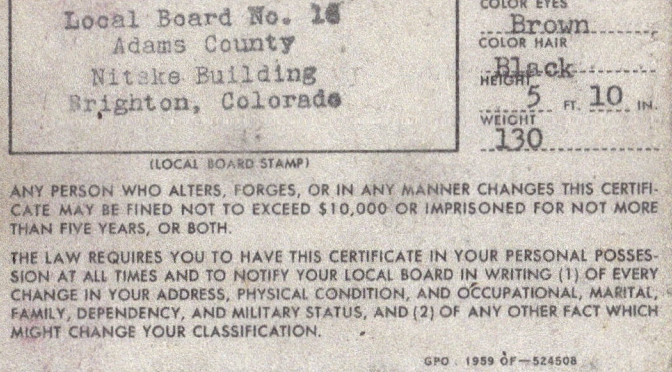

Local Board No. 18
Adams County
Nitske Building
Brighton, Colorado

(LOCAL BOARD STAMP)

COLOR EYES Brown
COLOR HAIR Black
HEIGHT 5 FT. 10 IN.
WEIGHT 130

ANY PERSON WHO ALTERS, FORGES, OR IN ANY MANNER CHANGES THIS CERTIFI-
CATE MAY BE FINED NOT TO EXCEED $10,000 OR IMPRISONED FOR NOT MORE
THAN FIVE YEARS, OR BOTH.

THE LAW REQUIRES YOU TO HAVE THIS CERTIFICATE IN YOUR PERSONAL POSSES-
SION AT ALL TIMES AND TO NOTIFY YOUR LOCAL BOARD IN WRITING (1) OF EVERY
CHANGE IN YOUR ADDRESS, PHYSICAL CONDITION, AND OCCUPATIONAL, MARITAL,
FAMILY, DEPENDENCY, AND MILITARY STATUS, AND (2) OF ANY OTHER FACT WHICH
MIGHT CHANGE YOUR CLASSIFICATION.

GPO : 1959 OF—524508

Collision of two airplanes over New York City that killed my great grandfather.

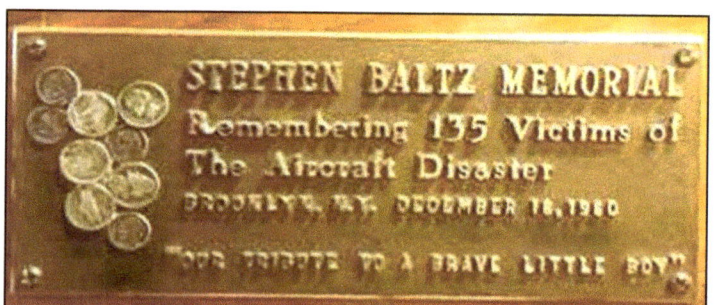

Stephen, age 11, was flying unaccompanied, was thrown from the same airplane that hit the church, and landed in a snowbank. He was taken to the hospital and died the next day. The memorial, including the coins in his pocket, is still displayed in the Methodist Hospital lobby.

On December 16, 1960, at 10:33 am, two airliners collided over New York City killing all 128 aboard and killing six on the ground, including my 90-year-old great grandfather, Wallace E. Lewis, who was the caretaker of the Pillar of Fire Church that was completely destroyed. Above, the New York City Fire Department on the scene of the United Airlines Flight 826 crash that destroyed the Pillar of Fire church located at 123 Sterling Place and Seventh Avenue, Brooklyn. The second plane, TWA Flight 800, exploded in the air killing all on board and its debris fell on Staten Island.

4. ENLISTED SERVICE

SEPTEMBER 1961 TO OCTOBER 1963

BASIC TRAINING, FORT ORD, CA
(SEPTEMBER 7, 1961 TO NOVEMBER 22, 1961)

UPON ARRIVING AT FORT ORD ON SEPTEMBER 7, 1961, WE WENT THROUGH IN-PROCESSING. WE were issued three uniforms, a hat and scarf, two belts, two pairs of boots, and combat gear including a helmet, a pistol belt, a canteen, and a poncho. We were also issued a duffle bag to carry everything from one assignment to another. We marched to our unit, Company C, 4th Battle Group, First Brigade.

Our company commander was 1LT Michael J. Beirne, our first sergeant was 1st Sgt H. M. Adams, and my platoon sergeant in the 3rd platoon was SFC R. L. Lewis. In our unit we were assigned a rifle that we had to sign in and out of a company arms room for training and for cleaning. I quickly found out that even though I had come from Colorado where there was snow on the ground, it seemed much colder at Fort Ord. I later found out it was because of the difference in humidity. In Colorado humidity was often only 40% or 50% while at Fort Ord it was almost always 90% or higher. Combine that with the wind coming off the ocean, and I could never seem to stay warm outside. We spent the first week at Fort Ord getting our uniforms and equipment, taking a physical, and getting oriented on what we would do in basic training.

In a letter to Mom and Dad (I guess I was at least prepared to add Dad to my letters), during our second week at Fort Ord I wrote,

September 17, 1961

Well, our training gets into full swing tomorrow morning. I'm not too sure what is going to take place, but I understand they intend to keep us mighty busy the next few weeks. We had our first physical training class Thursday during which we learned the 12 exercises we will go through every day, referred to always as the Army daily dozen. This life sure has done wonders for my appetite. I think I am eating about three or four times as much as I usually do, and I am still not getting enough. Every morning this week I have had at least two eggs, which I never used to touch; two pieces of toast smeared with butter, another thing I never used to eat, along with jelly; a bowl of dry cereal; bacon or sausage; fruit juice; and two glasses of milk. Whenever our platoon eats last, we get to go back and help ourselves to whatever is left, and I usually always fill my tray a second time. I also find myself eating things I never liked before such as fish, cabbage, and cold cuts.

Later in the letter I described my first week at Fort Ord, before basic actually started. I wrote,

Actually we did quite a bit last week. The first two days our company of 300 men supplied the post with men for details (any odd job you are sent to do is called a detail.) On Monday morning about 20 of us were sent over to the road repair and maintenance section and while we were lined up, one of the sergeants asked if anyone had ever worked with a survey transit, so I held up my hand. (Reverend Ray Sharpe had taught me the basics of using one while I was in high school.) I was sent to help this man survey a piece of ground for tennis courts. As it turned out I really almost had the day off, for after we surveyed for a while we went and had coffee and donuts and then finished just before lunch with the computation coming out perfectly square. When we started, the surveyor said he would be happy if it came out within three or four inches.

After lunch we went to Army Language School in Monterey to survey another field. After we finished, the man drove the long way back, showing me where some of the top army men lived along the coast, and we went by Fisherman's Wharf and watched the boats coming in from a long day at work.

Tuesday I was sent with four men to the base hospital. The other four were told to mop and wax floors while I rode around in a truck with a sergeant delivering papers and moving a mattress from one building to another. We had lunch with the doctors and nurses which included steak and apple pie, the best meal I have had in a long time. In the afternoon we sat around and had coffee until someone called and we went and picked up a bed that had been repaired. We then straightened up the medical warehouse a little bit and were sent back to the barracks at 4:00 o'clock. We had the rest of the day off.

For the rest of the week, we had classes in military courtesy, company tradition, moral guidance, and army history. We had lectures by the company commander, brigade commander, and the chaplain. We had physical training each morning and we were shown about ten movies which tried to show how terrible the Russians were and how much we owe to our country and our forefathers.

One of our first tasks was to learn how to march and take commands, both with and without a weapon. We had to right face, left face, and about face in a specific way. While marching, we had to learn how to turn left or right in formation or turn and go in the opposite direction.

Our second task was to qualify with our rifles. We first learned how to take them apart and put them back together and how to properly clean them, particularly the bore after firing. We were each issued a rifle cleaning kit that we carried in a pouch on our pistol belt any time we were carrying our weapons.

The firing ranges were right along the ocean. We took three open-topped tractor trailers to training. They had benches on both sides and one down the middle. The back wall also served as a ramp to walk up onto the truck. We would file up into the trailer in two ranks and take a seat. As soon as the last person sat down, one of the sergeants would call us to attention and have us face the rear of the truck. That was hard for those on the middle bench, because we sat facing the rear with one leg on each side of the bench. Once we were standing, we were told to back up and move toward the front of the trailer. This allowed for more soldiers to get on the trailer. There was a specified number and once that number was reached, the tail gate would be closed. We were then told to sit. As tightly packed as we were and carrying our rifles, we were either on someone's lap or someone was on ours.

In a letter to Mom and Dad I wrote,

November 1, 1961,

Basic training is almost over. We just had our final physical fitness test, and I improved my score quite a bit from the one I took six weeks ago. We will have our graded test next Monday which tests us on the 16 phases of our training to see if we have mastered them. This test is more or less what we have worked toward during basic training. We are tested on such things as the Code of Conduct, land navigation, military intelligence, first aid, gas mask drill, firing rifle grenades, night movement, military courtesy, guard duty, throwing hand grenades, cover and movement, rifle drill, rifle assembly and disassembly, rifle firing positions, range estimation, and squad tactics. Passing score is 70 but I hope to get much higher than that, in the 90s if possible.

I have been moved up to the position of assistant squad leader of the fourth squad (15 men) in our platoon. The private who was our assistant squad leader was put in charge of the first squad when the squad leader there was demoted. Being assistant squad leader does not mean too much, just leading the squad when the squad leader is absent. We've been having reviews for Monday's test and the squad leaders have been designated as assistant instructors. This means that I must supervise the cleaning of our 2nd floor in the barracks so that it is ready to pass inspection.

Sometimes we have to sweep as many as four times and mop the floors twice to get these 1940 vintage barracks clean. You can imagine how hard it is to get 12 or 13 lively young men, some three or four years older than I am, to go through the same routine every morning before and after breakfast.

Last Saturday they broke the company down into 16 groups to go through the different stations to prepare for Monday's test. Since they took out all four squad leaders from our platoon, I was designated to lead the entire platoon. What an experience! It always looked easy when someone else was doing it, but it sure wasn't. You must repeat all the commands from the lieutenant platoon leader as well as keep all the men perfectly in line and in step when we march. What makes it so tough really is that you must yell at different people to get them to straighten up and they often do not like it and do not listen, so you have to go in the ranks and kick their butts into place. Some try their best while others do not care a crap if the whole company looks terrible. I have concluded I am very happy I was not chosen as a squad leader.

It took me about three weeks to figure out that joining the army as an enlisted man might not have been the smartest move I had ever made. I began talking to Mother by phone in week one but refused to speak with Dad even though I added him in addressing my letters. On the weekend of October 21, 40 of us were selected to attend the Pacific Grand Prix at the Laguna Seca Racetrack on Fort Ord. It was great to have a free day and it was the first time I had ever seen a Formula 1 race event.

The following weekend I was one of about 20 soldiers who did not receive a single deficiency during inspections on our three-day field training exercise. I was told on Friday that I could have a two-day pass starting at 7:00 am Saturday and had to be back by 6:00 pm Sunday.

I called Reverend Reuben and Irel Truitt and asked if they would mind if I visited over the weekend. I told them I would take a bus to Oakland, and they agreed to pick me up. In the same November 1 letter to Mom and Dad I wrote,

My visit to Oakland was very refreshing as we spent a quiet evening on Saturday discussing the world situation as well as many other things. Everyone was very kind, especially the Frenkiels (Reverend Edward and Lillian who would later spend many years in England) who went more than out of their way to be hospitable. On Sunday Truitts took me sightseeing in Santa Cruz (about 90 miles south of San Francisco) and I left from there on a bus back to Fort Ord (About another 50-mile or so trip south) about 3:00 pm.

On the weekend before we graduated the unit held an "Open House" and a large number of soldiers had their families attend. We had to clean our barracks spotless as they were open for inspection by families until lunchtime. There were weapons and equipment displays set up in the street in front of our barracks, and all visitors were issued a meal ticket so they could eat lunch in the dining facility. We had a pass on post from 1:00 pm until 6:00 pm and we got to go to the post exchange, snack bar, and movie theater for the first and only time. Unfortunately, none of my family was able to attend my graduation, but I did get to go home on leave before my next assignment.

ADVANCED INDIVIDUAL TRAINING (AIT), FORT CARSON, CO (NOVEMBER 23, 1961 TO JANUARY 29, 1962)

Following our eight weeks of basic training I was sent right back to Colorado and was assigned to Fort Carson near Colorado Springs. This is the post where my daughter Shelley's husband Pete worked in the property disposal office until he retired in 2021. The post is near where they continue to live, and Shelley has a Counseling Business.

I departed Fort Ord, CA on November 11, 1961, and was granted ten days of home leave. I reported in to Company S, 4th Battalion, 2nd Training Regiment on November 23rd for my eight weeks of advanced Individual Training. My battery commander was CPT Clyde R. Miller, and my first sergeant was MSG John L. Brown.

I was selected for cannoneer training, which was military occupational specialty 140.00. This meant that we trained on towed, split trail, 105mm howitzers and had to learn how to put them into position, dig in the trails, set firing data on the instruments, and load and fire the weapons.

Fort Carson had the worst winter on record in about 80 years as I recall. We had snow on every field training exercise. Sometimes the ground was so frozen that we had to use picks to dig in the trails. Going to the field was simply no fun, as we were freezing all the time. Even though we were issued long underwear (shirts and pants), mittens, gloves, field jacket liners, and hoods, It was still hard to stay warm. I sometimes thought I was getting frostbite in my feet, but I never did.

About one third of the soldiers in my platoon were of Puerto Rican descent. They liked to speak their own language whenever they were out of sight of any of the cadre. Our second week there, our platoon sergeant came into the barracks before breakfast one morning and had everyone assemble around him. He told everyone they were in the United States Army, so they would speak English. He said if he caught anyone speaking any language other than English, they would be turned back to the next class and spend their time washing pots and pans in the dining facility until the next class started. This lecture changed things only slightly as once the training day was over and we were back in the barracks, they spoke their own language almost exclusively.

I do not remember the exact date, but I think it was sometime in December that my brother Stanley had an accident in a chemistry class. He mixed some chemicals in a glass test tube and carried the test tube out of the classroom and down the hall to the drinking fountain. As he bent over the fountain to get a drink, the test tube exploded in his left hand, blowing off the first two joints of his ring finger. Fortunately, the rest of his body was shielded by the metal water fountain. He was rushed to the hospital with his finger wrapped in a cloth, but the doctors quickly determined that it was too damaged to try to reattach. Stan kept the finger in a jar of formaldehyde for years and loved to show it off.

I was never sure exactly how it happened, but Dad called and talked to one of the commanders in the chain of command. I was called to the company commander's office on a Friday and informed that my brother had been in an accident that blew off part of his hand, so I was going to be allowed a two-day pass on Saturday and Sunday to visit him in the hospital.

Dad picked me up early Saturday morning and took me to the hospital both days. We had gotten back on speaking terms when I was home on leave after basic, but never in the 28 years I was in the Army did he have anything good to say about the service. He would often ask, "When are you going to get out of that stupid organization?" He drove me back to Fort Carson on Sunday afternoon as I had to report in by 5:00 pm.

I was given an automatic promotion to private E-2 on December 31, 1961, which resulted in a small pay raise. The day before graduation in January, I was informed that I had been selected as the outstanding trainee in MOS 140. Our platoon of 60 soldiers and two other platoons were all trained as

MOS 140 cannoneers. At least one platoon was trained as fire direction specialists, MOS 152.00, and a second private was chosen as the outstanding trainee of that MOS.

Graduation was held on Thursday, January 25, and Mom, Dad and my four brothers and sister attended. We were not given any leave after AIT and the next day we were put on a charted bus for the two-day trip to Fort Dix, NJ where we would process in preparation for our trip to Germany.

TRIP TO GERMANY, USS PATCH (JANUARY 30, 1962 TO FEBRUARY 8, 1962)

Since I had joined the army with an option of "Germany Unassigned," I was sent by bus with a few others to Fort Dix, NJ following graduation from AIT at Fort Carson. From there, after doing two days of processing, we were bused to the military port at Bayonne, NJ and boarded a small two smokestack ship named the *USS Patch*. It was a Navy transport ship for the nine-day ocean crossing to Bremerhaven, Germany. We left port at 2:00 pm on the 30th of January. The ship had a crew of about 400 and had accommodations for more than 4,000 enlisted soldiers and 280 officers. I believe that less than a thousand of us went aboard. I was assigned to a huge room about four levels down in the ship that had four levels of bunk beds in long rows. There was about 120 of us in the room, and I later found out there were eight compartments. Our room was at the very rear of the ship, so we got the full brunt of the ship going up and down over the waves. Each bunk had a locker that held a duffle bag and a small handbag.

We had been in the room about 15 or 20 minutes when a chief petty officer (E7) appeared at the door and shouted, "At ease!" When everyone got quiet, he shouted, "I need two volunteers!" I immediately raised my hand and headed for the door. No one else spoke up, so he said something to the effect that if he did not get a second volunteer, he was going to just pick someone. It was kind of tradition to NEVER volunteer for anything, but I figured how bad could it be?

He ended up pointing to a soldier and said, "You two come with me." We followed him up four flights of stairs and down long hallways to the front of the ship and then up about five more flights of stairs until we entered a carpeted area that had couches and chairs. He told us to sit down, and that he would be back in a few minutes. A few minutes later, he returned with the captain of the ship, and we were told we would be the captain's orderlies for the trip, working 12 hours on and 12 hours off. We would wear red tabs on our uniforms, and we were to sit outside the captain's door in a chair except when we were told we could go eat. We were to do whatever the captain asked, but primarily, we would pick up items for the captain from the Navy Exchange or from the officer's mess. We were told we could go anywhere on the ship when off duty and we would eat in the crew dining hall.

When off duty, we could go to the crew movie theater or to the crew's enlisted club. I thought I had died and gone to heaven—I could not believe my good luck. The only downside was that we had to sleep in our assigned bunks. We were crossing the Atlantic in January in the winter storm season. The ship was small, and since we were in the rear compartment, it felt like riding a roller coaster. I am proud to say I never got seasick even once, but hundreds of the soldiers did, sometimes while in their bunks. As hard as they tried to clean it up, by day three our compartment stunk like puke.

One of my experiences on board has stuck with me to this day. The stainless-steel tables in the dining hall had two-inch bumpers around them, so the metal trays we used as plates would not slide off as the ship pitched and rolled. One evening I was eating some kind of steak. Sitting next to me was an older grizzled looking chief petty officer. Since the meat was fairly tender, I was cutting off pieces with my fork rather than with the steak knife we were given. I noticed the chief watching me, and he suddenly exploded into a tirade of curse words. In essence he was saying to me something like, "No wonder our Go_ damn forks are bent up, are you too stupid to use a knife? If you are going to eat in this damn mess soldier, I had better never see you try to cut anything with a Go_ damn fork again or I will have you banned from eating here!" Needless to say, I used a knife to cut things from that day forward. Even today when I am sometimes tempted to cut something with a fork, my brain flashes back to that scene.

I regret to say that I enjoyed watching multiple movies in the evenings (I was on the day shift, 7am to 7pm.) and I also enjoyed going to the small lounge and reading the many magazines that they had on hand. On just the third evening I went to bed much later than I should have, and I overslept. I was awakened by the same Chief Petty Officer that had escorted us to the captain's quarters on the first day. He simply said, "Give me your red badge, you are relieved!"

I felt bad about it, and I feared that there would be some record of my screw-up. Fortunately, I did not hear about it when we were out-processed. We went through out-processing aboard ship on our last full day aboard, and I found out that I was being assigned as a track driver to the 1st Battalion, 27th Field Artillery located on Ray Barracks in Friedberg, Germany, about an hour's drive north of Frankfurt. We took a train to Frankfurt and were picked up at the station and taken to our base.

In a letter home to Mom and Dad I shared my experience living aboard ship, now as a simple soldier like everyone else. I wrote,

February 4, 1962

This trip has been miserable so far as we have been hitting rough seas. I would estimate that 2/3rd of the men have been seasick. So far, I have managed all right. Life aboard ship has started to fall into a regular routine. We get up at 5:30 am, eat at 6:00 and must have the bay ready for inspection at 11:00 am. There are 120 men in my compartment. Everyone is forced to go up on deck while the compartments are being cleaned. So, if you are not on the cleaning detail, the morning just drags by as all you do is await word the 11:00 am inspection is complete. We can then go below to get out of the wind and cold. After lunch everyone sleeps unless they are on detail somewhere. This ship is outfitted very well with a PX, a barber shop, a movie theater, a library, a hospital, and rooms with tables where one can play cards or games or read or write letters.

1ST BATTALION, 27TH FIELD ARTILLERY, 3RD ARMORED DIVISION, FRIEDBERG, GERMANY (FEBRUARY 9, 1962 TO OCTOBER 30, 1963)

Once I arrived in the battalion, I was assigned to Headquarters Battery as a personal carrier driver for a tracked vehicle used by the battalion fire direction center (FDC). It was wintertime in Germany, during the cold war with the Soviet Union. We were required to participate in exercises in which we marched toward the border, and the entire battalion had to clear the "Kaserne" (the German word for a military base) within two hours. To be ready, we had to start up all of our vehicles twice a night during winter months and let them run for 15 minutes. If the vehicle would not start, the maintenance section had to work on getting it started. If the vehicle wouldn't start during an exercise, it would be towed off the kaserne to meet the exercise requirements.

Within a few weeks of arriving in the battalion a soldier put up a sign on the barracks bulletin board that he had a Grundig console radio and record player for sale for $30.00. While this was about a third of a month's pay, I had never owned a stereo and I purchased it, used it for 2 ½ years, and sold it for more than I paid for it.

We were issued a liberty pass that had to be signed in and out each time you received a pass. Only so many were given each weekend and even though you applied for a pass, you did not always get one.

I quickly began hearing stories about Elvis Presley. He was assigned to the 32nd Armored Battalion also located on Ray Barracks from October 1, 1958, to March 2, 1960, meaning he had been gone for almost two years when I arrived in January. It was reported that he had been a good soldier, but he was already famous when he arrived to do his duty. Stories were told about him, like the first time he was asked to mow the grass. He was given the key to the shed where the lawnmower was kept, and as the story goes, he came back a few minutes later and asked the first sergeant if he could take a short break before mowing the grass. He was told he could, and about 20 minutes later the sound of a gasoline powered lawn mower could be heard. Everyone knew the lawn mower was a push mower so seeing Elvis walking behind a brand-new motorized lawn mower came as quite a surprise. Turned out that as soon as Elvis saw it was a push mower and he had asked to take a break, he caught a taxi to town, about a five-minute drive, and bought the best self-propelled lawn mower the hardware store had in stock, along with a gas can, and gasoline. The lawnmower was still in use when I arrived, and it had become a real souvenir of his time in Germany.

It was reported he would sometimes play his guitar and sing in the enlisted men's club free of charge. It was also reported that one Christmas the manager of the officers' club asked him if he would play and sing a few songs for the annual officers Christmas Ball. Elvis reportedly said that he would be happy to for $250,000. The manager apparently declined.

Elvis loved to play basketball and was on his unit's team. He became good friends with the manager of the gymnasium who reportedly let him use the weight room even when the gym had closed for the evening. When Elvis completed his tour, he had just been promoted to Sergeant E-5 and he gave brand new baby blue Cadillacs to the managers of the gym and the enlisted men's club, to the owner of a night club in Friedberg where he partied on occasion, and to the owner of a club in Bad Nauheim, a resort town about eight kilometers directly north, where Elvis rented a large apartment. Elvis met his future wife in Bad Nauheim on September 13, 1959, at a party at his apartment and married her in 1967 just after she turned 21. Pricilla Beaulieu was just 14 years old when they met, and her father was a US Air Force officer. The two civilian managers were still assigned to Ray Barracks when I arrived, and they were still driving their Cadillacs.

Early on in my tour I learned that all junior enlisted men in the battalion were on two duty rosters. One was for "Kitchen Police" (or KP) and another for guard duty. The first time my name came up for KP I asked what time we had to report, and I was told 5:30 am. I arrived a little before 5:30 am only to find a long line of soldiers sitting in front of the door that would be unlocked at 5:30 am or a little later if the mess sergeant was late. I learned that there were many jobs on KP, assigned based upon one's place in line. Turned out I was the last to arrive. The first soldier in line got to be the dining room orderly (DRO) for the officers who had a separate room, second in line was the DRO for the NCOs (Both cleared tables and would get things requested), third was two DROs for the enlisted dining room, then two servers, four soldiers for washing dishes with a large commercial washer, and finally four to wash pots and pans by hand. I washed pots and pans that first time and never did it again. From that day forward I would get up at 3:30 am, go to the dining hall, and sit with my back against the door. I think there was only one time I got there second. Being the officers DRO was a very simple job and required no hard work whatsoever.

After arriving in the battalion, I did not have any job other than driving one of the self-propelled vehicles that was part of a three-vehicle tactical operations center for the battalion. When I was not driving, I would sit in the FDC and observe other jobs being performed. Chart operator looked like a simple job, so I asked if I could learn how to be a chart operator. Both the captain fire direction officer and the section chief, a master sergeant, said it was good to have more people qualified.

Over the period of a month or more I joined the fire direction center for training any time I did not have to be in the motor pool doing maintenance on my assigned vehicle. Sometime in March of 1962, we went on a three-day training exercise. One of the chart operators was on leave and I was asked to fill in for him when I was not working on my vehicle. Apparently, I did well because when we returned to garrison, I was told that I was being reassigned as a chart operator full time and a new driver was being assigned to my vehicle. From that time forward I spent all my time trying to learn every job in the FDC. I found that "computer" was the most challenging job. This was before the time of digital computers, so all the computations were done on a slide rule. It was later noted in my application for Officer Candidate School that I had received "a 7th Army Training Certificate as a member of the Battalion Fire Direct Center after having worked in the Fire Direction Center (as a fire direction computer operator) for less than six months."

On March 29, 1962, I was invited to go with the battalion fire direction officer and his wife (I cannot remember the captain's name) to a concert in Frankfurt put on by the world-famous singer Ella Fitzgerald and the Oscar Peterson Trio, who played jazz. It was a very memorable experience.

In August of 1962, it was announced that our battalion had won the Division Artillery Gunners Trophy, given to the unit that had the highest scores on the annual live fire training exercise held in the large training area named Grafenwoehr. This award was won in no small part by the outstanding performance of the battalion fire direction center. The previous year the battalion had come in second out of the four battalions in the division artillery. We were the direct support battalion of the 3rd Brigade of the 3rd Armored Division.

On April 27, 1962, I was promoted to PFC. Several members of the battalion fire direction section snow skied. The army had two recreation areas in southern Germany where you could go for a weekend, spend one night in an army hotel, and purchase ski lift tickets all for about 10 dollars. As the winter of 1962/63 approached, I purchased a pair of ski boots, a ski jacket, gloves, a scarf, and a hat. We were able to rent very good skis at the recreation center for $2.00 a day so it was kind of a waste of money to buy skis, although later I was given a pair by a soldier rotating back to the states. I probably went with groups three or four times over the winter. On one of our trips, we were told about a British Armed Forces Recreation Center in Winterberg, a little less than a two-hour drive north of us as compared to more than a six-hour drive to Garnish-Partenkirchen or Berchtesgaden Army Recreation Centers. One weekend, four of us decided to go and try it out. We had not eaten breakfast before we left so we went to the British snack bar that sat right at the bottom of two slopes with two rope tows—not chair lifts—running up the middle of the mountain.

As we got our breakfast, we saw two busloads of British soldiers pull up. They were issued skis, poles, and boots and then they were taken to the bottom of the rope tow. A senior sergeant spent a few minutes explaining how the rope tow worked as they watched other skiers grab the moving rope, lean back slightly, and get pulled up to a flat place where they would let go of the rope and then move to the second tow rope to get to the top. We quickly got the idea that not one of these soldiers had ever even seen a set of skis. After the soldiers were briefed, there began what I can best describe as a day-long comedy show. Of the first 20 or so soldiers that grabbed the tow rope, not one of them made it even halfway up the first lift. As they fell, they would often let go of the rope after having been dragged a few feet and they would slide down the slope knocking two or three others off the rope. The senior sergeant was screaming at them and forcing them to get back in line. The soldiers were on one side of the tow lined up and other skiers were on the other side. The sergeant would stop the soldiers for a couple of minutes and allow the rope to run empty and then would direct other skiers to get on the rope. He would then stop letting other skiers on and again put soldiers on the rope. We were laughing so hard we could hardly eat our breakfast. A crowd gathered at the bottom of each of the two slopes to watch the soldiers and they would let out a cheer when one made it to the first level.

Watching the soldiers go up the tow rope was nothing compared to watching them try to get down the slopes without killing themselves. After about an hour of failed attempts, the majority of the soldiers seemed to grasp the concept of how to let the rope pull them up the mountain. We quickly confirmed that no one had thought it might be a good idea to have a bit of a ski lesson before sending them up the mountain. No such luck for them so we spent the day watching soldiers come down the mountain tumbling, sliding, rolling, headfirst, feet first, and every other way possible rather than skiing down. I am not sure we saw a single soldier ski down the mountain. Some were getting lots of hazing by young children and pretty girls who were zipping all around them like they had been born on skis. We did not even put our skis on that day, but it was one of my most enjoyable days ever. I am not sure I thought it would be possible to laugh for more than six hours straight.

The second duty roster was for guard duty. Once again, I knew nothing about the system, but I was simply told when and where to show up. It was the dead of winter and I found out quickly that walking guard in the snow was no fun whatsoever. I learned that before we went to the guard house there was an inspection of the 31 soldiers by the officer of the guard. He inspected our uniforms, our rifles, and the shine on our boots. He also asked basic military questions. As deficiencies were found the officer directed us to go to the guardhouse and wait. The last five received special treatment. Number five through two got to serve as gate guards and had a small building to stand inside and keep out of the wind and cold. There were 13 designated walking posts, two guards to a post that rotated during the night. They walked back and forth along the fence line from one designated point to another. The last man standing became the driver of the Officer of the Guard and sat behind a desk and answered the phone. Much like KP, I was a quick learner and after walking guard that first night, I never did it again. The next time I was one of the final five and quickly learned what the deficiencies were that kept me from being number one. I stood guard on my first inspection, was in the top five on the second and third time and later I was number two once when the lieutenant picked his own driver as number one. I remember breaking my roommate's record of 21 consecutive picks as number 1.

Sometime in early 1962 we received a new battalion commander, LTC Clarence F. Ax. At his change of command ceremony speech, he stated that he planned to show the battalion how to cut the time it takes to go into a firing position from a road march and be prepared to fire the first round. We had trained up to that point to try to do it in two minutes and we seldom achieved that standard. We soon found out that he felt the new standard should be one minute. Everyone felt this was an impossible standard. I learned a valuable lesson in this instance: if a leader can set higher standards and then show his subordinates how to achieve them, everyone will be better off. Within four months LTC Ax did just that, and we cut the time in half by following the guidance we were given, and practicing in the large training area behind our kaserne until we got it right.

My most memorable experience on guard duty was the night of October 7, 1962, when the Officer of the Guard was from our own battalion. This lieutenant, by the name of Domert, had a reputation for having a very high IQ—off the charts—but he had zero common sense. He had been nicknamed "Dumb Domert" by the enlisted men in the battalion. About midnight one quite night, the phone rang. I answered it and learned from the 32nd Armor Battalion duty officer that a drunk enlisted man from their unit had started up an M60 tank in the motor pool and had told a friend he was going to town to blow up a bar where some soldier had "stolen" his girlfriend.

I asked the duty officer where the tank was now, and he said it was still in the motor pool, but the man was backing the tank out of its parking space. I told 1LT Domert what was happening, and he told me to go get in the jeep. We both ran outside, and I drove toward the back gate of the 32nd Armor Battalion motor pool. Just as we were approaching the gate, the tank came rolling through the closed gate, tearing it to shreds.

Between the gate and town was about a two-mile square area of bare dirt that all the units used for training and driving practice. The lieutenant told me to step on the gas and pass the tank. I did so, and as soon as we were a little ahead of the tank, he told me to get further ahead, which I did. He then told me to turn right and stop in front of the path of the tank. I initially kept driving and told the lieutenant, "No sir, that is a crazy idea." He said, "I am giving you a direct order. Do it now." I immediately turned right and stopped the jeep in front of the tank's path as it was coming toward us fast but still perhaps a quarter mile back toward the motor pool. I jumped out of the jeep and ran well away from the jeep. The lieutenant got out of the jeep, put his hand up in a signal to stop, and stood there for a few seconds until he apparently realized *damn, I don't think the tank is going to stop.* Fortunately, he was able to get a few steps away from the jeep before the tank crushed it flat as a pancake.

As I watched the tank go by, another jeep came racing up with two people in it. The driver pulled his bumper right up against the back of the tank where blazing hot air blew out the back from cooling the engine. Braving the heat and the bumpy terrain, the passenger stood up in his seat and grabbed the tank turret that stuck out the back about 6 feet. He proceeded to shimmy down the turret until he could climb onto the tank. I watched as he went around the front of the tank, opened the driver's hatch, grabbed the driver, and dragged him up out of the tank. As soon as the driver's feet left the pedals, the tank stopped. It was just short of the second gate going into town.

The lieutenant and I ran over to the tank, and it turned out it was the soldier's first sergeant that had gotten the tank stopped. The sergeant's driver called their battalion duty officer with the radio in his jeep and a second jeep was sent to take us back to the guard house. In the weeks following that night, the first sergeant was presented an Army Commendation Medal for risking his life to stop the tank and Lieutenant Domert was given disciplinary punishment by the battalion commander, was required to pay for the jeep, never made first lieutenant, and left the army when his three years were up.

My roommate was Bob Waters, a farmer from Illinois. During the last week of October 1962, Bob and I purchased a light blue Volkswagen beetle from a soldier who was returning to the states. We split the $600 he was asking, and the car was registered in my name. In return I agreed to pay Bob $50 a month for six months as he was getting out of the army in about eight months. I borrowed the money from Dad to pay for my share of the car and actually paid off both early. Within a few weeks of purchasing the car, Bob and I each had earned a three-day pass that we were allowed to combine with a weekend. During the week of November 4, 1962, we put in for a three-day pass to go to Zurich, Switzerland and drive through the Alps, Austria, and back through southern Germany. We left on Saturday morning, November 8, and planned to return on Monday.

We marveled at the scenery as we drove the six hours to reach the city of Zurich, at the head of Lake Zurich. We had both been paid as E4s before we left Friedberg, so we were feeling flush with money. I had been promoted to PFC on April 27, 1962, and got a raise from $88 to $99 a month and was then promoted to SP4 on September 29 with a raise to $122 a month. I was promoted with a waiver of both time in service and time in grade requirements. We pulled up in front of a large hotel that looked out over the lake and decided we would spend one night in a great hotel. We walked up the steps, went into the hotel, and went to the reception desk. We said we would like a room for one night. We were given a form to fill out and we were asked for our passports. As we were filling out the form, I casually asked, "Oh, by the way, how much is a room for one night?" The clerk said, "$360 dollars." This was more than the two of us made in a month and twice what we had in our pockets. We quickly said, "Oh, I think we have changed our mind." We retrieved our passports and walked back out to our car breathing a big sigh of relief that we had asked before checking in.

We drove back down the lake about five miles and found a small bed and breakfast for $20 for the night. We ate at a restaurant with a pier out onto the lake where we could see the city at night. We both had movie cameras and took a few movies. I have no recollection of what happened to my camera or the movies I took. The next day we spent a long day driving into the mountains and over to Innsbruck, Austria. We took a number of pictures along the way.

I noted in a letter home that:

The mountains between Zurich and Innsbruck were magnificent, being much different from those in Colorado. These seem to rise right straight out of the ground up to great heights and the peaks seem to be much more rugged and jagged that those at home.

We arrived in Innsbruck about 4:00 pm, did a little sightseeing and drove on to Garmisch, Germany where we stayed in the armed forces recreation hotel, the Green Arrow Inn. It cost us a whole 75 cents a night (prices were based on rank).

On Monday morning, we first drove the 14 miles to Oberammergau, the town where a large passion play is performed every ten years. Legend has it that the first play was performed in 1634 and the residents promised God that if they were spared from the bubonic plague, they would perform the play every ten years. They continue doing so in years ending in zero with a cast of more than 2,000. The 2020 play was postponed to 2022 because of covid. The town is also famous for its wood carvings.

After a brief sightseeing stop, we drove on to Munich, and decided to visit the Nazi Concentration Camp at Dacau, just northeast of Munich. The camp was left much as it was when liberated by US Army soldiers of the 45th Division on April 29, 1945, the day before Hitler committed suicide. The headquarters buildings had been turned into a museum with hundreds of gruesome photos taken immediately after liberation. There were photos of the hundreds of dead bodies found, including a train with 30 rail cars containing many bodies. There were photos of some of the nearly 30,000 starving and emaciated prisoners. Some rooms were filled with shoes, clothing, and suitcases that had been taken from prisoners. The soldiers were so horrified over what they found that they executed a number of guards who had been caught trying to escape the area.

The commander of the 45 Division, the one in charge of the units around the camp, directed that the towns people of Dacau be required to dig graves and bury 9,000 dead. We were able to walk around the camp and see the 128 wooden barracks in 16 rows of eight that were still standing open with their crude multi-level bunk beds four high inside still in place. (By the time I returned in the 1970s with my family all the barracks had been torn down, save one that has been restored as a museum.) The crematoriums had signs describing how they were used and how prisoners were required to operate the ovens. This visit left a lasting impression on me as it is a living testament to the horrors of the Holocaust!

This was the first concentration camp built by the Nazis, and it opened on March 22, 1933. At first it housed Hitler's political opponents, communists, and other dissidents. It is estimated that during its 12 years of operation, it housed about 188,000 prisoners and that about 41,500 Jews were executed during its later years. I was overwhelmed by the size of the facility, about five acres. And the crematorium ovens used to burn bodies are a stark reminder that this was real. This is in direct contravention to the Holocaust deniers who continue to spout lies.

I remember riding on a train later sitting in a compartment with an elderly German man, who tried to tell me the Holocaust was all a big lie, that Dachau and other camps had been created as propaganda by the Americans to make Germans seem like butchers. I tried to argue with him, but it was like talking to a rock. After telling him I had seen the hundreds of photos taken the day the camp was liberated that could not possibly have been faked, I simply had to get up and leave the compartment.

My first battery commander, 1LT Richard J. Ozga, rotated back to the states about four or five months after I arrived. My second battery commander was 1LT David C. Hogan, later promoted to captain. He taught me everything I ever wanted to know about how *not* to be a good officer. I used him as an example throughout my career in talking to lieutenants and captains.

1LT Hogan would sometimes come through the barracks in the evenings while we were supposedly on our free time and inspect our wall lockers or check to see how well our boots were shined. He had the entire battery fall out on Thanksgiving afternoon and stand in formation in the snow while he harangued us about one thing or another. On Christmas day in 1962 he came through the barracks wanting to inspect our living areas. We were all mad as hell! In my 1963 diary, I wrote several entries about 1LT Hogan. January 4, "Hogan had PT in heavy snow and ice running the men around the parade field like a herd of horses." That night, "Hogan scheduled a GI party to clean the barracks starting at 7:00 pm. He inspected us at 10:30 pm. We passed." January 10, "Had PT again in the snow as Hogan got carried away running us all over the post on very icy roads." January 14, "It was bitter cold this morning. Hogan had PT for us when all other batteries cancelled." And finally on the Saturday after arriving at the Grafenwoehr Training Area the day before, and staying in temporary barracks, I wrote, "Hogan had us hold a GI party. We had to clean and wax the floors and wash windows inside and out in the dark. We understood we only needed to sweep out the barracks. Hogan inspected at 1930 hours (7:30 pm) and was not happy—he said we could just do it again tomorrow evening—Sunday, supposedly our day off."

In December 1962, I competed to represent the battalion in the division artillery soldier of the quarter competition. I won at the battalion and then appeared before a division artillery board and competed against 14 other soldiers and was selected. The battalion commander, LTC Ax, presented me with an engraved watch as the winner. I still treasure that watch today.

In late December 1962, I again appeared before both a battalion board and a division artillery board to compete for the semi-annual General Maurice Rose Spearheader Award, which was given to the best enlisted soldier in the entire division of 16,000 soldiers. I was selected at the battalion board and the division artillery board, but only came in second among the seven appearing before the division board.

Later in June and July of 1963 I would take the same route to be selected for the semi-annual General Doyle Hickey Award as the most outstanding noncommissioned officer in the division. In that case I was selected by all three boards and won the award. First was a battalion selection board from which I was selected the winner and was named the battalion NCO of the Quarter. I received a small transistor radio with an engraved brass plaque on the front. I still display it in my army memorabilia cabinet. I was then sent to a division artillery selection board where once again I was the winner. Finally, I had to appear before the division board along with soldiers from each of the division's seven major commands. In the last week of August 1963, the battalion was notified that I was the winner and a few days later the division commander, MG John R. Pugh, presented me with a large hand drawn certificate that has somehow disappeared over the years. I still have the certificate I received for being the monthly General Doyle Hickey Award winner as the division NCO of the month for April following my graduation from the 7th Army NCO Academy.

On January 16, 1963, we went on a week-long training exercise to one of the artillery firing ranges knows as Wildflecken. It was about 90 square miles in size and all types of artillery could conduct live fire as could tanks and weapons of all kinds. We convoyed up on about a seven-hour trip (290 miles) with about 30 vehicles keeping 100 meters apart. I was riding in the back of a 2 ½ ton truck that had a large metal prefabricated body on the back with a small window in front and two larger ones one on each side. Our radios were installed on a rack running across the inside of the truck just under the front window. I would lean on the radios and stare out the front window watching the scenery. Each radio had a 20-foot-long antenna attached to the front of the truck. If we did not require radio contact, we would not install the antennas, but on this trip I was also monitoring the radios, so we simply tied the antennas down to the back of the van. This way there was no danger of them contacting high voltage lines that crisscrossed the streets in small towns.

We had been warned that failing to properly tie down an antenna could be life threatening because if an antenna struck a high voltage line, the radio would burn up and likely explode inside the metal case. As we were passing through one small town, I observed a very good-looking young lady walking on the sidewalk. I immediately went to the side window to get a closer look as we were driving very slowly through town.

Just as I bent down to look out the side window, the radio I had been leaning on only moments before virtually exploded inside the case and burned up with smoke pouring out of it. It made a loud noise and the driver immediately pulled off the road and came around and opened the back door so I could get out of the smoke-filled van. As we looked behind us, we could see a piece of antenna dangling off an electrical line at the railroad crossing we had just passed and we observed that—sure enough—one of the antennas was standing straight up with the top burned off it.

I was very happy to, A. not be injured, B. not have been the one who tied down the antennas improperly, and C. thank God that he saw fit to have a young lady attract my attention at exactly the right time. I must say that girl watching has never ceased to be a favorite pastime of mine, as it likely saved my life once. I still have the burned antenna connection where the antenna cord simply melted off the burned radio (see photo section).

During a trip to "Graf" in late January, I was one of two full-time chart operators in the battalion fire direction center. We would often have a break of an hour or two while batteries were repositioning from one firing point to another. I should note that charts were kept on a large sheet of heavy-duty graph paper. Its large squares were the same size as a grid on a map. When the chart was created for a particular exercise, I would number the squares down the sides and across the top and bottom to match the map used by the second chart operator, and then with a number 2 pencil kept very sharp, I would plot the locations of each firing battery and other units. Finally, I would outline in pencil the impact area so that I could ensure that any fire mission that was called in would land inside the authorized impact area.

One very important criterion was that we were to keep the firing chart absolutely clean. We were issued large soft rubber erasers that could completely remove a pencil mark without smudging. Our FDC was in the back of a built-up truck with plywood sides, and a top and a door in the back with a set of about ten metal steps. It had a bench and a table down one side for the two computers (remember, these were people, not machines). The NCO would sit on the bench at the end by the door, and we stood behind the two firing charts on the opposite side. My chart was closest to the back door. Finally, there was a chair for the fire direction officer at the front of the van in the aisle. They could both walk back and forth observing what we were doing.

It was winter and there was snow on the ground, but one afternoon the sun came out, so we opened the back door to let in some fresh air. I am not sure if there was garbage nearby, but we soon had a number of flies inside the van. Out of boredom I started using my right hand to catch flies, squeezing my hand killing the fly I caught, and then contrary to all rules, lining them up on the top of my firing chart. I would track a fly as it went in front of me, swing my arm and hand, and catch it in mid-flight. I was not paying attention to anyone coming up the steps but was tracking a fly that I had been unsuccessful in snagging. The fly started across the front of my chart a little above eye level, and I swung my arm intent on catching the fly. Instead my arm knocked the hat off our Division Artillery Commander, Brigadier General Walter T. Kirwin, Jr., who had presented me an award earlier. He turned to me and said something like, "Nice to see you again, Sharpe. You are apparently quite the fly killer, but they do NOT belong on that firing chart." I quickly brushed them off onto the floor as someone handed him back his one-star hat. We would meet again a few times during my career, including at a briefing I gave to the 1st Cavalry Division CG in Vietnam. He also visited his son, Bruce, who was an officer candidate in the battery next door to the one I commanded. He was kind enough to stop by my battery and say hello. General Kirwin went on to become the Vice Chief of Staff of the Army and retired as a four-star general in 1978.

After having been selected as Soldier of the Month for the battalion, having been selected as Soldier of the Quarter for the Division Artillery, and having competed for the best enlisted soldier in the division for the last half of 1962 (I came in second), the battalion commander asked to see me in early-February 1963. He said he had one quota to the 7th Army Noncommissioned Officer Academy in Bad Tolz in Southern Germany for a class that would start on April 1 and graduate on April 26. He asked if I would be willing to attend even though I was not yet eligible to be a sergeant. I said I would and he told me that he was going to assign two sergeants (previous graduates) to help me prepare for the challenging course. For the next few weeks, I spent part of my days and my nights studying materials and being quizzed by the two sergeants. I reported on the last day of March and the classes started the next day.

The academic course of study included map reading and land navigation, instructor training, and leadership. There was daily physical training conducted by students and graded by instructors. There were daily inspections of our uniform and basic military knowledge each morning and our cubical was also inspected daily while we were in class. I did well on morning inspections as my guard duty experience had trained me well. On the last Saturday we had the "Commandant's Inspection." The entire class was in a class A dress uniform with bloused boots. Three senior staff officers did the initial inspection.

Just like on guard duty, students were told to "fall out" as a deficiency was noted or a question was answered incorrectly.

When there were ten of us left, we were told to take a ten-minute break and have our uniforms checked or rechecked. After ten minutes we were lined up in one rank at double arms-length apart (so the inspector could walk around us unobstructed) and the Commandant himself began the final inspection. He used the same process. Finally, only two of us remained, a sergeant first class and myself, a junior enlisted man.

The commandant went around each of us a time or two more and asked us each a few more questions. He had not found anything wrong. He went behind us and told us to lift one leg so he could see the bottom of our boot. He immediately said, "Sharpe, fall out." It turned out the sergeant first class had shined the soles of his boots. I had left mine untouched. He received an award at graduation for winning the Commandant's Inspection.

As FDC chart operators, we used two charts. One chart was on heavy graph paper and the second one was an actual 1:50,000 map. I became very good at map reading in my job, and the two NCOs who worked with me in preparation for attending put me through two or three land navigation exercises, during which I had to walk and locate three or four points marked on the map using only a compass. We had four exams, two written with map problems to solve, and two practical exercise exams. We were told that if we made a perfect score, our name would be added to a large board outside the map reading classroom titled "THE MAP READING HALL OF FAME." As I recall there were 12 or 13 names out of more than 100 classes and 40,000 students that had attended the course. I am proud to say my name was added to that board before graduation.

The afternoon before graduation I went to see the Commandant, Lieutenant Colonel Simmons, who had failed me in the Commandant's Inspection. He said he had some good news and some bad news. The good news was that I was number one in my class, and I would receive two awards at graduation. The first was one for being the Distinguished Honor Graduate and the second was the General Patton Award for Excellence, the school's leadership award. He said the bad news was that I could not be promoted automatically to SGT E5 as was the custom, because I was nearly two years short of the minimum time in grade and minimum time in service.

LTC Simmons also informed me that as the distinguished graduate I would be expected to give a short two-minute speech following the awarding of the graduation certificates. I see my name is included in the graduation program in the photo section. I went straight to the library following my meeting and asked the librarian if she had any historical information on previous graduations. She referred me to a shelf with some bound paper files where I found an article about a speech given by General George C. Patton on Christmas day 1944 to a graduating class in this academy. I used the article as a lead in to the following speech I gave.

Good morning,

It was on Christmas day in 1944, after many long months of fighting, that General George C. Patton spoke to the graduates of this very NCO Academy, and he told his men: "You have done a grand job, but I expect more of you now. So, our superiors will also expect more of us now as we return to our respective units. We have been given a challenge here today, a challenge to go forth and show others not only by words, but by deeds that we have truly benefited from the superior training and experience we have received from this, the oldest and most respected Noncommissioned Officer Academy in the Army."

So today, our success in completing this course of instruction shows our desire to accept this same challenge and to uphold the responsibilities that will be placed on us as Non-commissioned Officers.

Upon our arrival here four short weeks ago, many of us may have had doubts as to our ability to meet the high standards of this academy. Others were perhaps somewhat overconfident, but as time progressed, we all found that hard work and sincere effort could overcome almost any problem or difficulty. We learned that even good leaders may have a few weak traits, but that we must strive to improve on them, and even perfect these traits while at the same time concentrating on our strong points. If we do this, we can approach any task with confidence.

Let us be sure that our actions and orders continually uphold the highest ideals and traditions of the Non-commissioned Officer Corp. If we accept the challenge of leadership, and successfully carry out our responsibilities, we will, hopefully, someday stand together in a free world united in peace.

Thank you.

Our speaker for graduation was three-star LTG Hugh P. Harris, the Commanding General of Seventh Army. When I came up on the stage to receive my graduation certificate, he asked me if I knew what the best thing was about being the Commanding General of Seventh Army. I told him I did not. He said I can do anything I want to do and almost no one can tell me I cannot. He then pulled out SGT E5 stripes, and he and the commandant pinned them on my sleeves.

He told me that when I got back to my unit someone could figure out how to make this official. He also presented me with an engraved swagger stick for the General Patton Award and an engraved cigarette lighter as the Distinguished Graduate. I still have both today and they are pictured in the photo section. Upon my return to Friedberg, I was promoted out of the battalion fire direction center to section chief of the Battery C Fire Direction Center where I served until the end of my German tour. By then I had become an expert in every job in the FDC and thus it was easy for me to be the chief of section. I had a good lieutenant as the battery fire direction officer as well.

Sometime in late May of 1963 we were notified that the President of the United States, John F. Kennedy, would be coming to Germany. Plans were made for the entire division to put on a parade on the airfield at Fliegerhorst Kaserne in Hanau east of Frankfurt. In preparation for the parade, all the vehicles were to be painted NATO green, instead of camouflage. This meant we spent more than three weeks repainting all of our howitzers and other equipment. A BG, ADC, oversaw the project and he apparently did not coordinate the repainting with higher headquarters. He was later relieved of duty and forced to retire for wasting two or three years' worth of NATO green paint. The division then had to use another couple years' worth of paint to get them back to green camouflage in the month following the parade. One other criticism of the general was that he requested that the air force fly in a plane load of grass sod from Spain to replace the grass on the helipad where the President's helicopter was to land.

As the President's arrival (June 25, 1963) grew nearer, the weather turned bad and for the two days before the scheduled parade it rained day and night. Everything around the airfield became a sea of mud. The entire 3rd Armored Division was to line up on the airfield for an 11:00 am "pass in review" by the president. We had to walk barefoot carrying our shined boots and clean socks from an assembly area through mud onto the airfield where washing stations were set up for 10,000 soldiers along with our big guns, tanks, and equipment. There was a schedule by unit, and unfortunately, we were up early. It was still dark when we arrived. We were all in formation by 9:00 am, but we had already spent more than a few miserable, hot, and muggy hours waiting for the president to arrive.

The President's helicopter landed a little before 11:00 am, and he was driven in his limousine past long lines of tanks and howitzers on his way to the reviewing stand (see photo section). The parade commenced and after appropriate ceremonies, the president came down off the reviewing stand and got in a specially outfitted jeep where he and the commanding general, MG John R. Pugh, stood in the back where the seat had been removed and a bar was installed for them to steady themselves. The jeep then proceeded to conduct a pass in review of all the units lined up on the airfield. After passing the soldiers the jeep went behind the line of soldiers where all the weapons and equipment of the division were displayed. The President and the CG dismounted and walked among the equipment and weapons talking to soldiers. There is an historical video of the President's entire visit to Germany online that includes the parade and the later lunch.

50 noncommissioned officers (sergeants) were selected to have lunch with the President following the parade. Each battalion was allowed to select one sergeant. I had recently graduated from the NCO academy and been promoted to sergeant, and I had just appeared before the Division Artillery board for the semiannual award for the best sergeant in the division and was selected as the Division Artillery representative, so I was selected by my battalion. I lined up on the airfield in the last rank of our battalion and as soon as the President passed our unit and before he started looking at equipment, I

left and went to join the others at the NCO club where the lunch was to be held. The cooks had made a large cake in the shape of the President's boat, PT 109.

Once the President arrived in the dining hall, he made a few remarks, went over to the cake and did a ceremonial slicing, and then shook hands with a few of the most senior noncommissioned officers. Once back at the head table, about 30 military policemen and several secret service agents came into the room. The MPs stood shoulder to shoulder holding a heavy rope about six feet in front of the table. Upon a signal, the door was opened, and a flood of reporters and cameramen filled the space in front of the President. They were given about three minutes to ask questions, snap photos, and fight for a space near the President. Upon another signal, the MPs began herding the reporters back out the door by ever shortening the rope until all the news people and the MPs were outside and the door was shut. The President got up and apologized for what he called the "circus" we had just seen and noted that unfortunately he had to deal with that everywhere he visited. We had a nice lunch and he again stood up and thanked us for attending before he departed for a motorcade that would take him to the Frankfurt Airport down roads lined with thousands of people waving small American flags. Upon arrival at the airport, he departed for Berlin aboard Air Force 1. The next day he gave his famous "Ich Bin Ein Berliner" (I am a Berliner) speech.

After I returned from the NCO academy and was officially promoted to sergeant E-5. I then had the pleasure of keeping my liberty pass in my possession and I was able to become a member of the Noncommissioned Officer Club. As I previously noted, I was now in Battery C as the NCOIC of the FDC section. One morning, I was in the first sergeant's office on the third floor of our building, which was next door to the battery commander's office. We had not heard anything about getting a new first sergeant but in walked a tall, lean, immaculately dressed first sergeant. He introduced himself to the two or three of us that were in the area as the newly assigned 1st Sgt of our battery. He asked where our battery commander, 1LT Rufus B. Rogers, might be. We told him we thought he was up at the battalion headquarters. He indicated that was no problem. He then told us to please empty everything out of the first sergeant's desk. There was very little in the desk because apparently the old first sergeant had already taken everything he wanted out. There were only five pieces of furniture in the room: the desk, an office chair, two armchairs, and a small table.

I will summarize the rest of the conversation as I remember it. The new first sergeant then asked someone to get the battery supply sergeant to come up to the office and a few minutes later he arrived. The supply sergeant was asked by the new first sergeant if he had a hand receipt for the furniture in the office and whether or not he would be signing for any other furniture. The supply sergeant told the first sergeant yes, he had the hand receipt signed by the old first sergeant and no, these were the only items on the hand receipt form.

The supply sergeant said he would prepare a new one to be signed. The supply sergeant was told

COL(RET) GERALD W. "JERRY" SHARPE

not to bother, to simply make up a statement of charges for the destruction and loss of the five items of furniture, and he would sign it. The supply sergeant said something to the effect that he was not sure he understood. The new first sergeant told me to please get whoever was in the area to come into the office because he had a job for us. There were three or four others besides the supply sergeant and me available. We were told to open the large back window that looked down on the grass behind the barracks building. He then told us he did not want any discussion or comment on what he was going to tell us to do. He then told us to throw all the furniture out the back window, saving the desk for last. We had a hard time getting the desk out, but it basically resulted in a pile of smashed furniture on the back lawn.

He then turned back to the supply sergeant and asked him if he now understood why he was told to prepare a statement of charges for the destroyed furniture. He said he understood. He said good, now I assume as the supply sergeant you have access to a truck. He said yes, he did. The new first sergeant then said, "Good, let's go get me some new furniture." We soon found out that the new first sergeant was part owner of the King Ranch in Texas. He was in the Army because he loved it, and he donated his pay every month to charity.

He was probably the most knowledgeable noncommissioned officer I ever met. He showed back up about two hours later with brand new furniture he had purchased in town that included a very large desk that he kept shined to the point you could almost see your face in it. No one was allowed to touch his desk. It took a few ass-chewings to finally get the word spread that you would be in big trouble if you so much as laid a piece of paper on his desk. One morning a new lieutenant was assigned to the battery.

I happened to be sitting in the first sergeant's office waiting to see the commander about my OCS application when the lieutenant strolled into the office and said, "Good morning first sergeant," then walked up to the desk, hiked his butt up, and sat on it while at the same time dropping his hat on the desk next to the first sergeant. Quick as a flash, the first sergeant grabbed his hat, threw it out into the hall, and told the lieutenant, "Get your damn ass off of my desk, NOW, and you had better never touch it again." He added you had best not say one word before you take your sorry ass next door and report to the battery commander. He will likely explain to you that you have gotten off to a very bad start in this unit. A minute or two later you could hear the battery commander all throughout the building loudly chewing the lieutenant's ass for the next ten minutes. I am not sure he ever stepped foot in the first sergeant's office again.

On the weekend of May 20/21, 1963, Bob Waters and I went to the 1,000-kilometer Formula 1 endurance race at the Nürburgring racetrack which is about two hours west of Friedberg. We had gone to the German Grand Prix at the same track in August of 1962 and it had gotten rained out. Both times we spent the night in a gastehaus in Koblenz about halfway to the racetrack. Since we had a car this time, we left early Saturday morning and drove straight to the racetrack to be sure we knew where to park and where to sit on Sunday morning. We drove back to Koblenz and checked into our room.

We departed at 5:00 am the next morning and found driving to the racetrack in a VW bug quite an experience as even though it was still dark outside it was like being on the racetrack as everyone wanted to pass everyone else. We were surprised there were no head-on collisions, but fortunately the traffic coming from the racetrack was almost nonexistent. We arrived early enough to get a seat in the front row of the bleachers. The race started at 9:00 am and finished at 4:30 pm. It takes 44 laps around the 22.9-kilometer track to complete the race. There were 90 cars that started the race in four groups. Only 18 finished. Previously I had attended my first Formula 1 race while in basic training.

After being selected for the General Maurice Rose Award as the best noncommissioned officer in the 3rd Armored Division for the first half of 1963, I received the award from the commanding general. At the awards ceremony the CG said he would like to schedule an appointment with me to talk about my future in the army. I received word a few days later of the time I was to meet with the CG. My battalion commander told me he thought the CG was going to try to get me to go to West Point to become an officer and had asked that his admin officer (S1) research the regulation to see if I was eligible. It turned out I was about five months too old to apply for West Point, as enlisted soldiers had to go to a one-year prep school before entering and there was a maximum age you could be on the date of entry. If I had not had to go to the prep school, I would have been OK, but as it was, I would simply have been too old upon graduation from the prep school. I kept my appointment with the CG and the first thing he said was I would like to send you to West Point to get a commission. I told him I was pretty sure that I was too old to meet the West Point criteria. He called his chief of staff in and asked him to check based on my date of birth. Just a few minutes later it was confirmed I was too old. He then asked me to apply for OCS at Fort Sill, OK as the next best alternative. I agreed to apply, and he assured me he would ensure my application was approved. I did not particularly like the army, but I figured since there was a six-month option for active duty after graduation, it would allow me to get out of the army as a commissioned officer rather than as an enlisted man, even though I would have to serve time in the reserves. I found out I was going to get out of the service about two months later if I went to OCS, but I considered that a good trade off.

I applied for OCS and was accepted. A few weeks before I was to leave, the noncommissioned officer in charge of the battalion fire direction center, MSG Bodenhamer, was returning to the states. He and his wife lived in leased quarters off post. He scheduled a farewell party at his home, and I was invited as I had spent almost a full year in his section before being promoted and transferred to Battery C. I agreed to act as bartender and even though I seldom had a drink, for some reason I thought it was a good idea to mix myself vodka and 7 Up drinks. I suspect I had five or six large drinks during the evening.

I am told that a little after 10:00 pm MSG Bodenhamer had invited everyone to sit down in his living room so he could make a few remarks. I was sitting on the couch between two individuals and about the time the speeches were over, I apparently passed out, pitched forward, and threw up in a potato chip bowl on the coffee table in front of me. I was carried out to the back porch where there was an old couch. I was cleaned up a little, and then allowed to sleep. I woke up the next morning totally embarrassed over my actions and swore it would never happen again. I can say with some pride that I have never been drunk since and I have simply refused to drink alcohol from that day forward. This was not an easy position to take in the officer corps as drinking is a big part of military officer social activities. Years later, as a major, I was walking through the Dallas, TX airport when a loud voice rang out! "Sharpe!" It was all that was said and as I turned to look, striding toward me was MSG Bodenhamer. He was wearing green pants, a pink shirt, multicolored shoes, and a hat with peacock feathers. I found out he ran a clothing store for black customers in downtown Memphis, TN. He had retired as a Command Sergeant Major at Fort Hood, TX. We visited for a few minutes, and I never saw him again.

The last week in July 1963, Bob Waters and I took another trip, this time a six-day trip to London, England. We drove to Rhein-Main airbase in Frankfurt on Thursday and caught a free flight to England that evening landing at RAF Base, Mildenhall about 9:00 pm. We stayed at the on-post guest house and on Friday morning caught a train for the 80-mile trip southeast to London. There was an Air Force hotel in London called the Douglas House where we were able to stay for $2.00 a night. We spent all day Friday, the 26th of July, sightseeing and visiting Buckingham Palace, the House of Parliament, Big Ben, Westminster Abby, Hyde Park, Kensington Palace, the London Museum, Queens Art Gallery, Piccadilly, and Trafalgar Square. It was a very full day in which we passed up eating in favor of seeing more sites.

Saturday, Sunday, and Monday Bob had some friends to spend time with, so I took a train and bus to the Pillar of Fire Church in Hendon in northeast London. Two of the ladies who had been assigned to Oakland, CA when I was a child were there—Miss Keller and Mrs. Clark. I had not seen either of them for many years and enjoyed visiting with them. I met Harvey and Pauline Brandt, who were also assigned to the Hendon church. They were just a bit older than I was, and they spent Saturday, Sunday, and Monday showing me around Hendon and parts of London. On Saturday evening Harvey drove me back to my hotel showing me the American Embassy, the American Ambassador's quarters, and Old Baily, where many famous trials had been held. The trial of Stephen Ward was ongoing in what was called the John Profumo Affair. Mr. Profumo was the British Secretary of State, and one of the girls Mr. Ward provided to Mr. Profumo had also been sleeping with a Russian agent. Mr. Ward was charged with living off the earnings of girls he provided for Mr. Profumo. On July 30, the day before the verdict was to be read and the day we departed England, Mr. Ward took an overdose of barbiturates and died on the 3rd of August. He was found guilty, but this verdict has been a subject of controversy ever since.

On Sunday morning I again took a train and bus back to Hendon and had lunch with Harvey and Pauline and their two children. They took me sightseeing Sunday afternoon and again Harvey returned me to the hotel. There were two missionaries from Liberia staying at the church home and

Harvey had promised to take them to Windsor Castle on Monday. Harvey picked me up at the hotel and the four of us toured Windsor Castle and its surrounding grounds. The castle was the weekend home of Queen Elizabeth and is now the home of King Charles. We toured the Eaton School for Boys, one of the most exclusive boys' schools in the world. We also drove through the campus of the Harlow School for Boys, the chief rival of Eaton. The campus was closed, but we were able to tour the chapel which was said to be about 200 years old. I was taken back to the hotel, and Bob and I had supper together at a nearby pub. We departed for RAF Mildenhall about 5:00 am and our plane departed for Germany at 8:30 Tuesday morning, July 30.

After having been accepted for OCS and before being given a class starting date, I advertised my Volkswagen beetle for sale for the same price I had paid, $600.00. I thought I was going to have a week or two to sell the car, but I received orders to depart in just three days. I lived in a two-room NCO quarters in Charlie Battery and my roommate was a SGT William Parrish. Sgt Parrish offered to buy my car for $450 but only had $30 to make a down payment. I had no other offers and I trusted him, so I agreed to allow him to make monthly payments and I would put a lien on the car for the $420 he owed. He agreed. The day before I was to depart, we drove to the auto registration office, about a 30-minute drive north in the town of Giessen, arriving at 1:00 pm and stood in a line of about eight or nine people. It appeared the German lady was taking about 20 minutes to do each vehicle, but the line was moving along. We were the last in line and she finished the person ahead of us a few minutes before the 3:00 pm closing time.

She simply stood up and closed the window in front of us and told us we would have to come back the next day. I told her I was leaving for America the next morning and could not come back. A rather loud argument ensued, and we ended up in the provost marshal's office. The LTC listened to what had happened and agreed I would get waited on. While we were talking, the German lady left the office and was nowhere to be found. It turned out she had gone to the bank. When she returned about 3:45 she insisted upon arguing with the PM for about ten minutes. I had to be in Frankfurt, an hour's drive south, before 6:00 pm to get paid, and when she opened her window and took my paperwork, she stated she had NOT agreed to do a lien, so I said just forget it, I do not have any more time to argue.

She signed the title over to Sgt Parrish without a lien and we made it to Frankfurt in time for me to get paid. I had prepared a box with my ski boots, ski jacket, and some clothing to send home and I did not have time to go to the post office and send it. I gave Sgt Parrish the $30.00 he had given me, and he said he would send the box the next day. Long story short, I never received a penny and my box never arrived. Sgt Parrish wrote me a letter simply saying he had paid me everything I was owed and had shipped the box. I wrote the car off as a total loss, but Dad wrote to the battery commander twice trying to get them to do a more thorough investigation. Their answer was that without a lien or some written agreement it was his word against mine.

The members of the HHB Fire Direction Center became friends for life. We took a group picture of eight of us that is in the photo section. There are many other pictures of members of our section.

I already talked about Bob Waters, a farmer from Steward, IL. He and his wife Donna still live on their farm. Albert Lark arrived after the photo and was the only member who had his wife in the country. We have a couple of pictures that include Ruth on our adventures. They moved to CA where Albert worked for Gestetner Corporation selling copy machines until his retirement and he passed away in January 2018. Ruth has moved to Montana. Herbert Tabraham moved to Dallas, TX where he worked for a utility company until his retirement. He still lives with his wife Glenda. Raymond Gaglone and William Miracle both returned to Chicago, IL. William was a weightlifter who loved big women. Virgil Hill, Stanley Kulikowski, Milus Lewis, and Jackie Kimbell made up the enlisted men in the section. Over the years we talked on the phone, exchanged Christmas cards, and years later starting emailing back and forth. I began publishing a roster of addresses and phone numbers and sent out updated copies any time there was a change.

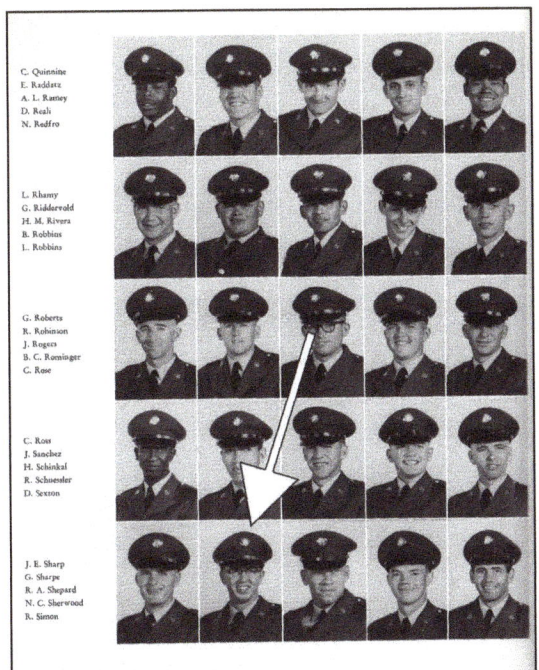

C. Quinnine
E. Raddatz
A. L. Rattey
D. Reali
N. Redfro

L. Rhamy
G. Ridderrold
H. M. Rivera
B. Robbins
L. Robbins

G. Roberts
R. Robinson
J. Rogers
B. C. Ruminger
G. Rose

C. Ross
J. Sanchez
H. Schinkal
R. Schneider
D. Sexton

J. E. Sharp
G. Sharpe
R. A. Shepard
N. C. Sherwood
R. Simon

Basic Training, Fort Ord, California.

Photos from the souvenir book published for our basic training class.

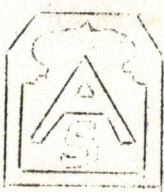

Gerald Sharpe Finishing Basic

U.S. Army Private Gerald W. Sharpe is presently undergoing basic Infantry training at the Monterey Peninsula training center, Fort Ord, Calif.

The eight-week program consists of such military subjects as rifle marksmanship, Infantry tactics, first aid, military justice and the conservaton and maximum utilization of military supplies. Upon completion of his basic training, the soldier will go on to either advance Infantry training or to one of the many Army specialist schools.

Private Sharpe is the son of Mr. and Mrs. William S. Sharpe, 8300 Irving. He graduated from Bellview Prep. School in 1959 and attended Western State College for two years prior to entering the Army.

Lowell Shissler In Joint Services Pacific Exercise

Participating in "Exercise Sea Wall", a joint Army, Navy and Air Force amphibious maneuver Sept. 15-20 on San Juan Island, Wash., while serving with the staff of Commander Amphibious Squadron Five, was Lowell E. Schissler, radioman seaman, USN, son of Mr. and Mrs. Paul H. Shissler of 8350 Green Ct.

Training Photo, September 61.

BATTALION STAFF

Major Paul A. Roberts   Battalion Commander

Captain Bernard N. Stout   Battalion Executive

Captain William F. Stecher Jr.   Battalion S-3

M/Sgt James E. Jubert   Battalion Sergeant Major

SFC Earnest Wilson   Operations Sergeant

BATTERY STAFF

Captain Clyde R. Miller   Battery Commander

1st Lt. Elmer Naber   Battery Executive

M/Sgt John L. Brown   First Sergeant

GRADUATION EXERCISES
ADVANCED INDIVIDUAL TRAINING

"S" BATTERY
4TH BATTALION
2D TRAINING REGIMENT
Fort Carson, Colorado

25 January 1962
1500 Hours
Mountaineer Theater

OUTSTANDING TRAINEES

Overall Outstanding Trainee of the Cycle

Palowski, William A. PVT E-1
NG25878014

Outstanding Trainee in MOS 140.00

Sharpe, Gerald PVT E-1
RA17607074

Outstanding Trainee in MOS 152.00

Roth, Ernest G. PVT E-1
US52516928

PROGRAM

March Music	179th Army Band
Invocation	Chaplain Blunt
Introductory Comments	Capt Clyde R. Miller
Introduction of Guest Speaker	Major Paul A. Roberts
Graduation Address	Lt. Col Henry J. Schroeder
Presentation of Awards	Lt. Col Henry J. Schroeder
Benediction	Chaplain Blunt
National Anthem	179th Army Band
Caisson Song	179th Army Band
March Music	179th Army Band

Graduation program from Fort Carson showing I was the Outstanding Trainee in MOS 140.00

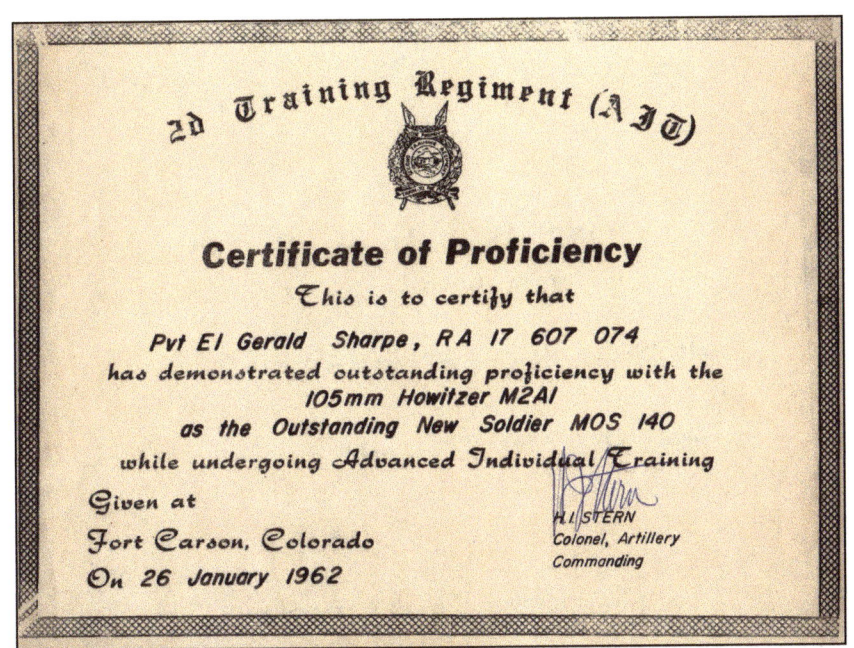

2d Training Regiment (AIT)

Certificate of Proficiency

This is to certify that

Pvt E1 Gerald Sharpe, RA 17 607 074

has demonstrated outstanding proficiency with the
105mm Howitzer M2A1
as the Outstanding New Soldier MOS 140
while undergoing Advanced Individual Training

Given at

Fort Carson, Colorado

On 26 January 1962

H.I. STERN
Colonel, Artillery
Commanding

Advanced Individual Training (AIT) at Fort Carson, Colorado.

ROYAL ORDER OF ATLANTIC VOYAGEURS

KNOW ALL YE ATLANTIC VOYAGEURS THAT ON THIS _____ DAY OF _____ 19__

THERE APPEARED WITHIN MY DOMAIN THE
USNS GEN. A. M. PATCH (T-AP 122)
BOUND ACROSS THE BRINY DEEP FOR THE PORT OF
BREMERHAVEN, GERMANY
AND KNOW ALL YE THAT

WAS DULY INITIATED INTO THE ROYAL ORDER OF
ATLANTIC VOYAGEURS

THE DOMAIN OF NEPTUNE — RULER OF THE RAGING MAIN

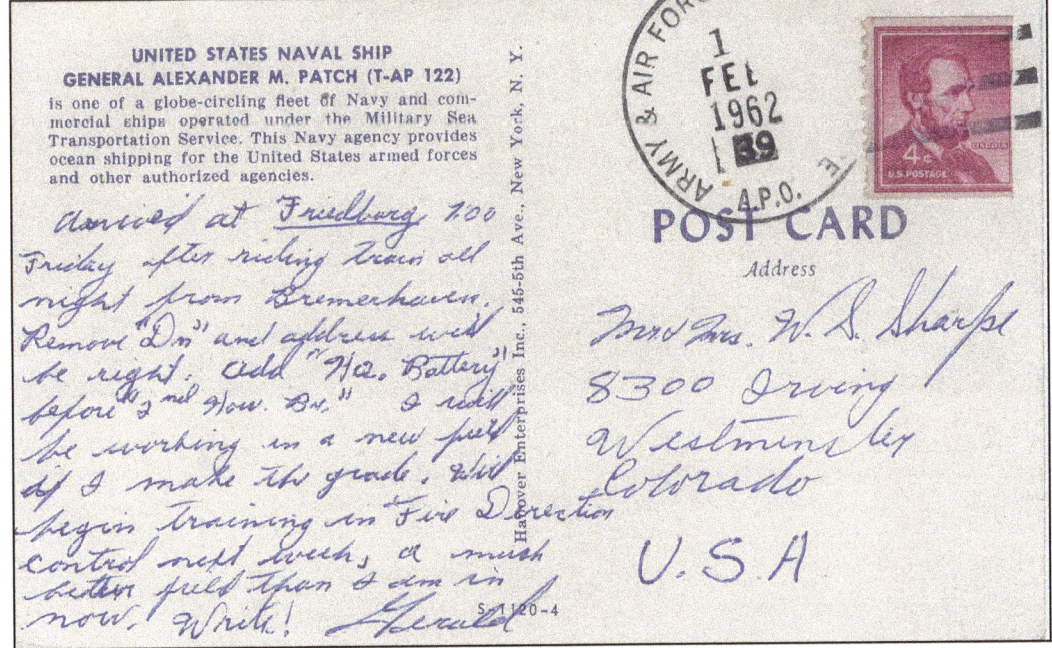

UNITED STATES NAVAL SHIP
GENERAL ALEXANDER M. PATCH (T-AP 122)
is one of a globe-circling fleet of Navy and commercial ships operated under the Military Sea Transportation Service. This Navy agency provides ocean shipping for the United States armed forces and other authorized agencies.

Arrived at *Friedberg* 100 Friday after riding trains all night from Bremerhaven. Remove "Dn" and address will be right. Add "HQ. Battery" before 2nd How. Bn." I will be working in a new post if I make the grade. Will begin training in Fire Direction control next week, a much better field than I am in now. Write! *Gerald*

POST CARD

Address

Mr. & Mrs. W. S. Sharpe
8300 Irving
Westminster
Colorado
U.S.A

HQ BTRY
27
2nd HOW BN

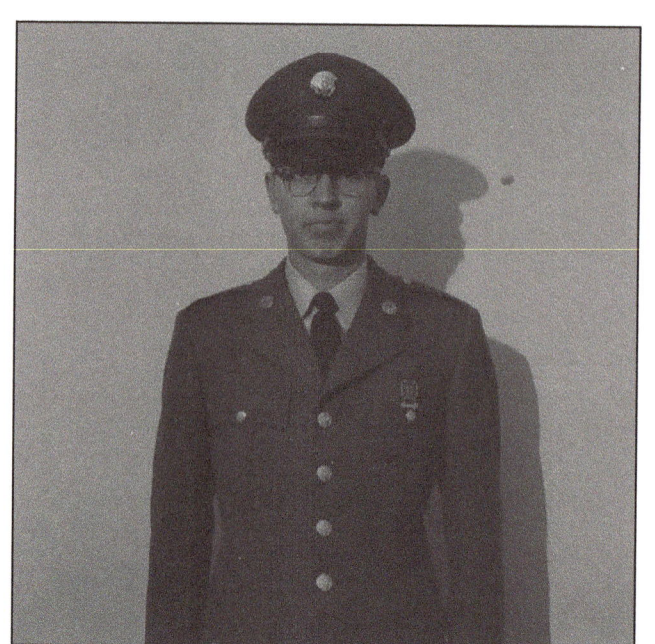

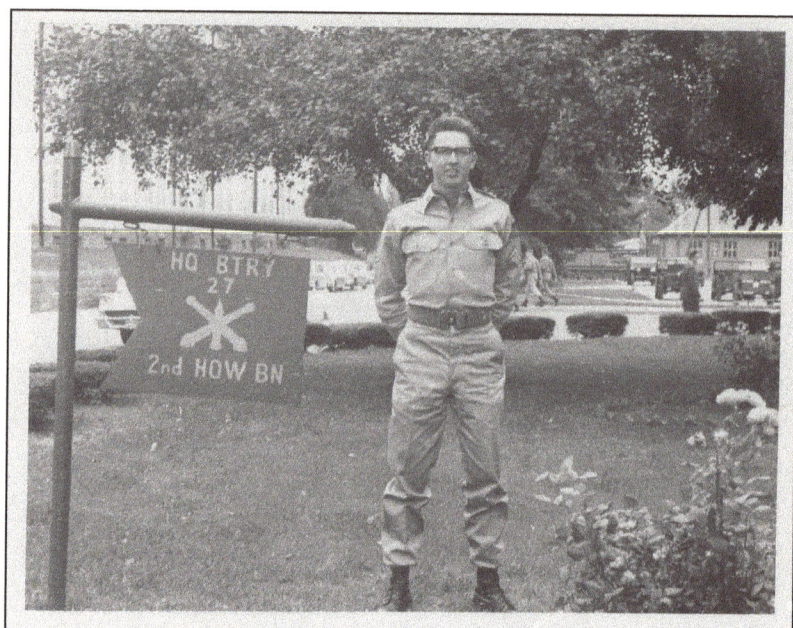

Photos from my enlisted service in the 2nd Howitzer Battalion, 27th Field Artillery on Ray Barracks, in Friedburg, Germany. left—from February 1962, right—from October 1962

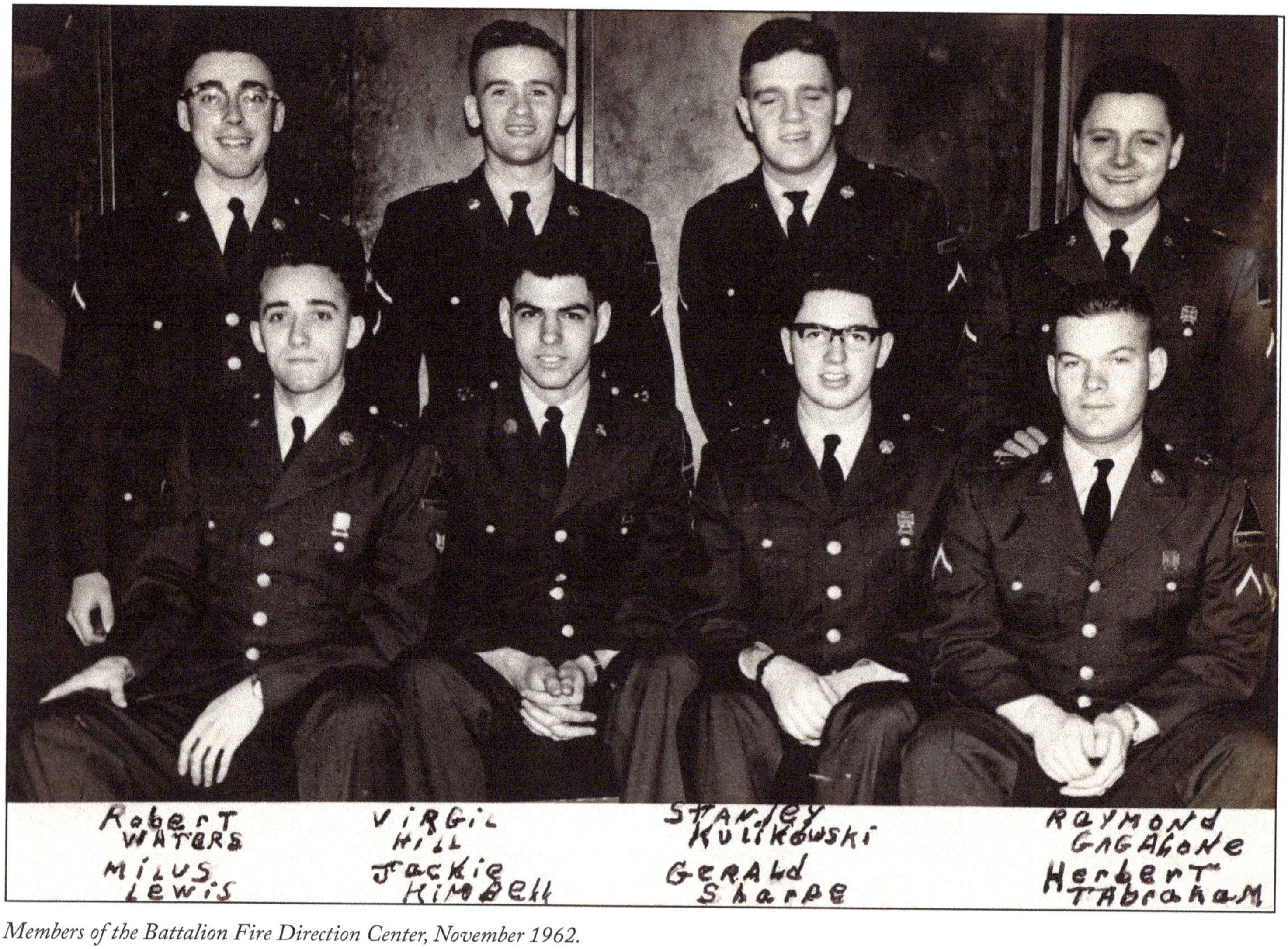
Robert Waters
Milus Lewis
Virgil Hill
Jackie Kimbell
Stanley Kulikowski
Gerald Sharpe
Raymond Gagalone
Herbert Tabraham

Members of the Battalion Fire Direction Center, November 1962.

May 3, 1962

Gerald W. Sharpe Participant in Army Exercise in Germany

Army Pvt. Gerald W. Sharpe, son of William S. Sharpe, 8300 Irving, participated in Command Post Exercise Grand Slam I, a five-day Central Army Group (CENTAG) exercise in Germany which ended April 13.

Grand Slam I involved headquarters units from the German, French and U. S. Armed Forces which are assigned to CENTAG, and was designed to test operational plans and procedures of these forces. CENTAG is a major element of NATO in Europe.

The 20-year-old soldier is a 1958 graduate of Bellview Preparatory School and attended Western State College of Colorado in Gunnison.

Grundig Stereo Console

Volkswagen purchased by Bob Waters and me. Bob is in the left photo, I am in the right one.

Ski equipment I purchased in Germany.

I was assigned to the vehicle circled in red.

2nd Bn, 27th FA Motor Pool, Friedburg, Germany.

Equipment layout & inspection in Motor Pool. 2 of the 3 vans used for the battalion headquarters.

Shooting to gain rifle and pistol qualification.

Photo from rear of the FDC van showing convoy.

Individual photos of members of the battalion Fire Direction Center.

My best friend, roommate, and fellow chart operator was Bob Waters. Shown bending over an FDC map, my chart was next to his; skiing; and across from the famous Colone Cathedral – we climbed to the top.

Milus Lewis

Jackie Kimbell

Bob Waters, Albert Lark, and Milus Lewis.

Virgil Hill and Ruth & Albert Lark.

More individual photos of members of the battalion Fire Direction Center.

Raymond Gagalone (Both from Chicago) William Miracle

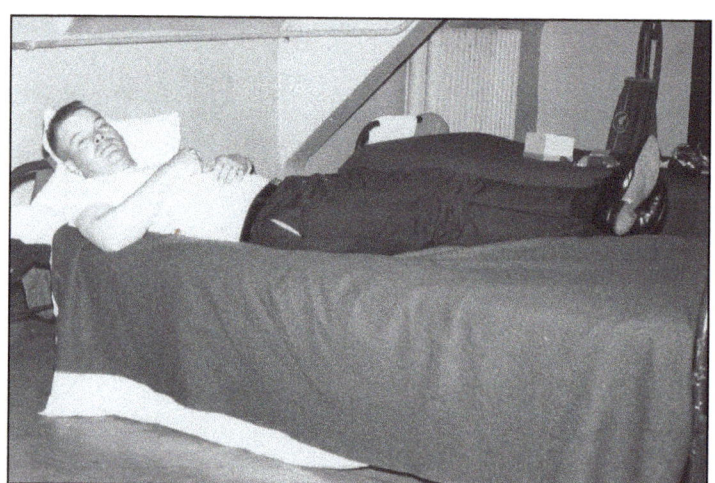 

Herbert Tabraham, driver for the battalion commander.

Front row seats, Nürburgring 24-Hour Endurance Race. Three views, connector from the burned radio.

Dacau, Nazi Concentration Camp, NE of Munich. Bob Waters and I visited in October 1962 Below are the ovens used to cremate dead bodies.

Above left: Bob Waters on exterior road. Above right: Catholic Mortal Agony of Christ Memorial Chapel. Below, a 1946 aerial photo of the camp. Headquarters at bottom/cremation ovens at top.

LTC Clarence F. Ax, Battalion Commander, presented this watch as a reward for winning the 3rd Armored Division Artillery Soldier of the Quarter for the period October 1, 1962 to December 30, 1962.

LTC Clarence Ax, Commander, 2nd Bn, 27th FA presenting me with the above watch.

Bob Waters and I with the battalion commander's antique Mercedes.

Lieutenant General Hugh P. Harris, CG of 7th Army, presenting my 7th Army NCO Academy diploma: immediately following he pinned on sergeant stripes.

MISSION

The mission of the Seventh US Army Noncommissioned Officers' Academy is to develop within the Noncommissioned Officer:

A. An ability to recognize his responsibilities.

B. A willingness to assume his responsibilities.

C. The confidence to apply his technical knowledge.

D. The leadership techniques that apply to his office.

E. The high personal and professional standards that must be maintained within the noncommissioned officers corps.

THE ACADEMY SHIELD

White and blue represent the state colors of Bavaria.

The figure is the "Munich Kindl", a child dressed as a monk. This city symbol of Munich represents the birthplace of the Academy.

The Seventh US Army insignia denotes the parent organization of the Academy.

The shield was officially adopted as the Academy Insignia on "Organization Day", 17 October 1956.

SEVENTH
UNITED STATES
ARMY
NON-COMMISSIONED OFFICERS' ACADEMY

GRADUATION PROGRAM

SEVENTH US ARMY NONCOMMISSIONED OFFICERS' ACADEMY

LT COLONEL WESLEY J. SIMMONS . . . . .COMMANDANT

LT COLONEL DON B. CLARK . . . . . . .ASSISTANT COMMANDANT

MAJOR THEODORE S. RIGGS, JR., . . DIRECTOR OF INSTRUCTION

CAPTAIN HAROLD D. WHITE . . . . .CHAPLAIN

GUEST SPEAKER

LIEUTENANT GENERAL HUGH P. HARRIS
Commanding General
Seventh Army

Class 63 - 18
Student Detachment "A"

Entered: 1 April 1963
Graduated: 26 April 1963

*Invocation. . . . . .Chap(Capt) White
Introduction of Guest Speaker . . . . . .Lt Col Simmons
Address by Guest Speaker. . . . . .Lt Gen Harris
Introduction of Award Winners . . . . . .Capt Smith
Presentation of Diplomas. . . . . .Lt Gen Harris
Address by Class Speaker. . . . . .Sp Sharpe
*Benediction . . . . .Chap(Capt) White
*British & American National Anthems . . .30th Army Band

*Please stand

R E C I P I E N T S   O F   A C A D E M Y   A W A R D S

DISTINGUISHED GRADUATE

GENERAL PATTON AWARD FOR EXCELLENCE

*SP4 Gerald W. Sharpe, Hq&Hq Btry 2d How Bn 27th Arty

HONOR GRADUATES

*SP5 Richard H. Capman, 251st USASA Proc Co

*SFC George R. Barnes, 54th Sig Co 504th Sig Bn

Sgt Jesse L. Johnson, Co B 2d Bn 509th Abn Inf

DISTINGUISHED LEADERSHIP GRADUATES

GENERAL DOUGLAS MAC ARTHUR AWARD

*SP5 Richard H. Capman, 251st USASA Proc Co

Sgt Robert N. Law, Btry A 2d How Bn 18th Arty

*SFC George R. Barnes, 54th Sig Co 504th Sig Bn

ASSOCIATION OF THE U. S. ARMY AWARD

Sp5 Edward A. Kieta, Hq Co 320th USASA Bn

GENERAL BRUCE C. CLARKE AWARD

*COMMANDANT'S INSPECTION

SFC Charles L. Booth, Fir Btry 3d Msl Bn 21st Arty

*Indicates double award winner

552P CST APR 25 63 KA701

SSD240 K CDU241 259 24 PD INTL CD BADTOELZ VIA MACKAY 25 1720

LT MR AND MRS WILLIAM S SHARPE

8300 IRVING WESTMINISTER (COLO)

WILL GRADUATE NUMBER ONE TOMORROW OUT OF 144 STUDENTS HOPE

TO BE SERGEANT TOMORROW

CFM 8300 144

(41).

Telegram I sent my parents when I found out I would graduate number 1 in the class.

L to R: (Back Row): SP4 Gerald W. Sharpe (Promotion to Sergeant E5, Distinguished Graduate & General Patch Award for Excellence; Sgt Robert Law, Distinguished Leadership Graduate and General Douglas McArthur Awards, and & SFC George Barnes, Honor Graduate, Distinguished Leadership Graduate, and General Douglas McArthur Award. (Front Row) Sgt Jesse Johnson, Honor Graduate, SP5 Richard Capman, Distinguished Leadership Graduate and General Douglas McArthur Award; SP5 Richard Kieta, Association of the Army Award; and SFC Charles Booth, winner of Commandant's Inspection (Shined the bottom of his boots) and the General Bruce C. Clark Award.

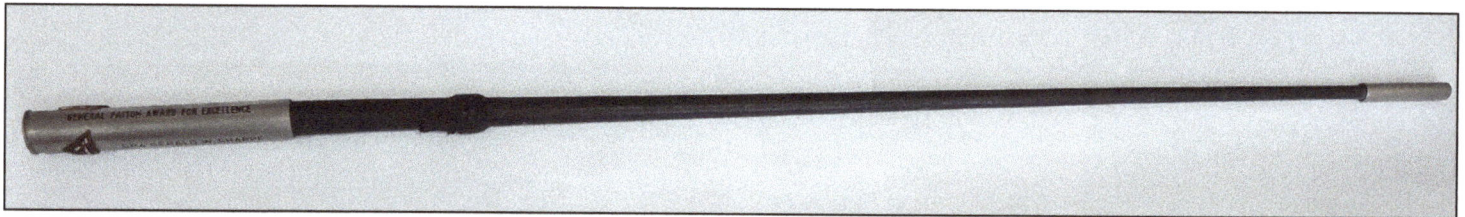

Cigarette Lighter and Swager Stick given to the number one graduate.

At Bad Toelz
Div Man Tops Class

BAD TOELZ — Top man in his Seventh Army NCO Academy graduating class is Sgt Gerald W. Sharpe of Hqs Btry 2d How Bn, 27th Arty.

In the class of 135, Sharpe was named Distinguished Graduate and won the General Patton Award for Excellence.

This award was presented for his superior ability shown during the course of instruction.

Sgt Sharpe

At the same time Sharpe received his new sergeant stripes from Lt Gen Hugh P. Harris, Seventh Army CG and Academy Commandant Lt Col Wesley J. Simmons.

In addition, Sgt Sharpe received a letter of commendation from Col Simmons. The letter was endorsed by Spearhead CG, Maj Gen John R. Pugh, who added further congratulations.

Col Simmons cited Sharpe for "having achieved two of the highest goals a noncommissioned officer can attain." He stated his confidence that the new sergeant would "continue to set the example by maintaining the the same standards of leadership" in his Army career.

27th Arty Sgt Gets 7th Army NCO Honors

BAD TOELZ, Germany (Special) Placing first in a graduating class of 135, Sgt Gerald W. Sharpe, Hq Btry, 2nd Howitzer Bn, 27th Arty, won the 7th Army NCO Academy's distinguished graduate and winner of the General Patton award for excellence.

On hand at Flint Casern to present the awards and address the graduating class was Lt Gen Hugh P. Harris, 7th Army CG. Speaking to the NCOs the general stated that, from his viewpoint, "The pay-off is that you go back to your unit and pass on the information and training procedures received here and thereby improve your unit."

Other honor graduates were: Sp5 Richard H. Capman, 251st USASA Processing Co; Sfc George R. Barnes, 54th Signal Co, 504th Signal Bn, and Sgt Jesse L. Johnson, Co B, 2nd Battle Gp, 509th Airborne Inf; who placed second, third and fourth, respectively.

Capman, Barnes and Sgt Robert N. Law, Btry A, 2nd Howitzer Bn, 18th Inf, were the distinguished leadership graduates. The Association of the United States Army award went to Sp5 Edward A. Kieta, Bad Aibling, Post Hq and the Commandant's Award was won by Sfc Charles L. Booth, Firing Btry, 3rd Missile Bn, 21st Arty.

Gerald W. Sharpe Officer Graduate

Army Specialist Four Gerald W. Sharpe, son of Mr. and Mrs. William S. Sharpe, 8300 Irving, graduated from the Seventh U. S. Army Non-Commissioned Officer (NCO) Academy at Bad Tolz, Germany, April 26.

Sharpe, a fire direction center computer in Headquarters Battery of the 27th Artillery's 2d Howitzer Battalion in Germany, entered the Army in August 1961 and completed basic combat training at Fort Ord, Calif.

The 21-year-old soldier is a 1959 graduate of Belleview Preparatory High School and attended Western State College in Gunnison, Colo.

Article from the Westminster & District 50 Journal dated May 30, 1963

SEVENTH UNITED STATES ARMY
NONCOMMISSIONED OFFICERS' ACADEMY
APO 108

OFFICE OF THE COMMANDANT

AETI   26 April 1963

Mr. & Mrs. William S. Sharpe
8300 Irving
Westminster, Colorado

Dear Mr. and Mrs. Sharpe:

Your son, Specialist Four Gerald W. Sharpe, was recently selected for attendance at the Seventh United States Army Noncommissioned Officers' Academy. This selection itself sets him apart from his fellow soldiers - putting him in the top category of our troops in Germany, since only the best come here as students.

At this time he has been graduated and further distinguished himself by earning two awards. He was designated as the Distinguished Graduate, in turn earning the "General George S. Patton, Jr. Award for Excellence". The ability to achieve distinction in competition with fellow soldiers is commendable, the ability to accomplish this in competition with the finest soldiers in Europe is an accomplishment from which you may draw real pride and satisfaction.

This Academy judges its graduates not only in the light of academics, but predominantly in the development of essential, practical characteristics vital for those who will lead other men. Your son has made a fine contribution to himself, his Army and his country. My congratulations to you.

Sincerely,

WESLEY J. SIMMONS
Lt. Col, Infantry
Commandant

SEVENTH UNITED STATES ARMY
NONCOMMISSIONED OFFICERS' ACADEMY
APO 108

OFFICE OF THE COMMANDANT

AETI   26 April 1963

SUBJECT: Commendation

THRU: Commanding General
          V Corps
          APO 79, US Forces

TO: Specialist Four Gerald W. Sharpe, RA 17 607 074
          Headquarters and Headquarters Batter
          2d Howitzer Battalion, 27th Artillery
          APO 39, United States Forces

    1. While attending the Seventh United States Army Noncommis-
sioned Officers' Academy, you distinguished yourself by attaining
the highest leadership and academic standing within your class.
Having been designated as the Distinguished Graduate, you have
earned the "General George S. Patton Jr. Award for Excellence".
You have brought great credit upon yourself in competition with
the finest noncommissioned officers in Europe. You have achieved
two of the highest goals a noncommissioned officer can attain.

    2. You are to be commended for your outstanding achievements
while a student at this Academy. Your performance gives me confidence that you will continue to set the example by maintaining the
same high standards of leadership and efficiency throughout your
Army career.

                              WESLEY J. SIMMONS
                              Lt. Col, Infantry
                              Commandant

67

United States Army Europe

3rd Armored Division

HEADQUARTERS 2D HOWITZER   BATTALION 27TH ARTILLERY

SPEARHEAD

This

Certificate of Achievement

is awarded to

SERGEANT E5 GERALD W SHARPE RA17607074

BATTERY C, 2D HOWITZER BATTALION, 27TH ARTILLERY, APO 39, US FORCES

for

Outstanding demonstration of leadership qualities and academic proficiency while attending the Seventh United States Army Noncommissioned Officers' Academy during the period 29 March 1963 through 26 April 1963. Throughout the length of the course, Sergeant Sharpe set the example for his classmates in appearance, discipline and all around military knowledge. His exemplary conduct, firm leadership, and scholastic attainment reflect great credit on Sergeant Sharpe, his organization, and the United States Army.

This 30th Day of   April   1963

CLARENCE F AX
Lt Col, Arty
Commanding

AR Form 2421A (Apr 61)   Distr: 1-TAG DA for DCSPER (Career Br) (Off); 1-TAG DA   AGL (1) 5-63-30M-50306
for AGPF (Off); 1-Field Mil File (Off & Enl); 1-File

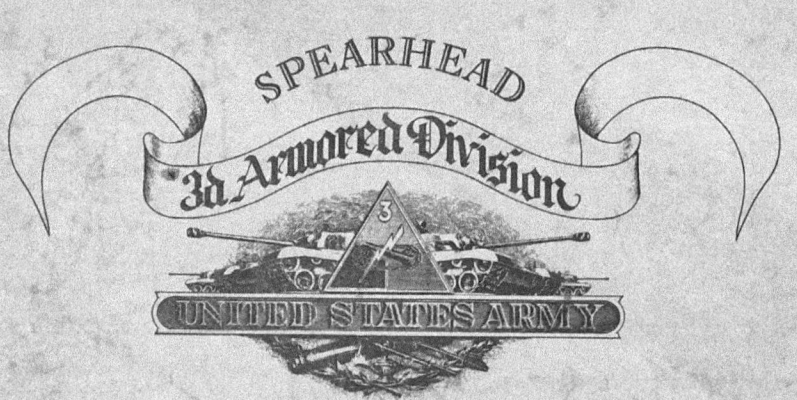

SPEARHEAD

3d Armored Division

UNITED STATES ARMY

be it known that:

Sgt (E-5) Gerald W. Sharpe RA 17607074
Hq & Hq Btry, 2d How Bn, 27th Arty

has demonstrated outstanding qualities in technical proficiencies, leadership, aggressive spirit, high moral standards and is awarded the

General Doyle O. Hickey Award

John A. Pugh
Major General, USA
Commanding

for the month of April 1963

NCO OF THE
QUARTER
APR - JUN

Visit to the US Army 3rd Armored Division on Fliegerhorst Kaserne in Hanau, Germany by President John F. Kennedy on June 25, 1963. The entire division and its equipment was on display.

I was one of the NCO's from the division selected to have lunch with the president in an Army dining facility. A cake was baked in the shape of the President's Navy ship, a torpedo boat. Below, the president cuts the first piece and NCOs can be seen in the background.

Press photographers were given five minutes to take picture. The president then spoke apologizing for the circus but thanking us for our service.

SPEARHEAD

NORMANDY ★ NORTHERN FRANCE ★ ARDENNES
RHINELAND ★ CENTRAL EUROPE

HEADQUARTERS
3D ARMORED DIVISION (Spearhead)
APO 39   US Forces

WEEKLY COMMAND BULLETIN   29 August 1963
NUMBER   25

EXPIRES 29 NOVEMBER 1963

OFFICIAL

---

20. AWARDS: The Commanding General is pleased to announce the following awards.

<u>AWARD OF THE SEMI-ANNUAL GENERAL MAURICE ROSE AWARD</u>

SSGT E6 Pl Robert Northcutt RA15445912 Co C, 1st ARB, 36th Inf Pd: 1 Jan 63-30 Jun 63

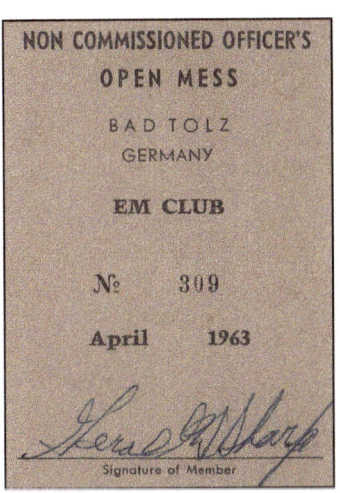

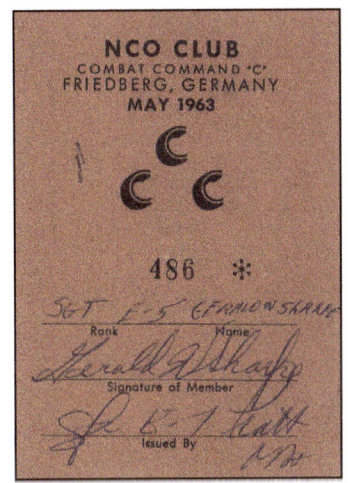

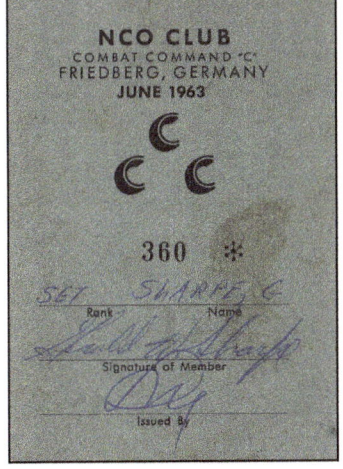

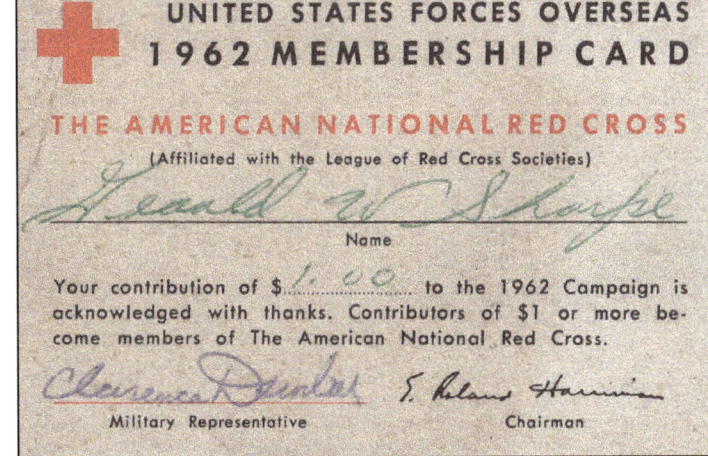

Card 1 (top left)

Diese Karte wurde ausgestellt für:

(NAME) **Gerald W Sharpe**

(Militärische Einheit oder Rechtsstellung) C Btry, 2 How Bn, 27 FA

Im Falle der Festnahme der vorgenannten Person wird gebeten, sofort die nächste US-Militärpolizeistelle (M.P., A.P. oder S.P.) und, wenn möglich, den für die festgenommene Person zuständigen kommandierenden Offizier zu benachrichtigen.

In diesem Zusammenhang sieht Absatz 2(a), Artikel 22 des Zusatzabkommens zu dem Abkommen zwischen den Parteien des Nordatlantikvertrages über die Rechtsstellung ihrer Truppen hinsichtlich der in der Bundesrepublik Deutschland stationierten ausländischen Truppen vor: "Haben die deutschen Behörden die Festnahme vorgenommen, so wird der Festgenommene auf Antrag den Behörden des betreffenden Entsendestaates übergeben."

AE FORM 9 APR 63 3317   AGL (1) 6-63-1M-94634

Card 2 (top right)

LEGAL STATUS OF US FORCES PERSONNEL IN THE FEDERAL REPUBLIC OF GERMANY

1. Please read this card carefully and carry it always.
2. As a member of the United States Forces, civilian component or as a dependent in the Federal Republic of Germany, this card reminds you of your legal obligations and will assist you if you get involved in difficulties with German police authorities.
3. While stationed in Germany, you are subject to, and must respect and obey, the laws of the Federal Republic of Germany, as well as the laws of the United States.
4. In addition to your being subject to the jurisdiction of US military authorities, you may be arrested by the German police authorities and may be tried by the criminal courts of Germany for violations of German law.
5. Here are some DO's and DON'Ts if you are arrested by German police.
   a. DON'T resist arrest or refuse to obey instructions.
   b. DO furnish German police immediately with your name, rank and organization and show them your identification card or passport.
   c. DO request that your Commanding Officer and the nearest US Military Police (M.P., A.P., or S.P.) be immediately notified of your apprehension. The reverse of this card contains a request in German that this notification be accomplished.
   d. DO request the assistance of a qualified interpreter. Remember that any statement you make may be used against you.
   e. DON'T leave the scene of an accident in which you become involved until authorized by the police.

SHOW THE REVERSE OF THIS CARD TO GERMAN POLICE IF ARRESTED

Card 3 (second row, left) — MEAL CARD

MEAL CARD

DATE ISSUED: 30 April 1963

ISSUED TO (Last name, first name, middle initial)
SHARPE, Gerald W.   G 665 479   Room 247

SERVICE NO. OR BADGE NO.
RA 17 607 074

TYPED NAME, GRADE, TITLE, AND ORGANIZATION
RUFUS B ___, 1st Lt., Arty
Commanding ___ C, 27th Arty

AUTHORIZING OFFICIAL SIGNATURE
Rufus B Rogers

DD FORM 1 May 53 714   AGL (1) 11-59-180M-73534

Card 4 (second row, right)

LEGEND

A. Issued to personnel entitled to meals without charge.
B. Issued to personnel who will pay for meals in cash.
C. Issued to personnel of other departments or agencies.

INSTRUCTIONS

This meal card is for the personal use of the person to whom it is issued. It shall not be loaned out, destroyed, or thrown away. When the person to whom it is issued leaves, he will return it to the authorizing official. Persons violating these instructions will be subject to disciplinary action.

CERTIFICATE

I CERTIFY THAT I HAVE READ AND UNDERSTAND THE INSTRUCTIONS SET FORTH ABOVE.

Gerald W Sharpe

Signature of Bearer

Card 5 (third row, left)

US ARMY, EUROPE

Operator's License / Führerschein   № 025938   G

Name of Operator / Name des Fahrers
SHARPE, Gerald W

Issue Date / Ausgabedatum
20 Nov 1962

Signature of Operator / Unterschrift des Fahrers
Gerald W Sharpe

Not valid unless dated and officially stamped / Ungültig ohne Datum und amtlichen Stempel

Grade / Grad   Sp-4
AGO or ASN Nr.   RA 17 607 074

Valid / Gültig: 4 years / 4 Jahre   Until / bis

Organization and APO
Hq Btry, 2nd How Bn, 27th Arty   APO 39

Valid only for Operation of / Gültig nur für:
(X) Passenger Vehicles / Personenkraftwagen
( ) Motorcycles / Motorräder
( ) Trucks / Lastkraftwagen
( ) Other — Specify / Andere — Bestimme

ORIGINAL: To Operator
DUPLICATE: To USAREUR Registrar
NOT VALID WITHOUT GLASSES
AE Form 208   Replaces VR Form 206
30 Dec 55   which is obsolete

Card 6 (third row, right)

RECORD OF ADMINISTRATIVE ACTION

a. All convictions for violations of maximum permissible speed, or for violations indicating disregard for traffic laws, or for negligent driving will be recorded below.
b. All suspensions will be recorded in appropriate columns below and reported immediately to Headquarters, USAREUR, Attn: Registrar of Motor Vehicles.

VIOLATION	ACTION TAKEN	DATE	RECORDING OFFICER	UNIT HQ

AGL (1) 10-61-125M-85214

Card 7 (bottom left)

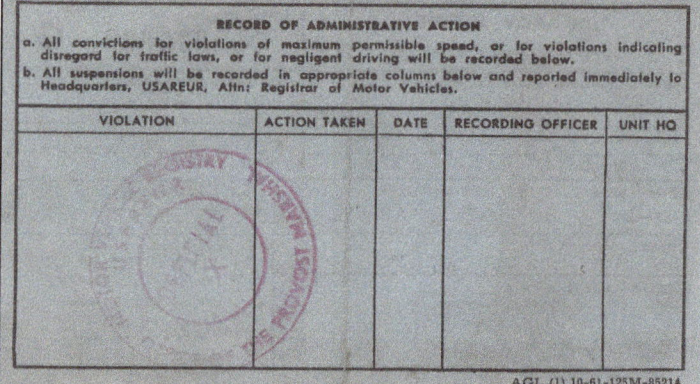

SHARPE

IF YOU SEE A LICENSE PLATE LIKE THIS

3 SOVIET MILITARY MISSION USAREUR

513th MI GROUP
North Mil (2311) 8006

PLEASE CALL ___ OR NOTIFY THE NEAREST MILITARY UNIT

AE FORM 3231, 5 OCT 60

Card 8 (bottom right)

WHAT HAVE YOU SEEN?
A SOVIET MILITARY LIAISON MISSION VEHICLE

REPORT:

1. Time and place of sighting.
2. Color and make of vehicle.
3. License plate number.
4. Number of occupants (military or civilian dress).
5. Grade or rank of occupants.
6. Direction of travel.
7. "SMLM vehicles will not be halted or pursued to obtain above information."

AGL (1) 3-61-150M-81634

Elvis Presley served on Ray Barracks from October 12, 1958, to March 3, 1960. I arrived in January 1962. Below, first day in 1st Bn, 32nd Armor. He often played guitar in the barracks.

Elvis was a good soldier. He was promoted to sergeant and served in a scout platoon. Above, left, with his father in 1959, the first photo of Elvis & Pricilla, and his departure from Ray Barracks

Family photo taken at our home at 8300 Irving on the Belleview College campus in late October, 1963 before I reported to the Artillery Officer Candidate School (OCS) at Fort Sill, Oklahoma on November 1, 1963. L to R Sitting: Thomas Thompson, Marie Thompson, Robert, Milton, Dad, Mother, Aunt Mary Lewis, (the unmarried sister of Mother's father) Eileen, and Renee Thornton. Standing: Theodore Thompson (the youngest brother of Grandmother Ritchie who changed his last name from Lewis for unknown reasons), Grandmother Marion Ritchie, Charlie (in his scout uniform, Stanley, Me, Aunt Irene Thornton (my mother's sister), her husband Fred Thornton, Jeff Thornton, and Fred's Mother, Ethel Thornton.

5. OFFICER CANDIDATE SCHOOL (OCS)

FORT SILL, OKLAHOMA
NOVEMBER 1,1963 TO APRIL 28, 1964

I REPORTED TO FORT SILL, OK FOR FIELD ARTILLERY OCS ON NOVEMBER 1, 1963. OCS WAS DE-signed to commission officers after 23 weeks of intensive physical and educational training. In WWII, OCS graduates were referred to as "90-day wonders." In 1963, there were only two OCS programs, one for Infantry at Fort Benning, GA, and one for Artillery at Fort Sill, OK.

OCS was organized with six batteries and four platoons in each battery. It was divided into three periods: lower class, middle class, and upper class. The lower class lasted for seven weeks including one week of in-processing and orientation. Middle and upper class were eight weeks each. We lived in two-story WWII-era wooden buildings with a fire escapes on one end of the 2nd floor of each building. The lower class lived on one side of the first floor of a platoon's barracks, and the middle class, who wore green felt bands on their epaulets, on the other side. There were two beds to a cubicle, and each bed had a metal hanging frame (see photos) and wooden footlocker. The first floor had an aisle down the middle no one was allowed to walk on. Each day, lower classmen would get on their hands and knees and shine it with old t-shirts, using black shoe polish. When they were done shining, they would buff it with a buffer, making perfectly aligned half circles from one end of the building to the other. I quickly became the platoon expert in using the buffer, and I do not think we ever received demerits for the circles NOT being aligned.

The upper class, who wore red felt bands on their epaulets and wore metal horseshoes called "clickers" on the heels of their boots and shoes, lived on the second floor with one bed to a cubicle and a study desk replacing the second bed. Everything had to be ready for inspection from 7:00 am to 5:00 pm.

Upon our enrollment, the upper class was Class 1-64 while the middle class was Class 3-64. The three classes were assigned to three different batteries: A, B, and C. Each Class had a controlling battery and ours was Battery A. Classes 2-64, 4-64, and 6-64 were assigned to Batteries D, E, and F and were four weeks ahead of us in the schedule. Our Alpha battery commander was CPT James E. O'Bryant, the XO was 1LT Richard S. Anderson, and the four platoon leaders were 1LT Germain H. Gershback (my 2nd Platoon Leader), 2LT Arthur C Atkins, 2LT Kenneth T. Legum, and 2LT Joe R. Younginer. Our 1st Sgt was Casimer F. Konopacki. The officers in Battery B and C are shown in the photo section listed in the graduation booklet.

The lower class was designed to make you quit. It often resulted in the loss of about 30% of the students who started. In the first week, the upper classmen took it upon themselves to give each candidate a nickname. Sometimes it was for something they did, and sometimes it was for how they looked.

On our third or fourth day, we were required to fall out of the barracks wearing our helmets with no liner. This was like having a steel cooking pot on your head with no way to keep it in place. We were standing in formation when one of the upper classmen came over to me, looked me up and down, and said, "From this day forward, your nickname is 'Turtle.' I swear your head virtually disappears under that steel pot." He then shouted out to his contemporaries that, "Another one has been named – Sharpe is Turtle."

Every lower classman was directed to decorate a 3" x 5" loose leaf notebook and always carry it. Your nickname went on your notebook, and I still have mine today.

Every minute of the day was filled. We were awakened at 5:00 am by the tactical officers and we had exactly five minutes to pee, dress in shorts, jock strap, t-shirt, socks, and tennis shoes; make our bed; and get into formation on the grass beside the building. The day sometimes started with physical training (PT) led by the upper classmen, who were called "Redbirds."

We did exercises called the "Daily dozen" (the same ones I learned in basic training and the ones used during my entire career) and we usually ran a mile or two in formation. Some mornings the lower classmen were told to wear their steel pot helmet without the liner, which signaled we were in for an hour of harassment by the tactical officers. On such mornings the middle and upper classes conducted PT on their own. We would be made to do such things as low crawl back and forth under the barracks which were built up on stilts and/or we would perform exercises designed to ensure our helmets bounced up and down on our heads or fell off. If your helmet fell off, you had to do 20 pushups.

We would have races around the barracks in teams of four and while the winner stood at parade rest, the three losers again did pushups. Some mornings, the tactical staff would create new diabolical games such as making us pretend we were inchworms. We were told to lie on our backs and "inchworm" head-first back and forth under the barracks. These activities were designed to make us unhappy, and if we so much as frowned, a lieutenant would ask if we wanted to quit. There were a few mornings someone did, and they would be told to go into the barracks, pack all their belonging in a duffle bag, and report to the headquarters for out processing.

Our class was designated as the "Elephant Joke Class" by our upper class and whenever an upper classman would stop us on the street, we were expected to tell an elephant joke that he had never heard before. My little 3x5 black notebook is filled with pages of elephant jokes, and I have a spiral notebook with many more pages of them as well. Just like other things, if the upper classman had heard the joke before, or if he failed to even smile, he would issue demerits. The following is a sample from my notebook.

ELEPHANT JOKES

Why did the elephant watch reruns on TV?
He did not like the shows the first time.

Why do elephants jump across small streams?
So they do not step on the fish.

Why do elephants have wrinkled knees?
From shooting marbles.

Why does an elephant never forget?
What do they have to remember?

Why does an elephant never forget?
Because no none tells them anything.

What did the banana say to the elephant?
Nothing, bananas cannot talk!

How does an elephant get out of a phone booth?
The same way he got in.

Where do you find elephants?
It depends on where you lost them.

Why are elephants not allowed on beaches?
They do not know how to keep their trunks up.

How do you get 4 elephants into a VW?
Two in the front - two in the back.

Why did the elephant and mouse get married?
They had to.

Why do elephants climb trees?
Do you expect them to climb bushes.

Why do elephants wear green tennis shoes?
So they can hide in the grass.

How to tell if a woman had elephant sex?
She is 16 months pregnant.

Game you should never play with an elephant?
Squash.

How do you raise a baby elephant?
With a forklift.

What is the biggest type of ant?
An eleph –ant?

How do you stop an elephant from charging?
Take away his credit cards.

From morning to night, candidates could get demerits. On Saturday morning, demerit totals were posted on the bulletin board for all to see. There were a thousand reasons you could get demerits, ranging from failing inspection of your cubicle, to having your uniform inspected, or simply failing to answer questions in a loud enough voice or not coming to a proper position of attention quickly enough when asked a question by senior candidates and/or tactical officers. There were certain questions that, when asked, had expected answers and if you gave any other answer, you received demerits.

Demerits were most often given in multiples of five. On weekends, there were two supervised runs called "Jarks," named after the first commandant of the school. These runs went from the OCS area up a small mountain called Medicine Bluff 4 (or simply MB4) and back. The runs were conducted in full tactical gear and as lower and middle classmen, we carried our rifles. It was a little more than four miles round trip. If you received less than 60 demerits, you only ran on Saturday morning. If you received 60 or more, you ran on both Saturday and Sunday mornings.

One of the favorite questions asked of lower classmen was, "You think you have this school made, don't you, candidate?" The standard answer was, "No sir, I do not think I have this school made, sir!" shouted as loud as you could while standing at rigid attention and never with a smile on your face.

From almost the first hour of the first day I decided I was never going to give the standard answer and so when asked I would shout, "Yes sir, you could not drive me out of this school with an atomic cannon, sir." Needless to say, this answer would draw a swarm of upper or middle classmen writing demerits as quickly as they could get pen to paper. I was sometimes asked this question several times a day, and I took great pleasure in putting a big smile on my face when answering. In my first seven weeks I got more demerits than I could count and went on two jarks every week as a lower classman.

Our battery held weekly latrine competitions. There were four platoons in each battery, each in a separate building. Each week, Monday through Friday, the four latrines in the platoon buildings had to be decorated with a theme decided on by the upper classman in charge. The tactical staff judged the results each Monday morning, and the winning platoon received a prize – I cannot remember what the prize was. I was in charge for one week and talked the sporting goods manager at the Lawton K-Mart into lending me archery and fishing items to hang on the walls. We won that week.

November 22, 1963, turned out to be a very memorable day in our lives as officer candidates. President John F. Kennedy was shot in Dallas, TX while riding in a motorcade at about 12:30 pm. He was declared dead about 30 minutes later at Parkland Memorial Hospital. It is said most people can remember where they were when momentous events happen. I was in a survey practical exercise very near the OCS area when the shooting and death were announced over the loudspeaker system. Others in our class have mentioned that they were allowed to watch television in Snow Hall that afternoon as the terrible news was covered. Snow Hall was the large three-story classroom building where classroom instruction was held. President Kennedy, as noted in my college chapter, was the youngest person ever elected to the presidency, was the first catholic elected, and was very popular among young people.

We ate in a huge dining hall that accommodated all the OCS candidates in one sitting. It was arranged with ten-person tables, and the food was served family style in serving dishes. Each table had one or two upper classmen on one end, three or four middle classmen next to them in the middle of the table, and four or five lower classmen on the opposite end. The dining period was exactly 45 minutes, including the time it took for everyone to be seated. We marched in formation to the dining hall and entered upon a signal. Tables were assigned by number each week. The food was served by the lower classmen, and the bowls were always handed first to the senior upper classmen at the end of the table, then passed down to the middle classmen, and if there was any food left in the bowls, to the lower classman. Lower classmen were only allowed to carry one bowl and the bowls were never filled up. It was a game the upper and middle classmen played to make it as hard as possible for the lower classmen to eat.

Some upper classmen were more diabolical than others and they would ensure you had only a few minutes to eat the small amount of food left in the bowls when they finally reached the end of the table. To make matters even more challenging, lower classmen could only sit on the front two inches

REMINISCENCES OF A LIFE WELL SPENT

of their chair, had to sit at a rigid attention, and food had to be brought to your mouth in a square motion, straight up from the plate, and then straight across to your mouth.

As if that was not enough of a challenge, the upper classmen always finished first and some took great delight in issuing demerits for not eating properly. Fortunately, most upper classmen ensured you got at least some food to eat. At the end of exactly 45 minutes a whistle was blown, and you were NOT allowed to touch another morsel of food. Things like dessert and bread and butter simply never made it to the lips of a lower classman. Our lower-class period ended on December 19, and we were allowed to go home for 17 days. I do not think my mother had ever seen me eat a piece of bread and butter. Growing up I would eat toast with butter and jam but never just bread and butter. On my first afternoon home I brought a loaf of bread and a dish of butter from the kitchen to the dining room table and ate five or six pieces of bread covered with butter. The family watched me like I was from another planet until I explained how the dining hall worked.

Because of our dining hall difficulties, we would ask our friends and family members to send us boxes of cookies and candy. This was called "pogey bait." The term originated in the military and meant anything not included in army ration or things bought at commercial establishments, but most often candy. In China, during WWII, marines were issued candy like Baby Ruth's and Tootsie Rolls. The Chinese term for a prostitute was a "pogey." Thus, the marines began calling this candy pogey bait. There was a standard rule for the receipt of pogey bait in OCS by a lower classman. Mail and packages were handed out after the evening meal. If a lower classman received a box, they had to share it with everyone on the first floor, middle classmen first, and it had to be all eaten before lights went out or it had to be thrown away. Some middle classmen took great delight in emptying the boxes before any lower classman, including the person receiving the box, got any of it to eat. Every lower classman had a big brother and some middle classmen had two little brothers. I was lucky in that my big brother, candidate David Haddock, struggled to pass gunnery and since I had been the chief of a Fire Direction Center, I knew gunnery inside and out. We quickly came to an unwritten agreement. In return for helping him study, when I received a box, I would give it to him, and he would distribute a cookie or a couple pieces of candy to each middle classman and he would then return the box to me to share with my classmates. As middle classmen we had to share with upper classmen.

I found the academic portion of OCS to be fairly easy, as I could concentrate my studies on subjects other than gunnery. I most enjoyed rotating through leadership positions. Middle classmen were assigned positions similar to noncommissioned officers, such as squad leader, while upper classmen rotated through officer positions at both the battery and battalion level such as battery commander, XO, and platoon leader and battalion commander, Battalion XO and S1, S2, S3, and S4. I did well in my tactical staff ratings and scored high on candidate rankings as we had to rank every person in our platoon from number one to number 30-something every two weeks. As I recall I was fortunate to be in the top ten in every rating period.

Following Christmas, we welcomed Class 7-64 on January 5 as the new lower class into our platoons and watched Class 1-64 graduate and become 2nd lieutenants. In middle class I became a part-time gunnery instructor for upper classmen in my barracks during our two hours of study hall every evening and as a result I was cut some slack in receiving demerits for silly stuff. I was still far too self-confident to suit many upper classmen, so even in middle class I managed to Jark twice every weekend.

I clearly remember my little brother, candidate Michael W. Totten, Jr. I was called in to see Captain O'Bryant and told that I was being assigned a little brother with some special challenges. I was told his father was BG Michael W. Totten, Sr., the Assistant Commandant of the Artillery School and that he had a number of famous relatives including his grandfather, GEN George Patton of WWII fame. CPT Bryant also explained that Candidate Totten had officially requested that he have no contact with his mother and father, but I was instructed to explain to him that he had to understand that while we would try our best to honor his wishes, we could not stop his parents if they were insistent on seeing him. In my first discussion with Candidate Totten, he explained that his parents had wanted him to go to West Point, but he refused. He told me he went to Williams College in Williamsburg, Massachusetts, and then got a master's degree from American University in Beirut, Lebanon. Following graduation, he signed up for a two-year archeology expedition in the middle east to, in his words, stay far away from his parents. He explained that when he returned to the USA, he had gone to see a recruiter to see if he qualified for the Infantry OCS at Fort Benning, GA. He said he told the recruiter he DID NOT WANT TO GO TO FORT SILL. Apparently, the recruiter said that yes, he was qualified, and they would ensure he went to Fort Benning.

But after signing all the papers, enlisting, and attending basic training — where he was selected as the outstanding trainee of his cycle — he was informed that someone saw fit to change his OCS assignment to Fort Sill. Candidate Totten was sure his father had something to do with the change and he was not at all happy about it. Turned out the battery did get a few calls from his mother, and she even showed up once demanding to see her son. Fortunately, he was out of the area. His mother spoke at an Officer Candidate Wives Club meeting and our classmate, Ted Doucette, was designated as her escort officer. Ted recently wrote that he, "was told she specifically asked for an escort by someone in the class from 'back home.' And I guess I was the closest thing to back home as you can get. I lived about 7 or 8 miles from their estate."

Candidate Totten turned out to be an outstanding student and leader. We were informed in early March that OCS had been tasked to put together a soccer team to play a German team on Sunday, March 22. Middle classman Totten came to my cubicle and told me he had been requested to play by name, and he was sure his parents both set up the game and had his name put forward as a team member. He asked me to get him taken off the team. I suggested we go see the battery commander. CPT O'Bryant asked if he was a good soccer player. He said he was, but he didn't want to perform for his parents. He was told, sorry, but OCS needs you to play as we have very little chance of beating this team. He reluctantly agreed and he played well, but we lost. I believe Candidate Totten was the distinguished graduate of his class.

As week sixteen approached, we were more than ready to ascend to upper class status on February 29. We traded in our green tabs for red ones that we received from Class 3-64 who was graduating, we had clickers put on our dress shoes and boots, and we looked forward to moving upstairs in the platoon buildings. We held a "Red Bird Dinner" at the officer club on Saturday evening, February 29. Each platoon also had an informal ceremony giving our green tabs to our little brothers in Class 7-64 and they in turned welcomed Class 9-64 to lower class status. We looked forward to now occupying officer positions and the most senior ones were assigned to those near the top of the class standings so we could be better evaluated on our leadership abilities.

The final exercise for every class, about two weeks before graduation, was an "escape and evasion" course that started midday and ran late into the night. We were briefed that the exercise would be in two parts. First, a day-time part in which we were all given a map that had two "safe" areas marked on it. We were all supposed to get to the first safe area during daylight without being captured by the aggressor force. We would have supper in the safe area and then we would have to try to get to a second safe area at night. The exercise started with our convoy of six vehicles loaded with the upper classmen in my class being ambushed by school support soldiers acting as the aggressor force. We jumped off the trucks and ran off into the woods and up a ridge. Some students were captured in the ambush, but after some harassment they were let go to try to complete the exercise.

As we escaped, Jim Chapmen and I, both of us prior service students, fell in together. Jim would become the distinguished graduate of our class while I came in fourth. We would later be classmates as majors at the Command and General Staff College at Leavenworth, KS. There was a high wooded ridge that ran along one side of the road where we were ambushed, and the safe area was on the other side of the road about three or four miles further up the road. We climbed to the top of the ridge and started following it up the road. We could see aggressors patrolling the road on foot and in jeeps. We were alone and we didn't know where our classmates were. After walking for about an hour and getting close to where we could see the safe area across the road, Jim said quietly. "Jerry, freeze, do not move. There is a copperhead snake behind you coiled up to strike." He continued, "I am going to get a big rock and when I throw it at the snake, I will say RUN. When I do, run forward as fast as you can." Less than a minute later he said "Run!" and I did so. About a second later I heard a rock hit the ground and I stopped and turned around. Jim had made a perfect throw and the large rock had smashed the snake's head, killing it instantly.

After taking several deep breaths over escaping a potentially deadly situation, we talked about what to do. We got a heavy forked stick and hung the snake on it. We snuck our way down the mountain to where we were right across from the safe area and waited until there were only two aggressor soldiers between us and the safe area. They were walking down the road and we waited until they got as far away from us as they were going to get and as they were turning around to come back, we burst out of our hiding place shouting, "Copperhead snake, copperhead snake." We were waving it in the air shaking it making it look like it was alive. The two aggressors stopped in their tracks and before they could recover, we were across the road and ran into the safe area. We were the only two that did NOT get captured in the first phase. We hung the snake on a tree as a souvenir of our class.

After supper, we had to wait until after dark for the night portion to begin. We were briefed on the course for the night. We were shown the safe area on our maps where we were supposed to get to. Between us and the safe area was a small mountain bounded on one side by a large lake that we were told was off limits. We were told it would be patrolled by two speed boats with flood lights. Anyone found in the lake would be considered captured. On the other side was a paved road, which was also off limits.

Four of us, all who would end up as distinguished graduates (top five), including Jim Chapmen, Mike Garcia, Bill Clark and myself, took our supper on paper plates and sat far away by the lake. We sat to eat on a partially submerged log. We looked at the map and concluded that the aggressors likely had enough people to create a line from the lake up over the mountain down to the road that would be nearly impossible to sneak past. We talked about what we might do to not get caught and the more we talked, using the lake seemed to be a possible solution. We talked about the lake being off limits, but the language that was used did not seem to indicate that we would get kicked out of school if we were caught using the lake, only that we would be considered "captured" if we were caught in the lake. Someone suggested that since it would be a long swim, about a mile, perhaps we might consider using the large log we were sitting on to both support ourselves and hide from the searchlights on the boats. One real disadvantage of the idea was that it was late March, and the water was still very cold.

We were told there would be a siren sound to start the night exercise about an hour after dark. Each of us had his poncho rolled up on his pistol belt and after discussing how to have dry clothes, we decided to strip naked, tie our boots and clothes inside our ponchos, using one boot lace to tie the poncho shut so it would float, and using the other boot lace to tie the poncho to the log. We then pushed the log into the water and all four of us stayed on the side of the log nearest the shore while keeping our left arm on the log to support ourselves. We then simply kicked to propel the log along the shoreline well enough away so that the hand-held flashlights being used by the aggressors could not reach us in the dark. It was bitterly cold, and we talked about pulling onto shore, getting dressed, and trying to get through the aggressors, but we agreed to keep going as we could see aggressor flashlights along the water ahead and figured if we could just get past the flashlights, we would be home free. Twice, speedboats came by shining their spotlight on the log, but we stopped moving and hid behind the log for the few seconds they kept the light on the log before they moved off to sweep the light along the shore. I suspect we were only in the water about 30 or 40 minutes by the time we pulled in to shore, but we were almost frozen. We detached our ponchos from the log, and the four of us pushed it back out in the water where it hopefully stayed. We quickly got dressed after using our socks to dry off and then put our ponchos on and huddled under them getting warm for about half an hour. We then rolled our ponchos up, put them back on our pistol belts, and we made our way around to the roadside of the safe area and walked into camp.

We reported to the exercise desk and were told there were still some 20 students not yet accounted for and we again were the only ones not captured. There were hot drinks and snacks, and we had no sooner gotten some than we were accosted by our battery commander. CPT O'Bryant wanted us to describe how we had evaded capture, but we said we had agreed we were not going to reveal how we did it and we were not going to answer any questions. He said OK, but I will see you four in my office tomorrow morning. The next day he called us in and told us he suspected we had broken the rules, but did not know exactly how. He asked if we were sticking to our position that we were refusing to answer any questions. We said we were. He told us that he had no desire to kick us out of school as we had been the best candidates he had ever had, so he was simply going to put us in charge of the last two Jarks and even though we were upper classmen we would carry our rifles. This was the last weekend before our graduation, and we were very happy to have gotten off with only two Jarks. Just as in lower and middle class, I made two Jarks every weekend of upper class. I recently learned from our classmate Stanley Ball, of Battery C, that he had the distinction of not going on a single jark during our 23-week course. I had already known that Dave Salmon had also escaped running up the mountain.

We graduated on April 28, 1964. Before the actual graduation ceremony, on Friday, April 24 at 1615 hours, we had a parade in which the class marched in a "Pass-in-review" for the school commandant. There were five Distinguished Graduates designated from the 55 students in the class and I was number four of the five. As previously mentioned, Jim Chapman was # 1, Mike Garcia # 2, Bill Clark #3, and Ira Reed # 5. All but Bill stayed in the Army and retired with more than 20 years of service. The night before graduation, April 27, we had a reception and dinner dance for the class and faculty at the Officers Club. I was the narrator for a medley of army songs sung by a group of classmates. We had our graduation ceremonies the next morning in the auditorium of Snow Hall, the large building that held classrooms and instructor offices. We received our second lieutenant bars and, by tradition, received our first salute as an office from our first sergeant. Afterwards, also by tradition we were told, we each handed the first sergeant a one-dollar bill. Once again, I do not believe any of my family traveled to Fort Sill for my graduation.

 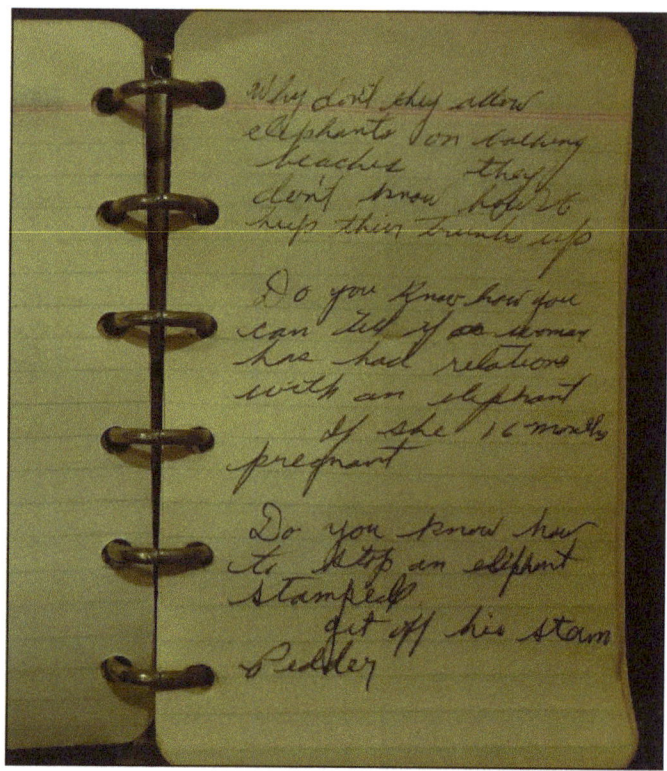

Black notebook carried by every candidate with your "nickname" clearly shown. In lower class there was no color behind the "US". In middle class it was green and red in upper class.

JARK:

The term "Jark" was coined by the OCS cadre to describe a fast-paced trip from Robinson Barracks to the top of Medicine Bluff 4 (Known as MB4) and back at port arms, a physically strenuous task. The events were held every Saturday and Sunday for those candidates who had accumulated a certain number of demerits. The step was expected to be 30 inches and the pace was 130 steps per minute. The prescribed uniform was baseball cap, fatigues, combat gear, and rifle. Total distance was 4.2 miles.

Thank you to my classmate, Richard Chricchio, who shared his class photos shown below.

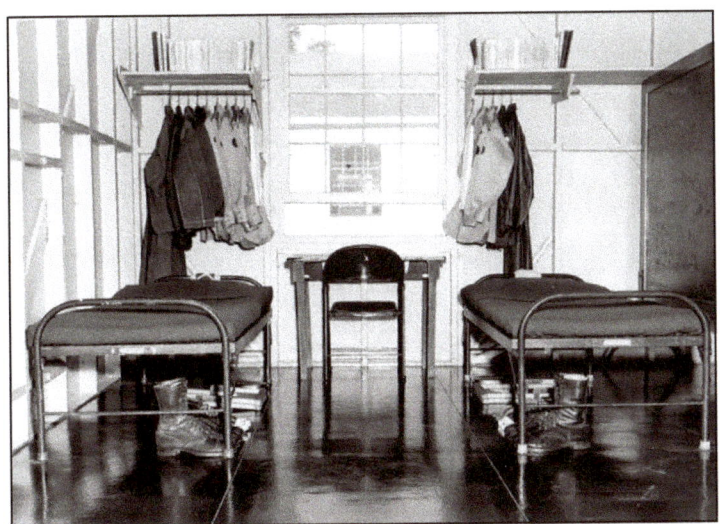

Typical OCS cubicle.

Physical training conducted daily.

Top left, picking dandelions. Top right, Marching in the parking lot. Bottom left, center back is Ed McCormick, my roommate after graduation. He became engaged to my first wife, as we lived next door to her. She changed her mind. Above right, aligning the ranks.

Standing at a "Brace."

Outside the OCS Operations Office

Proud new "Redbirds"

Rappelling down Medicine Bluffs

River crossing at bottom of Medicine Bluffs

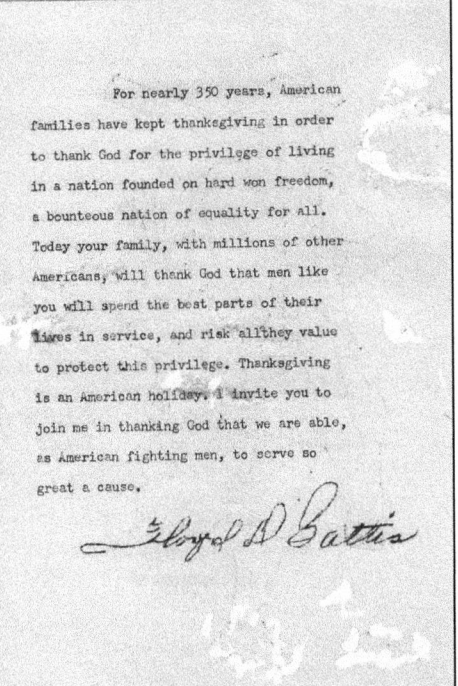

For nearly 350 years, American families have kept thanksgiving in order to thank God for the privilege of living in a nation founded on hard won freedom, a bounteous nation of equality for all. Today your family, with millions of other Americans, will thank God that men like you will spend the best parts of their lives in service, and risk all they value to protect this privilege. Thanksgiving is an American holiday. I invite you to join me in thanking God that we are able, as American fighting men, to serve so great a cause.

Floyd D. Battis

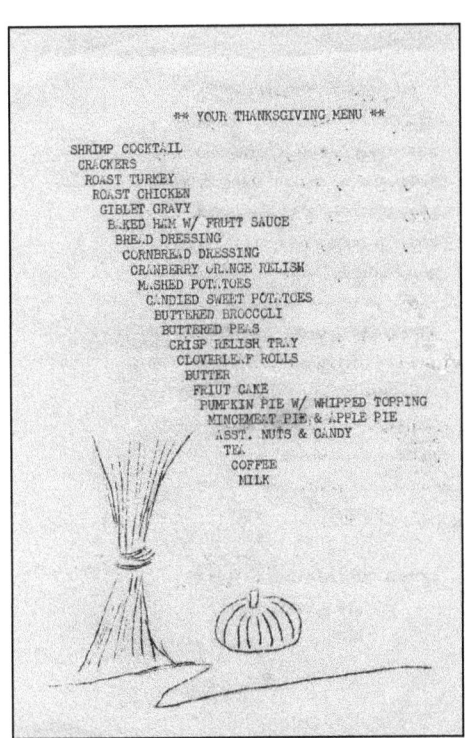

** YOUR THANKSGIVING MENU **

SHRIMP COCKTAIL
CRACKERS
ROAST TURKEY
ROAST CHICKEN
GIBLET GRAVY
BAKED HAM W/ FRUIT SAUCE
BREAD DRESSING
CORNBREAD DRESSING
CRANBERRY ORANGE RELISH
MASHED POTATOES
CANDIED SWEET POTATOES
BUTTERED BROCCOLI
BUTTERED PEAS
CRISP RELISH TRAY
CLOVERLEAF ROLLS
BUTTER
FRUIT CAKE
PUMPKIN PIE W/ WHIPPED TOPPING
MINCEMEAT PIE & APPLE PIE
ASST. NUTS & CANDY
TEA
COFFEE
MILK

Thanksgiving Day dinner menu from OCS on my birthday, November 28, 1963.

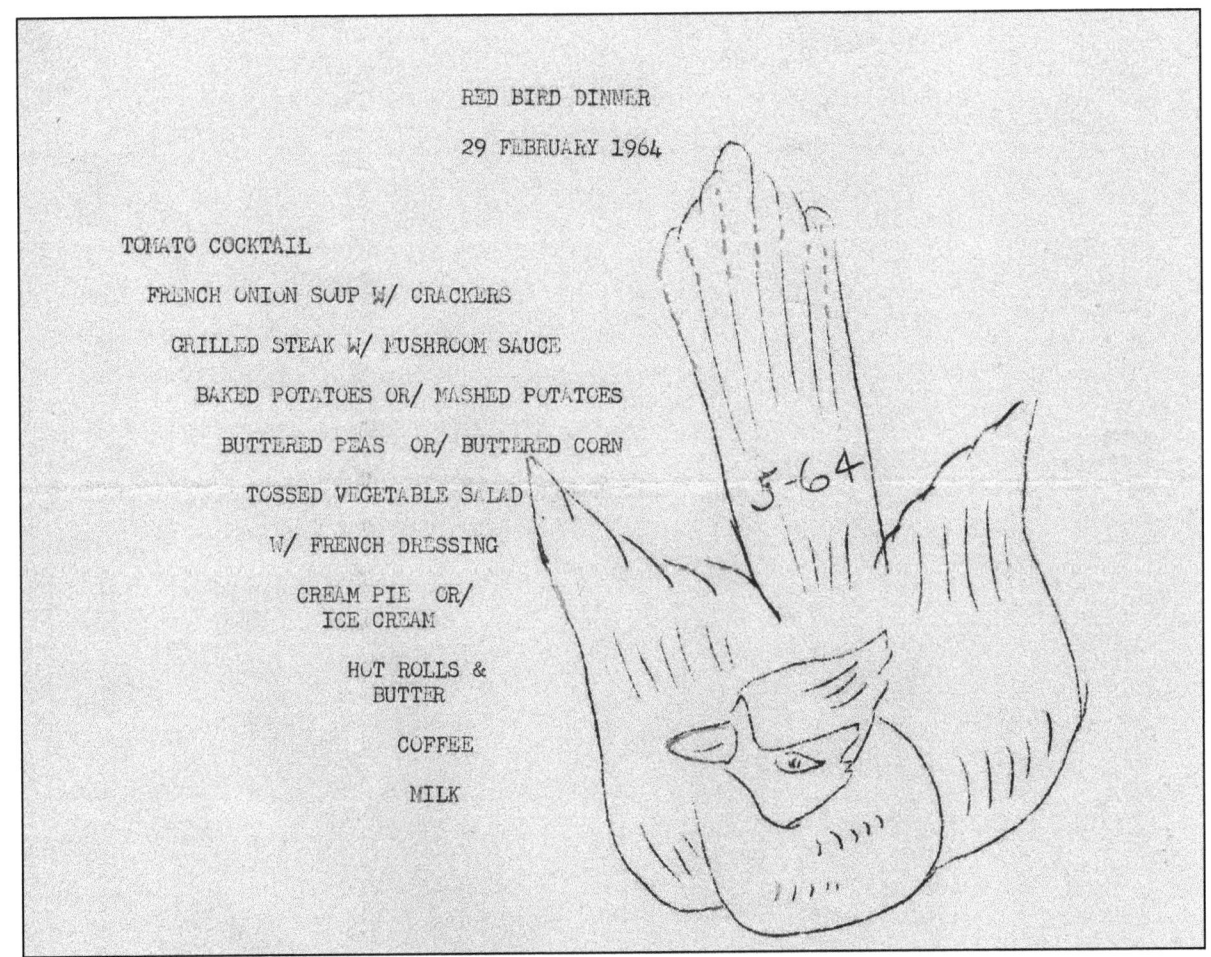

RED BIRD DINNER

29 FEBRUARY 1964

TOMATO COCKTAIL

FRENCH ONION SOUP W/ CRACKERS

GRILLED STEAK W/ MUSHROOM SAUCE

BAKED POTATOES OR/ MASHED POTATOES

BUTTERED PEAS OR/ BUTTERED CORN

TOSSED VEGETABLE SALAD
W/ FRENCH DRESSING

CREAM PIE OR/
ICE CREAM

HOT ROLLS &
BUTTER

COFFEE

MILK

Red Bird dinner to celebrate Class 5-64 reaching upper class status.

MEAL CARD	SERIAL NO.
ISSUED TO *(Last name - first - middle initial)* Sharpe, Gerald W	
SERVICE NO. OR BADGE NO. RA 17 607 074	DATE ISSUED 8 Nov 63
AUTHORIZING OFFICIAL	TYPED NAME, GRADE, TITLE, AND ORGANIZATION JAMES E. O'BRYANT, CAPT, ARTY SIGNATURE *James E. O'Bryant*
DD FORM 714 1 SEP 61	PREVIOUS EDITION IS OBSOLETE.

Meal card shown while entering mess hall.

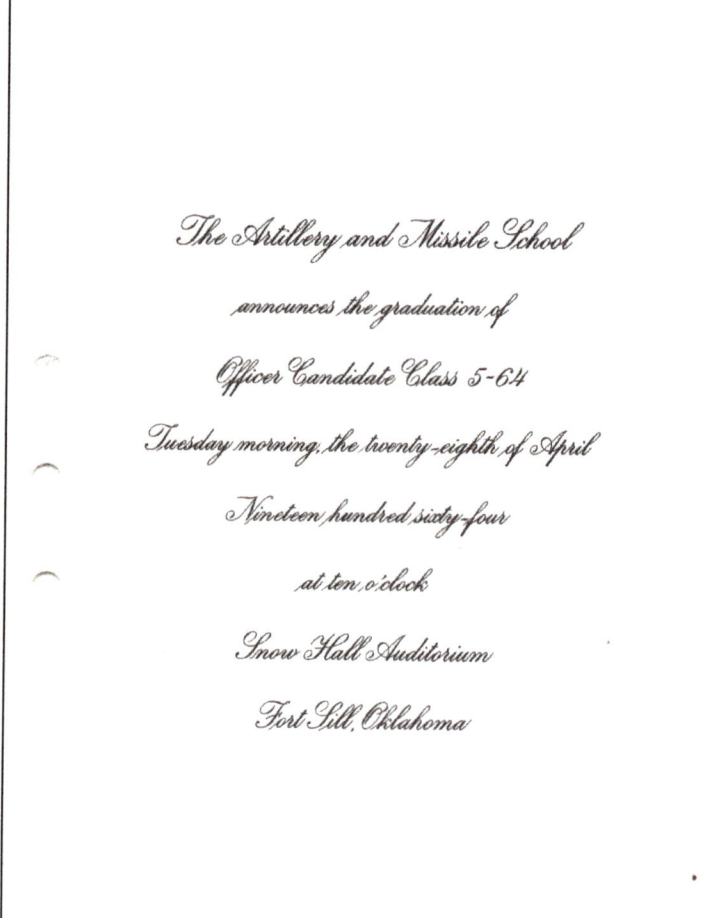

The Artillery and Missile School

announces the graduation of

Officer Candidate Class 5-64

Tuesday morning, the twenty-eighth of April

Nineteen hundred sixty-four

at ten o'clock

Snow Hall Auditorium

Fort Sill, Oklahoma

Engraved graduation invitation we were allowed to order to send to family and friends.

Gerald William Sharpe

Lieutenant
United States Army

2/LT GERALD W SHARPE
O 5406578
3rd TGT AQ BN, 26th ARTY
FORT SILL, OKLAHOMA

"Calling Cards." Each officer got them; protocol required you to leave a card at social events.

Metal horseshoe shaped "Clickers" worn by upper classmen on their shoes and boots.

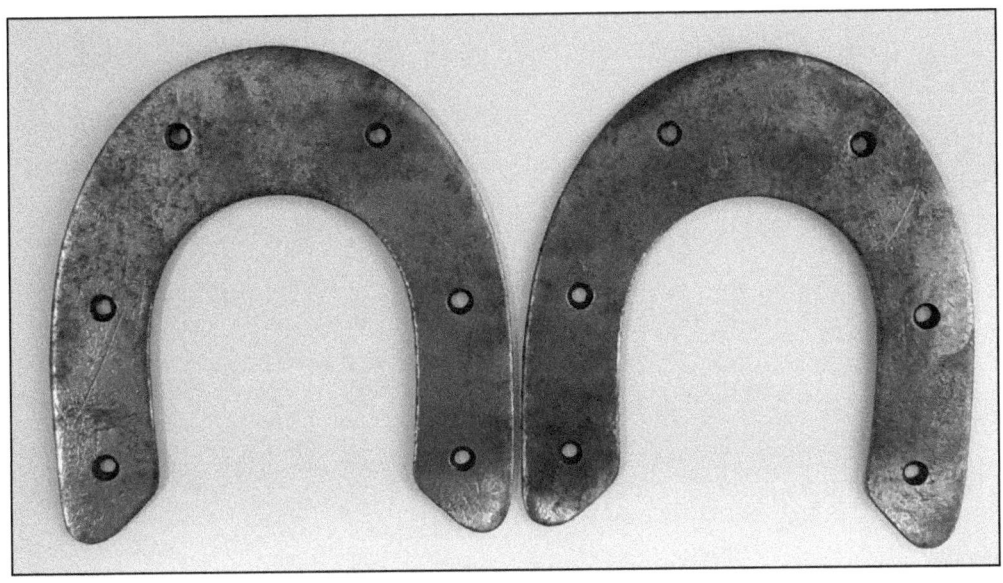

Graduation parade on Friday, 24 April on the OCS Parade Ground, before graduation on Tuesday.

Class was led by James Chapman, the Distinguished Graduate. Mike Garcia, standing right was #2.

Class 5-64 observing the other OCS classes passing in review. Cake cutting by James Chapman at our graduation dinner.

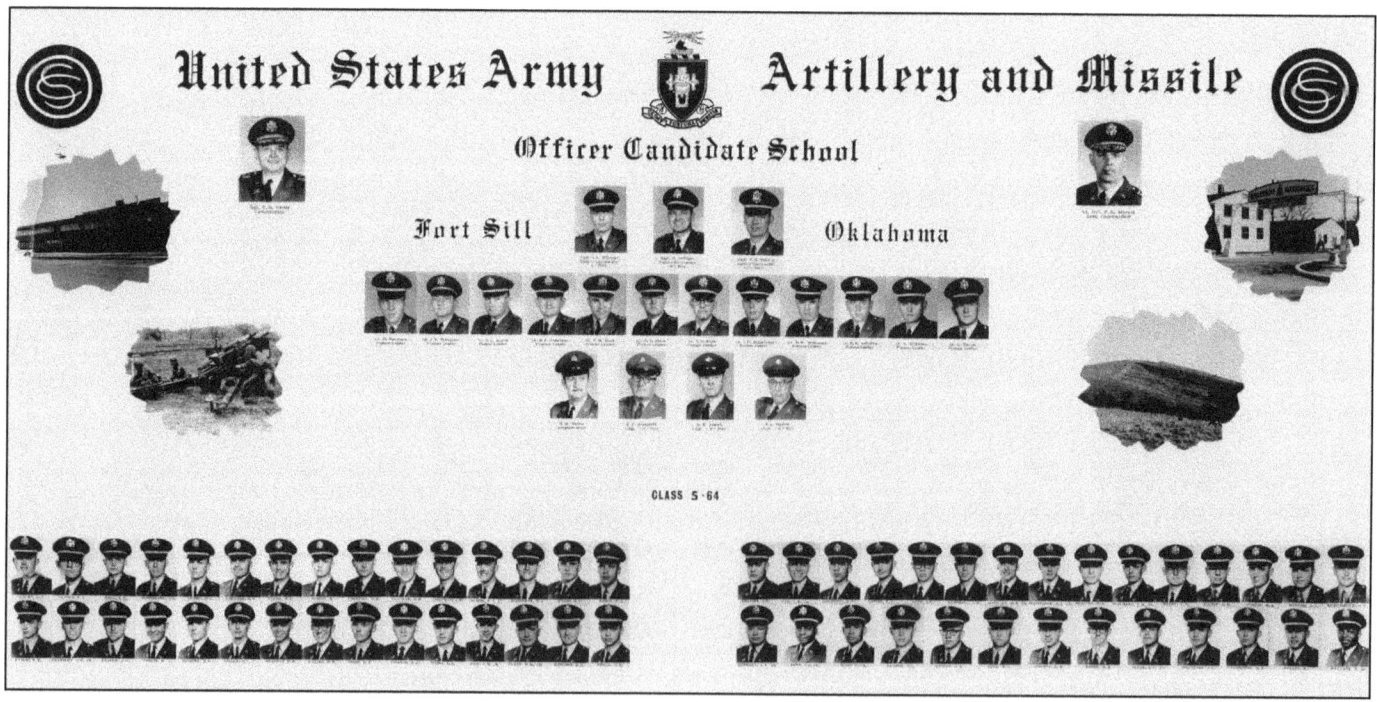

Receiving my diploma from Colonel C. A. Christin, Guest Speaker (Assisted by our battery commander, Cpt James O'Bryant.)

Class photo of OCS Class 5-64.

Graduation booklet for OCS Class 5-64.

U. S. ARMY
ARTILLERY AND MISSILE SCHOOL
FORT SILL, OKLAHOMA

Cedat Fortuna Peritis

OFFICER CANDIDATE CLASS 5-64
GRADUATION EXERCISES
24 April - 28 April 1964

THE MISSION OF THE ARTILLERY

The Artillery accepts with pride its mission as the primary supporting arm to the Infantry and Armor. It operates zealously to maintain its established reputation as the greatest killer on the battlefield, whether by conventional projectile, atomic projectile, free rocket or guided missile. It executes its missions with an esprit which is widely recognized as the Spirit of the Artillery.

THE MISSION OF THE U. S. ARMY ARTILLERY AND MISSILE OFFICER CANDIDATE SCHOOL

To develop selected personnel to be second lieutenants of the Army of the United States, capable of performing duties appropriate to their grade in artillery units and, with a minimum of additional branch training, prepared to serve as second lieutenants in other branches designated by the Department of the Army.

—1—

GRADUATION PARADE
1615 hours, 24 April 1964
OCS Parade Field

* * *

GRADUATION RECEPTION AND DANCE
Monday, 27 April 1964
Ball Room
Fort Sill Officers' Open Mess
Reception: 2000-2030 Hours
Dance: 2030-2300 Hours

—2—

GRADUATION CEREMONY
SNOW HALL AUDITORIUM
1000 hours, 28 April 1964

Opening Selection . . . 77th & 97th Army Bands

Invocation . . . Chaplain Raymond L. Kasper
(Audience stand)

Introduction . . . Major General Harry H. Critz

Address . . . Colonel C. A. Christin, Jr.

Oath of Office . . . Lt Colonel William H. Durand

Presentation of Honor Graduate
Award . . . Colonel C. A. Christin, Jr.

Presentation of Association of US Army
Award . . . Brig Gen James F. Brittingham (Ret)

Presentation of Diplomas and
Certificates . . . Colonel C. A. Christin, Jr.

National Anthem . . . 77th & 97th Army Bands
(Audience stand)

Closing Selection . . . 77th & 97th Army Bands

—3—

U. S. ARMY
ARTILLERY AND MISSILE SCHOOL

Commandant
Major General Harry H. Critz

Assistant Commandant
Brigadier General James W. Totten

Deputy Assistant Commandant
Colonel Joseph H. Harrison

Director of Instruction
Colonel Lewis A. Hall

Secretary
Colonel Raymond H. Lumry

—4—

OFFICER CANDIDATE SCHOOL

* * *

TACTICAL STAFF

"A"
Capt James E. O'Bryant
1st Lt Richard S. Anderson
1st Lt Germain H. Gersbach
2d Lt Arthur C. Atkins
2d Lt Kenneth T. Legum
2d Lt Joe R. Younginer
1st Sgt Casimer F. Konopacki

"D"
1st Lt Larry L. Langley
2d Lt Randolph L. Austin
2d Lt Edward C. Carter
2d Lt Guy L. Hovis, Jr.
2d Lt Francis E. Israel, Jr.
2d Lt Denton W. Thompson
1st Sgt Ralph Cheatwood

"B"
Capt Sam Jernigan
2d Lt Daniel T. Black
2d Lt John O. Kork
2d Lt John P. McDermott
2d Lt Carl M. Peck
2d Lt Roy S. Stephenson
1st Sgt Walden E. Powell

"E"
Capt Herbert D. Harris
2d Lt Richard J. Diamond
2d Lt Clifton M. Gatehouse, Jr.
2d Lt Alton D. Morris
2d Lt James A. Wueste
1st Sgt Edward C. Bryant

"C"
Capt William T. White, Jr.
2d Lt Robert B. Andrews
2d Lt Arthur L. McFatter
2d Lt Neal W. McDonnell, Jr.
2d Lt Clyde D. Taylor
1st Sgt Richard L. Heston

"F"
Capt Kenneth L. Morrison
1st Lt Hugh W. Holden
1st Lt Gary L. Schneider
2d Lt Robert D. Caputi
2d Lt Dayne L. Davis
1st Sgt Jack J. Oates

—7—

U. S. ARMY
ARTILLERY AND MISSILE
OFFICER CANDIDATE SCHOOL
FORT SILL, OKLAHOMA
Graduates of U. S. Army Artillery and
Missile Officer Candidate School
Class Number 5-64

Ahart, Louis C. — Dow City, Iowa
Ball, Stanley E. — Evansville, Ind.
Bondurant, James A. — Paducah, Ky.
Butler, Edward F., Jr. — Aurora, Colo.
Chapman, James H. — Rifle, Colo.
Chartier, Larry M. — Tillamook, Ore.
Clark, William L. — Saratoga, Calif.
Clutter, Howard H. — Avella, Pa.
Comiso, Richard — Ridgefield Park, N. J.
Cricchio, Robert J. — New Orleans, La.
Crump, Raymond T. — Danville, Ind.
Damon, William F., III — San Antonio, Texas
Doucette, Theodore J., Jr. — Beverley, Mass.
Ervin, William J. — McAlester, Okla.
Evans, David L. — Lawton, Okla.
Fawns, William E. — Lexington, Ky.
Fearnley, John W., Jr. — Newport Beach, Calif.
Feyrer, David A. — Bethlehem, Pa.
Garcia, Michael A. — Albuquerque, N. Mex
Gilmore, Kenneth D. — Austin, Texas
Herbold, Paul E., Jr. — Olympia, Wash.
Hiller, Robert L. — Wichita, Kans.
Holden, Parker W. — St. Louis, Mo.
Jackson, Edward J. — North Bay, N. Y.
Jones, Francis R. — Midland, Texas

—8—

Kidd, Vernon N., Jr. — Tulsa, Okla.
Klaus, Walter M., Jr. — Montague, Mich.
Koska, Robert L. — Cleveland, Ohio
Kosonen, Richard H. — Oronville, Wash.
Kroll, Geoffrey T. — Lake Zurich, Ill.
Lackman, Thomas W. — Newark, Del.
Landis, Richard C. — Sable Forks, N. Y.
Ledy, James C. — Phoenix, Ariz.
Lee, Clarence E., Jr. — Pensacola, Fla.
Marshall, John W. — Livonia, Mich.
Masella, Alphonse A., Jr. — North Bergen, N. J.
McBride, Reid A. — Brookings, S. D.
McCormick, Edmund J., Jr. — Yonkers, N. Y.
McDonald, Joseph N., Jr. — Hazlehurst, Ga.
Nahay, Yaroslow W. — Savannah, Ga.
Pierce, Oliver D. — Anita, Iowa
Pincoe, William A. — Battle Creek, Mich.
Reed, Ira M., Jr. — Kansas City, Mo.
Richtsmeier, Ronald C. — Hampton, Iowa
Robbins, Charles L. — Redondo Beach, Calif.
Rodwell, James A., Jr. — Allen Park, Mich.
Salmon, David L. — Ft. Benning, Ga.
Seek, Walter G. — Silver Spring, Md.
Sharpe, Gerald W. — Westminster, Colo.
Solis, Pete, Jr. — Azusa, Calif.
Soukup, Howard R. — Palatine, Calif.
Tabor, Woody R. — Wickett, Texas
Traub, Arthur C., Jr. — Springfield, Ohio
Urquhart, John C. — Phoenix, Ariz.
Valdez, Robert — Douglas, Wyo.
Voorhees, Merlin K. — Hitchcock, Colo.
Wallace, Clarence, Jr. — Lockport, La.

—9—

AKPSIOC   *28 April 1964*

SUBJECT: *Distinguished Graduate of Army Officer Candidate Course*

TO: Officer Candidate Gerald W. Sharpe
        Class Number 5-64, Officer Candidate School
        Fort Sill, Oklahoma

1. You are designated a Distinguished Graduate of Class Number 5-64, U. S. Army Artillery and Missile Officer Candidate School.

2. I congratulate you for your outstanding achievement in completing this course of instruction with such distinction. In doing so, you have not only shown yourself to possess the mental and physical qualifications for leadership but, more important, you have shown that you recognize and accept a moral obligation to your country to qualify yourself to serve in the highest capacity possible. This reflects great credit on yourself and your family.

3. Selection is made from those members of the class in the upper ten percent in final class standing who have demonstrated that they possess outstanding qualities of leadership and personal attributes which indicate a successful career as a Regular Army officer.

4. As a Distinguished Graduate, you have an excellent opportunity to obtain a commission in the Regular Army. A career in the Regular Army has many desirable aspects, chief of which is a continuing opportunity to devote yourself to the service of your country. In addition, there are varied and interesting assignments, membership in a select and elite profession, cultural development through travel and the satisfaction of leading men.

5. Inclosed is information which explains the method of applying for a Regular Army commission.

AKPSIOC   *28 April 1964*
SUBJECT: *Distinguished Graduate of Army Officer Candidate Course*

6. I sincerely hope that you will consider favorably the Army as your profession and that you will continue the career in which you have, thus far, advanced so admirably.

1 Incl   FLOYD D. GATTIS
  Ltr re Application for   Colonel, Artillery
  apmt in RA with 1 incl   Commandant

Honorable Discharge

from the Armed Forces of the United States of America

This is to certify that

GERALD WILLIAM SHARPE RA17607074 SGT E5 Artillery Regular Army

was Honorably Discharged from the

Army of the United States

on the 27th day of April 1964 This certificate is awarded

as a testimonial of Honest and Faithful Service

Floyd D. Gattis

FLOYD D. GATTIS
Colonel, Artillery

DD FORM No. 256 A
1 MAY 50

Discharge Certificate as an enlisted sergeant to become an officer.

OCS entrance sign and brass cannons, shined daily, outside OCS headquarters.

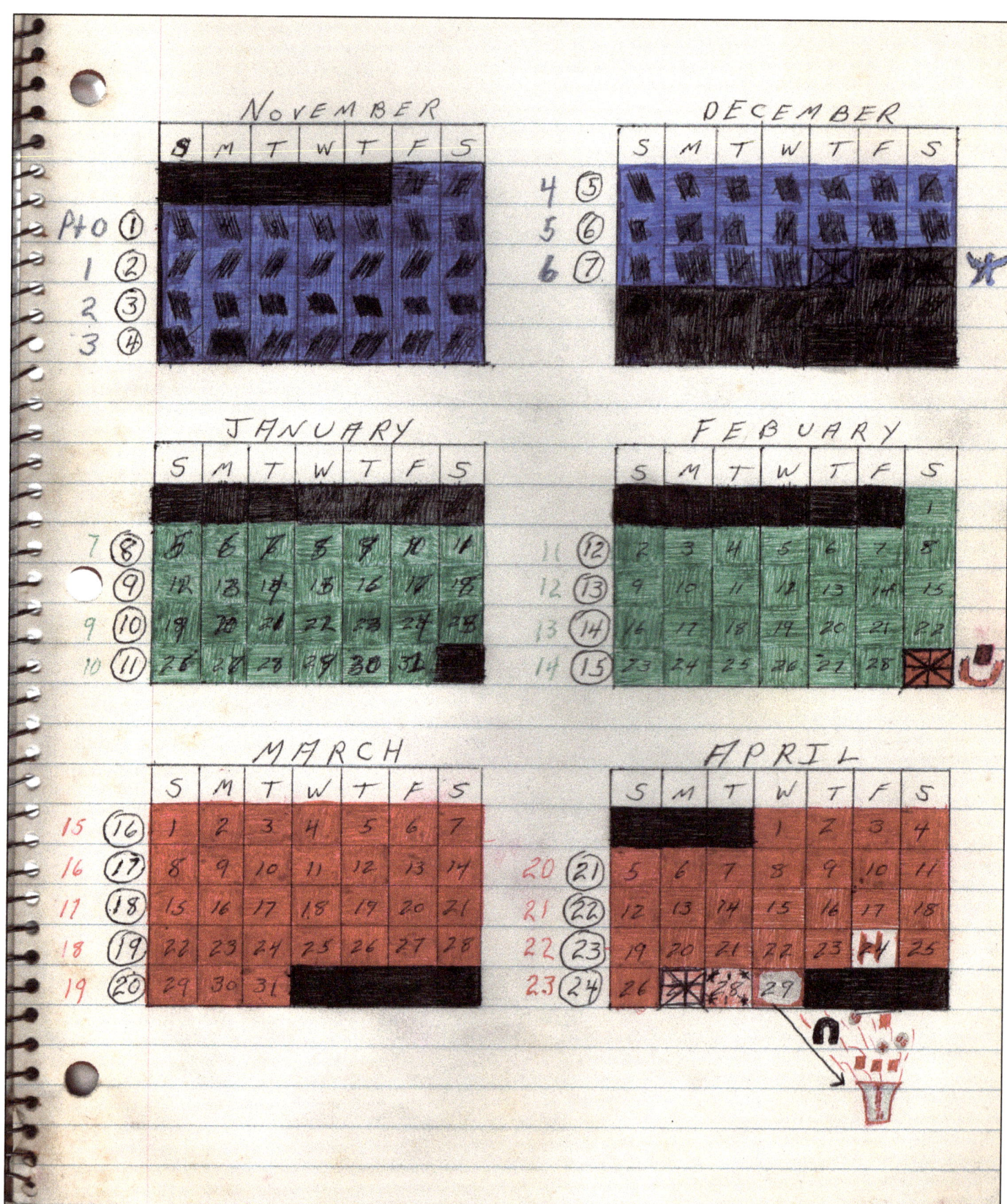

Short Timer Calendar I kept in OCS.

OCS area consisted of World War II wooden, two-story barracks and office buildings.

Battery F

Battery E

Battery D

Battery C

Battery B

Battery A

Dining Hall

Operations

Jark Assembly Area

OCS Sign

OCS Hq, Durham Hall

Photo as a 2nd lieutenant following graduation.

6. FORT SILL, LIEUTENANT AND CAPTAIN

APRIL 29, 1964 TO MARCH 6, 1967

SURVEY PLATOON COMMANDER
HHB, 3RD TARGET ACQUISITION BATTALION
(APRIL 29, 1964 TO JUNE 7, 1965)

MY FIRST ASSIGNMENT AS A COMMISSIONED OFFICER AT FORT SILL WAS TO THE HHB, 3RD TAR-get Acquisition Battalion, 26th Field Artillery as the Survey Platoon Commander. I suspect this assignment was a result of my final OCS target acquisition course score—a 93.85 as compared to my gunnery score of 91.66. I had attended OCS with the idea that I would get out of the army after six months and then spend six years in the reserves as an officer rather than an enlisted man. My scheduled discharge date was October 27, 1964. I was prepared to dislike the army just as I had as an enlisted soldier, but I quickly determined that I really liked being an officer. And what's more, I was good at it!

I met my first wife, Mary Katharine "Kathy" Kennedy while living with my OCS Classmate, Ed McCormick, Jr., in a house in Lawton, OK that we rented immediately following graduation. She lived next door to us with her uncle, Henry Sabine. She had come to Lawton from her home in El Paso, TX to help care for her uncle while he healed from a disc fusion surgery. Kathy and Ed became engaged over our first six months living there. Like me, Ed had signed up to serve only six months of active duty as a lieutenant. He got out of the Army in October and returned to New York while I applied for and received a Reserve Indefinite Army Commission on the 27th day of October, 1965.

I began a life-long habit of being the first to arrive and the last to leave my unit. This resulted in 18-hour workdays that left virtually no time for socializing, going out to dinner, or partying. We worked six days a week, and I would often come home on Saturday evening, go to bed, and get up at noon or later on Sunday. Kathy, Ed, and I fell into the habit of going out together for dinner on Sunday evenings.

Ed got a yappy little dog that never stopped barking and he named him "Nahumpkah," which he said was an American Indian name. I have been unable to find the word on the internet, but I digress. The dog always raced around the house, peed on the carpet, and even though I was not home a lot, drove me nuts. One Sunday afternoon in late September, Ed and Kathy got in the front seat and I was in the back seat of Ed's two door sedan. As I recall the dog was in the small backyard behind the house but apparently the gate was not closed as it was supposed to be. As Ed went to close his car door—which had to be slammed to close properly—Nahumpkah tried to jump in the car. He only got his head in as the door closed, and in an instant, he was dead as a doornail. I do *not* regret to say I did *not* mourn the dogs passing, but we didn't go out to dinner that evening.

On Friday, October 21, 1964, our battalion commander decided to conduct a competition between the batteries of the unit in the way of a timed 10-mile run. Soldiers would wear combat gear and would carry a rifle and a gas mask. There were a number of criteria published about how many soldiers must start the run (80) and there would be a penalty of ten seconds per man that dropped out. A small bonus in seconds would be added if fewer than five soldiers failed to complete the run. I was asked to lead the HHB team. We discussed a strategy and decided that we would take the 10 second penalty for 54 soldiers at the start and try to win with 26 very fit ones. We won by seven minutes (see the pictures at the end of this chapter to read more about the run).

Following Ed's departure, Kathy and I went to dinner often and she would come over to the house to tell me how much she missed Ed. She agreed to spend Christmas 1964 in New York with Ed and meet his family. I sent her flowers as a Christmas present, we talked on the phone a few times, and in the middle of her second week she called and told me she wanted to come home as she was breaking off her engagement. I sent her money for an airplane ticket by Western Union and picked her up at the airport in Lawton. A month later I asked her to marry me, and she said yes.

One of the things that surprised me the most as a brand-new 2nd lieutenant was the number of additional duties I was assigned. I became the battery motor officer responsible for overseeing the battalion maintenance section and the maintenance of headquarters vehicles; the battalion mess officer, which meant I oversaw the battalion mess sergeant and cooks; the A&R (athletics & recreation) officer where I had to oversee organizing sporting events for the battalion; and the battalion training officer responsible for overseeing the training of all National Guard and Reserve enlisted personnel as well as reporting on their training status. In addition I was appointed as the postal officer, West Point/OCS advisory officer, member of the battalion safety council, custodian of the headquarters battery unit fund, secretary of the unit fund council, member of the battalion character guidance council, alternate fire marshal, CBR (chemical, biological, radiation) officer, education officer, and several one-time appointments like being the inventory officer for a deserter's personal effects, investigating injuries sustained by a soldier, witnessing officer for the destruction of classified documents, and verification officer for the daily count of dining facility funds.

One of my most unusual additional duties was assistant defense counsel for court-martial cases. It was not too many years later that the army decided that only lawyers could perform the duties of trial and defense counsel. Generally, the most junior officers served as defense counsels and the more senior lieutenants and captains served as trial counsels or prosecutors. This meant that the deck was generally stacked against soldiers who were charged with crimes serious enough to warrant jail time or discharge. In my first two cases, I was the assistant defense counsel to more senior lieutenants who seemed more than willing to let me prepare the cases. The first case in July 1964 charged a soldier with desertion. I found out that that the soldier had made a written request for an extension of his leave, but it had not been acted upon by his commander. The soldier had been gone a month before voluntarily returning. I argued that while he met the technical definition of desertion, he should be shown leniency if found guilty, because he had tried to get his leave extended and had returned voluntarily. The maximum punishment for special court-martials was reduction to E-1, jail for six months at Fort Leavenworth, KS, and forfeiture of two thirds of ones pay for six months—often referred to as a 6 plus 6. This soldier was found guilty but was only sentenced to 2 plus 2 (Reduction to E1, two months confinement, and forfeiture of two thirds of his pay for two months).

Before I was assigned to my second case as a defense counsel, I was tasked to train a 12-person team to carry out a week-long assignment of raising and lowering the flag in front of the post headquarters as well as firing the cannon to mark the event. We started on Saturday, July 25, 1964, and finished on Friday, July 31. Nine men (one sergeant and eight enlisted men) did the flag raising and lowering, and two NCOs and one enlisted man fired the cannon. Each morning and evening I marched the team to the post headquarters, lined the soldiers up in front of the flagpole, and waited for the post duty officer to come inspect the detail and observe the process. The team was graded 14 times, and we received a "superior" rating each time, the best score possible.

My second case was in August 1964. It involved an SP4 who was accused of striking a noncommissioned officer and being disrespectful. When I interviewed him, he denied striking the sergeant. He said there were at least two other sergeants present when he was told they were going to get him kicked out of the army. He told me that the sergeant had threatened to frame him for something he had not done, so an argument ensued. He did not know the names of the other two sergeants present.

I told him that if a witness backed up his accuser and he could not produce a witness to refute the charge, he was likely in big trouble. On the morning of the court-martial, I was standing in line in the dining facility when I overheard a conversation between the sergeant who had accused my client and two other sergeants. I overheard the accuser say that by close of business he was going to be rid of that pain in the ass in his section as there was no way my client was going to be able to prove he had made the story up. They were laughing among themselves and once they had their food they went and sat at separate tables.

I got my food and sat as close to the two sergeants as I could, and later saw the accuser leave the dining facility. I immediately went and sat down at the table with the two sergeants and told them to give me their ID cards. They wanted to know why, and I said as soon as I look at them, I will tell you why. They both handed me their ID cards and I asked if they would like to go to jail. They said they did not understand. I told them that one of two things was going to happen: "You are either going to recount your previous conversation with Sergeant X or I am going to charge you with lying." At first, they denied having the conversation, but I told them, "I am sorry guys, but I was taking notes the entire time you guys were talking." I told them to come with me to the battalion headquarters. I had them sit in the conference room and asked the battalion S1 to join me as a witness.

I asked each man to give a statement while the other was out of the room, and had both recount what actually happened and how the section sergeant had made up the story of being assaulted.. I had both statements typed up. I then had each of them raise their right hand and swear that what was written was the truth, the whole truth, and nothing but the truth. They both signed their statements, witnessed by the S1 and me.

The court-martial was due to start in an hour, so I had the S1 notify their battery that the sergeants would not return to duty until about lunch time and then had them remain in the conference room. I said I did not know if they would be called to testify, but they were not to leave the conference room without my permission. Just before the trial started, I shared the two statements with my lead defense counsel and suggested we hold them until it was our turn to question the accuser. He was concerned that we should share the statements with the prosecution. I argued that if we did, they would just ask for a delay and redo their entire case. He agreed we should hold them. The trial got under way and the prosecution told the jury of eight officers and warrant officers that they would show that the accused was a poor soldier and that he had in fact struck his platoon sergeant and had been disrespectful.

In our opening argument I said that since no witness had been listed by the prosecution it appeared the case was going to hinge on two different stories. The prosecution called the accuser to the stand. They spent about 15 minutes going through what had happened, who said what to whom, and why the sergeant thought the accused should go to jail for what he had done. When the prosecution rested their case, I got up and handed the prosecution a copy of each of the statements made by the two sergeants, outlining how the story had been made up to get rid of the young soldier. I then gave copies to the judge and to the accuser. I reminded him he was still under oath and was obligated to tell the truth, the whole truth, and nothing but the truth. Before the accuser had a chance to finish reading the two statements, the captain (who was the senior trial counsel) stood up and said, "Judge, we would like to withdraw the charges against the accused." The judge asked if the defense had any objections and I said we did not, but I would like to recommend that the accuser be placed under arrest and charged with perjury. He was charged and was later convicted and sent to jail himself. I sent a messenger back to the headquarters telling the two sergeants they could go back to their unit.

That should be the end of the story on this case, but it was not. The next morning, I was told the battalion commander, LTC Leo D. Johns, wanted to see me. I had no idea why, but I reported in as trained. I walked to the front of his desk, came to attention, saluted, and said, "Sir, Lieutenant Sharpe reporting as ordered." He asked me to take a seat and then asked if I enjoyed being a defense counsel. I told him I had only done it twice, but it was very interesting work. He asked me if I knew why he assigned the most junior second lieutenants to work for the defense. I said I supposed it was so we could learn how the system works. He said, "No, it was because when I charge someone with a crime, I want them convicted." I was somewhat shocked by that answer, and all I could say was, "Yes Sir."

He then said, "With that in mind, you are never going to be a defense counsel again. I am going to start assigning you as a trial counsel. How about you see what you can do to get convictions and maximum sentences." Again, all I could again say was, "Yes Sir." After that I was an assistant trial counsel twice and then trial counsel six times—all eight cases were guilty with seven receiving maximum sentences. One soldier was court-martialed for one day of being AWOL. The defense counsel agreed with me that the young man did not deserve jail time, so I recommended to the judge that he be reduced one grade and fined $20.00. The commander was not happy with *that* case, he was with the other seven.

In December of 1964, our battalion was assigned to support a five-day exercise called "STRIKEX 64" at Fort Sill. It was conducted by Strike Command out of MacDill Air Force Base in Florida. Three three-story barracks buildings were emptied and surrounded by two rows of chain link fencing. Strike Command was the predecessor of the US Readiness Command and was commanded by a very well-known four-star general named Paul D. Adams. He had a reputation for having no mercy for other general officers who did not do exactly what they were told.

The exercise went on 24 hours a day, with more than 100 general officers from all four services along with a hundred or more colonels and support personnel participating. They were billeted in hotels and on-post accommodations. I was assigned as the "Officer of the Guard" and was responsible for commanding a 48-person guard force plus 24 senior NCOs and two duty officers. The force was divided in half with each working 24 hours on and 24 hours off. The senior NCOs maned two entrance gates and soldiers maned eight walking posts between the fences. Each gate and walking post had three eight-hour shifts. There was an extensive set up period to prepare the three buildings including putting in office furniture, setting up conference rooms, and building out a tactical operations center. Telephone communications were installed in every part of the three buildings, and I had direct lines to the two entrance guard posts and to each of the walking guards. There was an extensive security SOP published and a clearance roster published listing every person authorized inside the exclusion area. Unknown to me General Adams had a senior communications sergeant monitoring all telephone lines. For anyone to enter they had to present a special ID card issued for the exercise, and the name and number on the ID had to be checked against the clearance roster before personnel were allowed through the second gate.

On the second or third morning, a BG (one star) showed up at one of the gates and told the gate guard that he had forgotten his special pass at his hotel. He insisted that since he was on the access list and had his ID card, the gate guard had to let him in as he was supposed to be in a briefing with General Adams in 15 minutes. The sergeant told him he was sorry, but he was not authorized to let him in without his special pass, but he would have the second sergeant call the officer of the guard (me) who could perhaps make an exception to the rule. The general became even more irate, and I could hear him screaming at the sergeant as I was listening to the problem on the telephone. The guard headquarters was in the basement, about a two-minute walk from the gate. I told the guard I was on the way. I grabbed my helmet and dashed up the stairs and out the door. Before I could get to the gate, I heard someone screaming. At first, I thought it was the one star still yelling at my guard. But as I got closer, much to my shock and amazement it was General Adams himself. He had the BG standing at attention and he was berating him for having the nerve to yell at *his* gate guard and to try to get the gate guard to violate *his* orders. He ordered the general to go back to his hotel, get his badge, return, and give the badge to the gate guard, and then leave and find transportation back to his unit as his participation in the exercise was over. As soon as the one star left, General Adams told the guards he was proud of them for doing exactly what they were supposed to do, and he apologized to them for the one-star's screaming. General Adams turned to me and asked me to please send the names and information on the two senior sergeants to his office and he would see to it that they received a letter from him congratulating them on a job well done. They each received a very nice letter a few days later.

In early February, I was called into my battery commander's office and told that the battalion had received a request to nominate someone to represent Fort Sill on the Board of Directors of a new Boy's Club that had been in operation in the Lawton community since September 1964. I agreed to have my name submitted, and on February 26 the post commander, MG Harry H. Critz, notified the President of the Board of Directors of the Lawton Boy's Club, Incorporated of my nomination. I received a call from the board president, Mr. Lew Johnson, inviting me to attend a board meeting for the election of officers for 1965 the second week of March.

Upon attending, I found that the board was made up of some very influential people, including a state senator, a county judge, and the Lawton Assistant Police Chief. Over the following months, I became well acquainted with the board members, by both attending the monthly board meetings and working on special projects to improve the Boy's Club's programs. The Assistant Police Chief, Mr. Albert "Al" Hennessee, was elected as treasurer of the Boy's Club Board of Directors. In early May, Al invited me to attend a meeting of the Kiwanis Club of Lawton. Al introduced me to his fellow Kiwanis Clib members and made some complementary remarks about my participation in the Boy's Club Board. After the meeting, the club president introduced himself and gave me an application form saying the club would like me to consider becoming a member. I agreed to join, and thus began a civic club journey that has extended for years, as I will describe in later chapters.

On the 5th of March 1965, Kathy and I drove to El Paso on a three-day leave so I could meet her parents. I stayed at the Caballero Motel in El Paso, while Kathy stayed with her parents. Kathy's father was a big-time gun enthusiast. He had been a member of the US Army Rifle Team and knew weapons inside and out. He had many stories about his experiences, and I recall listening to him tell stories for hours on end. It was clear to me that her mom and dad accepted me as a future son-in-law, and we were pleased the visit went well.

After returning to Fort Sill on the 8th of March, I was notified that I had been selected to escort a nuclear weapons shipment from the USA to Germany. I told my commander I was getting married the first week of April, but he said I would be back in time. I had a staff sergeant and two enlisted men assigned to go with me to guard the warheads anytime we landed at an airfield. The four of us flew to Texarkana Regional Airport aboard a military plane on March 18, and we were picked up at the airport by a military vehicle and taken to the Red River Army Depot about a 30-minute drive west of Texarkana.

After having lunch, we were taken to the weapons storage area where there were two tractor trailers to be loaded with six nuclear warheads, three in each truck. We were to observe the loading of the trucks, which we were told would be sealed and remain in the exclusion area guarded 24 hours a day until our departure the following morning. I signed a hand receipt for the six weapons, and we were told we could take the trucks any time after 8:00 the next morning. We were given a place to stay in the visitors quarters, and I took the soldiers to supper at the officers club.

The following morning, we had breakfast, then met with the pilot who would fly the warheads to Germany. He had two staff cars; one was for himself, his co-pilot, and me; and the other was for the crew chief and my three soldiers. There were also two MP cars. I was told we would be driving back south about 90 miles from the Red River Army Depot to Barksdale Air Force Base where the weapons would be loaded on a C130 transport airplane. We departed about 9:00 am, and the trip to the outskirts of Shreveport, LA was uneventful, with the convoy being led by a military police car with flashing red lights, then our car, the two tractor trailer trucks, followed by the crew chief's car, and finally the second military police vehicle also with lights flashing.

We were instructed to pull into a large truck stop on the northern edge of Shreveport to await a police escort through the city as the airbase was on the opposite side of town. After a few minutes we saw a parade of 15 or 20 motorcycle policemen pull into the truck stop. The leader of the group asked that we meet him near the lead MP vehicle. I did not think this was more than a formality, but the policeman explained that it was important that the convoy stay closed up so no one could get a vehicle into our group. He also told us that he and one other policeman would lead the convoy, and no matter what happened, we were to follow them closely. Finally, he told us why there were so many policemen: they would ride ahead of us, blocking side streets and clearing the road ahead of our convoy. He instructed the two MP drivers to turn on their flashing lights and sirens. He stressed that if any of these instructions were not followed, we were risking a nuclear incident.

As we went back to our vehicles, we commented on the large number of policemen and wondered aloud if they were really necessary, as we had already driven 70 miles without an escort with no apparent problems. As all the motorcycles left the lot, the two leaders pulled in front of the lead MP car, and we slowly pulled out onto the main highway. Once we were all lined up on the highway, the leaders began increasing their speed. As we passed 40 mph, then 50 mph, and finally neared 60 mph, I was about to have a heart attack as it seemed there was no way we were going to get through town unscathed at such a high speed. We had no way to communicate with the police leading us or even with the lead MP vehicle ahead of us. We simply followed as best we could.

At one point, there was a motorcycle policeman right next to our car beating on the window of the passenger side of a vehicle next to us, shouting at them to move over as we were driving on the shoulder. It was an elderly couple who were likely scared to death, and they did not roll down their window or pull over into the other lane, but they did slow down, so we flashed by them and got back on the road. I have to say the drive was the most terrifying 30 minutes of my life, but by some miracle we arrived safely at the gate of Barksdale AFB where all the policemen went away. The pilot called his commander from the gate and asked to see him. After leading the two trucks to an area near his airplane and leaving my three solders to stand at the back of the two trucks for a short time, we drove to the commander's office, and we described the trip through the city. The commander said he would relay the information to the base commander, and we later learned the leader of the motorcycle group was suspended and disciplined for risking the lives of so many people. We were also later told the police department had a maximum speed of 40 mph for such convoys.

We drove back to where the airplane was parked, and I was introduced to the rest of the crew that would fly the C-130 transport plane to Germany and back. A large group of vehicles, equipment, and people were waiting to load the Honest John missile warheads onto the plane. After cutting the seals off the two vehicles, I observed the loading process, which took a little over three hours.

As we left the plane, the pilot showed me how they put wire seals on the door of the airplane and recorded the number. He gave me the number of the seal and said I should check it before it was cut off in the morning. I said I would schedule my soldiers to take shifts guarding the plane as I had been instructed to do before we left Fort Sill. This was to cover the time before we were scheduled to depart at 5:30 in the morning. The pilot said something like, "Lieutenant, I am a Lieutenant Colonel in charge of this flight, and you and your soldiers will get dinner and try to get eight hours of sleep." He then said, "I have armed MPs guarding this aircraft all night." All I could say was, "Yes sir!"

The crew chief took my three soldiers to the enlisted quarters, while the navigator took me to the bachelor officers' quarters (BOQ). On the way there, he told me he would be hosting a party at his house that evening. He had a large house with a pool and about 20 officers and their wives attended. They served steaks cooked on the grill and had lots of beer, liquor, and wine, but I did not drink any alcohol. Swimming and storytelling seemed to be the main activities.

The next morning, March 20, we had breakfast at 4:30 am. After the crew completed preflight operations, we departed at 5:30 am, as scheduled, headed to our first stop at the Seneca Army Depot in upstate New York. We arrived about 11:00 am and picked up some additional cargo bound for Germany. We departed about two hours later for McGuire AFB in New Jersey arriving about 3:00 pm. The pilot requested an isolated parking spot, and I later found out he also requested military police guards.

At 6:00 the next morning, after verifying that the seal on the aircraft door was intact, we boarded the plane and took off for the Azores Islands. We landed at Lajes Airfield with a few hours of daylight remaining. Once again, the pilot sealed the airplane and had MPs guard it until the next day. The crew took my two soldiers to supper and got them a place to stay. We checked in at the BOQ and immediately went to supper at the officers club. After eating, the navigator, who was six foot six inches tall and weighed well over 200 pounds, asked me if I would like to tour the island with him. He had a friend assigned at Lajes who let him borrow his car. I said yes, and we drove to the top of a small mountain overlooking the airfield. We saw a spectacular view of the sunset from there and then drove into a small town. He asked me if I might be interested in a lady for a short "roll in the hay." I said sorry, but I am getting married in two weeks. He said something like, "No problem, we will only stop for 30 or 40 minutes." He parked the car, and we walked down an alley and entered a nondescript building.

We were met by a very good-looking older lady who obviously knew him because she gave him a big hug. He introduced me to her. She also asked if I might be interested in a lady, and I gave her the same answer. She told our navigator that she had gotten some new ladies since he had been there last, and he could have his pick. We went into the next room and there were about eight ladies sitting around the room. She said something that caused them to stand up and lower their dress tops so their breasts were exposed. The co-pilot went to each lady, fondled their breasts a bit, and a couple he asked to lift their skirts as they were not wearing underwear. When he finished, he turned to the madam and said he would pass on these ladies and would pick her. She told me I could wait in a chair outside the room they went into and as it turned out it had no door and only a few beads hanging down. I had my first experience of a live sex show for about 40 minutes. I was impressed that they got into multiple positions, and neither were at all quiet. It was one of my most memorable experiences at that point in my young life.

The next day, we departed about noon and flew to Portugal for refueling and then on to Ramstein AFB outside of Frankfurt. We arrived around midnight and were met by an Army unit with trucks and soldiers lined up to take the warheads to their permanent storage location.

As soon as I got off the airplane, I was met by the battalion commander of the unit, an Army lieutenant colonel. He told me he would like to sign for the weapons tonight. He said they had been told we would arrive before dark, so they had been waiting most of the day. He said they would like to start unloading the airplane right away as they had a long drive back to his unit.

As we were talking, the rear ramp of the C130 aircraft started closing. The commander started cursing at me furiously. I told him I was sorry, but he would have to talk to the pilot. It was about ten minutes before the pilot came out and started calmly putting a seal on the door. The army commander shouted at him that he wanted to unload now. The pilot told everyone except me to go get on the waiting crew bus. He calmly waited until everyone was on the bus and out of earshot. He walked up to the army commander, got right in his face, and told him very quietly that he could stop being an asshole because Air Force regulations required him to allow his crew to get "crew rest," so they would not return until the morning. He further told him if he said one more "blankety-blank" word, he would exercise his prerogative and come back long after lunch. Apparently, the Army commander knew enough to walk away, because the pilot simply said to me, "Let's get on the bus and go get a good steak."

The next morning, March 23, we did not actually show up until about 11:00 am, and it took the unit about four hours to unload the plane and load the warheads on six trucks. The Army commander apparently had his ass chewed by someone for the way he screamed at the pilot, because he was very apologetic for the way he had acted. The pilot had reported him to the base commander who happened to be in the officers' club the previous evening while we were eating.

I reminded the pilot I was getting married in less than two weeks and wondered when we would be leaving. He told me not to worry and to try to enjoy the day by doing a little shopping at the PX, because we were *not* leaving until about 5:00 the next morning when everyone was well rested. As it turned out, my three enlisted soldiers did not have to guard the airplane a single time at the orders of our pilot. They were happy to be along for the ride as they said it was far better than being in their units for ten days.

We left Ramstein AFB as scheduled at 5:00 am and landed in the Azores at Lajes Field just before noon. We had lunch, refueled, and everyone got about eight hours sleep. We departed for Bermuda as scheduled at 5:00 am on March 25 and arrived at Kindly Air Force Base on Bermuda after lunch. Once again everyone had two good meals and got about eight hours of sleep.

We departed for Charleston Air Force Base the next morning at about 6:00 and arrived in Charleston about 10:00 am. After refueling and eating, everyone once again got a good night's sleep. We departed about 9:00 am for Standiford Airfield in Louisville, KY where the crew had another mission waiting for them. We said goodbye to the crew and caught a commercial flight from Louisville back to Will Rogers Field in Oklahoma City. We took a taxi back to Lawton and we were all home about 11:00 pm on March 27, which left me about a week to spare before my wedding. After my arrival back at Fort Sill, I applied for a ten day leave. On March 31, Kathy and I departed for El Paso.

When we arrived in El Paso, it was decided that Kathy would stay with her mother and father while

I was able to get a room at the BOQ on Fort Bliss. I was also able to arrange for Mom and Dad, Charles, Eileen, and Robert to stay at a duplex that the housing office maintained for visiting officers. They had decided to leave Milton in Colorado, and Stanley was off on his own. On Thursday, April 1, Kathy and I did some shopping and visited with her family some. On the next day, my family arrived, and I got them checked into their quarters. Charlie helped me check out of the BOQ and move over to the duplex. That evening, both families went to an early spaghetti dinner at the Caballero Motel Restaurant.

Kathy and I had a 7:00 pm conference with the priest who was going to perform the ceremony, and at 8:00 pm we had a short rehearsal at the chapel. Afterwards, we all went back to Kathy's parents' house where we opened presents our folks brought from my many friends in Westminster. On Saturday morning we decided to take a short sightseeing trip to Mexico. Just as we were about to leave, I got a call from Willard Staats saying he would be arriving within the hour. The folks went on to Mexico without me and I agreed to meet Willard at a large shopping center. After about ten phone calls, we were able to link up and we drove back to Fort Bliss where we changed clothes for the wedding, scheduled for 3:00 pm.

Mother wrote in her diary from April 3, "We got a late start to the church and Charlie said Gerald was very nervous, thinking he might not make it to the church on time. Charlie went in Gerald's car with Willard. We all got there on the stroke of 2:45 pm, just when we were supposed to be there. Kathy was lovely in her white, street length dress with a shoulder length veil. We all got corsages or boutonnieres, and the ceremony was short but nice. We discovered we had a flat tire as we left the church and had to stop and get it pumped up. Another couple waited for us to show us the way to the reception at the Officers Club at Fort Bliss, which was nice, and they received more gifts there. At six o'clock Gerald and Kathy left the reception amid a rain of rice and well wishes from everyone. Their car had been appropriately decorated with lots of tin cans and 'Just Married' written on their windows as they disappeared to the tune of tin cans banging along behind."

After a short honeymoon of which I have absolutely no recollection, I went back to work. We had decided I would stay in the Army, so immediately upon my return I applied for a Regular Army Commission to replace the Reserve one I held. I had to have two years of college work and fortunately I had been attending night school since getting commissioned. I was also given credit for some of my studies at OCS, and in early July I was notified my two-year college equivalency had been accepted.

Shortly after returning from our honeymoon, we also started looking for a house to purchase. We found a nice three-bedroom brick home for sale at 1638 NW 50th Avenue, about a 20-minute drive from Fort Sill. We made an offer a few thousand dollars under the asking price, and the sellers accepted. I was able to use my VA loan benefit and made a very small down payment. We moved into the first of many homes we would occupy over the next 20+ years and we really enjoyed having one of our own. Soon after moving in, we were overjoyed to learn Kathy was pregnant with our first child.

In early June of 1965, I was called in to see the battalion commander. I was told the battalion was required to send an officer to the higher headquarters of the battalion, the 1st Field Artillery Brigade. LTC Johns said I had been selected and I would report to the Brigade S-3 section for duty the following Tuesday, June 8. He thanked me for my service and indicated he thought I was going to be replacing a senior captain, but he was confident I could do the job.

ASSISTANT OPERATIONS OFFICER
1ST FIELD ARTILLERY BRIGADE
(JUNE 8, 1965 TO OCTOBER 10, 1965)

When I reported to the brigade S-3, LTC Charles K. Harris, told me I was going to be placed in charge of "School Support" for the brigade. I would be working for his assistant S-3, CPT Alan H Lubke. I had three sergeants working for me and the brigade received about 20 to 30 requests for school support daily. I had been vaguely aware that the eight battalions of the brigade supplied men and equipment to the Artillery School to support the many officer and enlisted classes being taught but had not realized the size and scope of the requirements.

My predecessor had left for Germany the previous week, so he didn't get a chance to explain the pitfalls of the position or what the job actually entailed. He also neglected to leave any written notes that might have been of help. When I asked my new boss, CPT Lubke, if he could brief me, he told me he was sorry, but he had only arrived in the brigade two weeks previously. I also found out that the brigade S3, LTC Harris, had only been on the job a few weeks. I immediately got my three sergeants together and asked them to explain what our section did and how it was being done now.

The three sergeants were very knowledgeable, but they said that the captain in charge had made all the decisions and they simply did what they were told. I asked who had been making decisions since the captain left, and they indicated they had been doing the best they could based on their experience. I asked them to show me what had been assigned to whom that morning. As they walked me through each request, it became clear that for some requests only one unit had the equipment while others could be provided by any one of the eight battalions in the brigade.

I asked if the section had a record of past requests, and they showed me three four-drawer file cabinets filled with requests and taskings. I also asked if the section had a list of what equipment each unit had, and they showed me a binder with the army table of organization published by the Department of the Army for each battalion. These documents show what is authorized but *not* what is on hand. I said something like surely you have a better way of knowing who has what and they said that they got many of the same requests over and over again so they would just find an old one and do the same tasking. I asked how they knew if this was fair and what they did to keep the units from complaining. They said that was the captain's job not theirs.

I told my three sergeants that I would do my best to figure out a better way of doing our taskings, but for the next couple of days I was going to have to depend on them to try to keep us out of trouble and continue to provide the support the school requested. I felt more than a bit overwhelmed as a relatively new 2nd lieutenant that had to deal with majors and lieutenant colonels in the eight battalions. I was surprised that there was not a more organized system in place. I went back to the three file cabinets and found the latest requests and took out all those from the previous week. There were perhaps 200 requests and taskings. I got a three-ring notebook and paper and began listing what was asked for with a separate page for each item of equipment, how many were requested, the date, and what unit had been assigned to provide the equipment. I quickly found that many requests were simply for vehicles and drivers while others were for soldiers only to act as aggressors, umpires, graders, and other tasks. I worked until 2:00 am and only got through about three weeks of requests. As I drove home, I came to the conclusion that what was being done was neither fair nor equitable.

The next day I sat with the three sergeants for about two hours, and I listened as they explained why they were giving each request that day to one or more units. I did not yet know enough to comment, so they completed the tasks without much input from me. I went back to reviewing past taskings and entered another three or four weeks in my notebook. It became clear that the majority of the taskings could be provided by any one of the eight battalions while maybe 20 percent could only be provided by either only one unit, or just a few, based upon the equipment being requested.

We had three artillery battalions with 105mm howitzers, two battalions with 155mm howitzers, one battalion with 8-inch howitzers, one target acquisition battalion (my old unit), and one maintenance battalion. Each of these had some specific equipment no one else could provide. I recall completing my second day beginning to understand the scope of what the brigade was providing on a daily basis.

I decided to prepare a tasking record with one or more pages for every item of equipment and one each for soldiers, NCOs, and officers requested. Every day we entered what was tasked and tracked the number of items and individuals provided each day per unit. Before tasking each morning, I had my three sergeants check the book and task the unit with the least number of items or soldiers provided since we began tracking items. In my third or fourth day, I received an angry call from a major who was the operations officer in one of the units complaining that he was being overtasked. I explained to him that it was very possible that was happening in the past and gave him a short explanation of the new system we were going to use. I also told him that if he stopped by my office, I would show him how we were doing the taskings going forward and he was welcome to make suggestions on how we could do it better. That seemed to satisfy him.

At the end of my first week, I asked my boss, CPT Lubke, to please come sit down with me and let me show him the system I had created to make taskings fairer and more equitable. He reviewed what we were doing and said it looked great. The next week, I was summoned to the S3s office with the comment that one of the battalion commanders was angry with the amount of taskings he was receiving. I took my book and asked the two colonels if I could have five minutes of their time without comment so I could explain the current situation, and if after five minutes they were not satisfied, they could chew my ass. I gave them the same short explanation I gave the major, then showed the battalion commander my tracking book, pointing out that he had *not* been tasked if it was not his turn. When I finished the S3 said I could leave and away I went. A bit later CPT Lubke came to my office and told me that the colonel said to "Keep up the good work."

We took a four-day leave over the 4th of July weekend to visit my family in Westminster, CO. It was the first time I had seen Milton in a long time, and he had grown considerably. We enjoyed our visit but it was simply too short.

The Board of Directors of the Lawton Boy's Club, Inc was faced with many challenges during 1965. An application for support from the Lawton-Fort Sill United Fund was denied and although we had served more than 1,000 boys, we were struggling financially. To make matters worse, the Salvation Army also operated a Boy's Club, and they were members of the Lawton-Fort Sill United Fund. They had also begun construction on a new larger boys' club facility that was scheduled to open in early 1966. The first anniversary of the Club's founding was September 5 and it was decided we should make the boards tenure run from September through August of each year. At the August board meeting elections were held and I was elected to fill a new position of second vice president. An awards banquet was held in mid-September in which the new officers were installed. Also in September, the board received a proposal from the Salvation Army to combine the two Boy's Clubs into one and close both current club facilities and occupy the new facility being built for the Salvation Army.

I held the 1st Brigade job of school support for only four months, but at my farewell ceremony the brigade S3, LTC Harris, commented that while he hated to lose me to the Officer Candidate School that was rapidly expanding to meet the requirements of the Vietnam War, he was pretty sure that the system I had set up would be used for a long time. He went on to say that he had not had a single complaint after my first month on the job, which was a big change from the in-briefing he received from his predecessor who had told him that school support taskings would be a major headache for him and he should expect many complaints every week.

In my efficiency report, LTC Harris wrote,

LT Sharpe did a superb job as the Assistant Operations Officer of the Brigade. With his detailed, up-to-date knowledge of the personnel and equipment capabilities of the units in the brigade, he made sound decisions in planning and committing the brigade to daily school support missions. Once committed, he reacted quickly and surely to any changes that arose. Further, he so allocated these requirements to subordinate units so as to cause minimum interference with readiness training and other essential activities. Gaining their confidence and esteem, he worked in close harmony with senior personnel from both higher and lower headquarters."

LTC Harris further rated me a "10" in potential (top block possible) and recommended that I be given command of a tactical unit. I was allowed to take a five-day leave starting on October 5 and once again we drove to Westminster to visit my family. When I returned from leave, I was informed that I had been awarded a Regular Army Commission by a Department of the Army General Order dated October 6, 1965.

OCS TACTICAL OFFICER AND BATTERY COMMANDER
BATTERY D, OFFICER CANDIDATE SCHOOL
(OCTOBER 11, 1965 TO APRIL 17, 1966)
CLASS 8-66, OCTOBER 31, 1965 TO APRIL 14, 1966

I reported to OCS on October 11,1965, and was assigned as a tactical officer responsible for a platoon of officer candidates in Battery D of the Student Brigade. The battery was two weeks away from graduating Class 11-65 and was scheduled to welcome Class 8-66 to lower class status on October 31, 1965.

The day after reporting in, our first daughter, Sherrin Louise, was born. I should say that while we had agreed on a name, we had not discussed how to spell her first name. I thought I knew how to spell Sharon or Sherin or even Sharyn, but Sherrin was unknown to me. I told Kathy that since the birth certificate had already been filled out, it was too late to change it, but I noted that she would likely spend her whole life saying, "My name is 'Sharon' (how it sounds when you say it) and you spell that, S H E R R I N!" According to my daughter, this turned out to be a true statement. I put in for a seven-day leave starting on the 16th of October and got to spend time with our new daughter, learning to change her diapers and burp her, and just enjoying holding her. A few months after Sherrin's birth, I completed my first "self-help" project in the back yard of our home by constructing a covered sandbox.

Before Class 8-66 began, we were notified that an agreement had been reached between the Information Office of the Artillery Center and School, in coordination with the OCS Commandants office, that a Fort Sill Public Affairs lieutenant and a photographer were going to follow the weekly progress of the class. They would write six monthly articles to be published in the January through June issues of the US Army Recruiting & Career Counseling Journal. 1LT Joseph A. Rollo was given a desk in our headquarters and spent considerable time observing activities, questioning the staff and candidates, and gathering information from other parts of the Artillery School. These articles and photos describe the training and activities undertaken during the 23-week period of OCS. 1LT Rollo did a great job of describing the OCS program and his articles are included as Appendix 4 of this book. I have decided not to spend any time describing day-to-day life at OCS during my tenure as these articles do that in great detail.

I returned to OCS 18 months after graduating to find it organized almost the same as the day I left, with six batteries, but quickly learned that the Army had directed a rapid expansion of OCS over the next two years to meet the ever-increasing need for lieutenants in Vietnam. Where I was a candidate in 2nd Platoon, Battery A; I was now a platoon leader in 2nd Platoon, Battery D. I was promoted to 1st Lieutenant on October 28 and Class 8-66 started on October 31, 1965, with about 130 candidates.

CPT Floyd D. Whitehead, my battery commander, was another example of how *not* to be a very good officer. He placed no trust in his subordinates, offered little in the way of productive counseling, and was in general not a shining example of good leadership! He wore heavily starched fatigues that he changed at least once a day, so he always looked like he did not have a wrinkle anywhere. Just sitting down in a starched uniform caused wrinkles. He carried a swagger stick and loved to tap soldiers on the chest with it to make a point while talking to them.

My Seventh Army NCO Academy Distinguished Graduate swagger stick was displayed on my desk, but I never felt inclined to carry it around. He would often ask questions in a loud voice so one of us could answer. He would ask something like, "What time is staff call?' The answer was posted in the OCS daily bulletin which everyone got every morning, but apparently, he did not care to check. I would sometimes answer, "Sir, it is at 10:00 am in the command conference room!" Without a pause or an acknowledgement of the answer, he would call the headquarters and ask the exact same question. We never heard the answer, but we always assumed it was the same or he would start shouting.

We had a daily staff call with him and he would often pick out something one of us did that he thought could have been done better or differently. He always started about the same way by making it seem we had all committed the same minor transgression. He would say something like, "I do not want to have to tell you again that (insert minor transgression here) is not going to cut it in this battery." It was almost never me, and I got in the habit of raising my hand when he was finished and, when called upon, would say something like, "Sir, I do not know who did this, but it was not me, but it sounds like you are accusing all of us of doing it!" He would always make some excuse, but I think we got our point across.

My primary job as a platoon leader was to evaluate the leadership potential and performance of each of the 30 or so candidates in my platoon. Each week, leadership positions were rotated among the candidates. Each candidate was inspected and questioned daily and received a weekly written evaluation and report. Each candidate was counseled one-on-one about his performance and was required to acknowledge receipt of a copy of the evaluation by signing at the bottom of the form. Some candidates did very well, responded well to counseling, and improved their performance week over week. Others did not. Candidates were also evaluated by the instructional staff on participation, knowledge, improvement, and test results. After six weeks or so in the 23-week course, it became clear that some candidates were not progressing along with their contemporaries. The battery XO held a meeting every Friday afternoon, and each of the platoon leaders reviewed the performance of the best and the poorest candidates in our platoon. We each had to make a recommendation on whether or not to keep the poorer candidates another week, recommend in writing which candidates should be sent to an evaluation board to be turned back to another class, or which ones should be referred to an evaluation board to be released from OCS as not having the potential to become an officer. This was a heavy responsibility that required keeping detailed written records. I enjoyed the work and did my best to help every candidate graduate.

During October and November of 1965, the Board of Directors of the Lawton Boy's Club, Inc. held weekly evening meetings to discuss the possible merger of our Club with that of the Salvation Army. In late November we held a long but productive meeting with the Salvation Army Advisory Board and decided combining the two clubs would be a good idea. In an article in the *Lawton Constitution* newspaper in early December, it was announced that, "The new Boy's Club will be known as the Salvation Army Boy's Club and will be housed in the new structure being built by the Salvation Army with matching funds from the McMahan Foundation. A Boy's Club Council, to be headed by Alford Hennessee, assistant police chief, has been organized to operate the merged programs of the two clubs. Twelve members from the governing bodies of each organization will make up the new council. The new Boy's Club building will be dedicated, and the newly formed council will be installed on January 17, 1966." I was one of the 12 members selected from our board. To closeout our independent existence, the Board of Directors of the Lawton Boy's Club, Inc. coordinated an awards banquet with the Southwest Regional Director of Boy's Clubs of America. The awards banquet was held on Wednesday evening, December 7, 1965, and three of us received Distinguished Service Awards from Boy's Clubs of America.

In November of 1965 I met with the candidate leaders of Class 8-66 and requested that they and their wives consider sponsoring a Christmas party for the Boy's Club before they left for Christmas vacation. The Class agreed and on Saturday, December 17, they gathered at the Boy's Club facility and welcomed nearly 400 boys to the second annual Christmas party. They brought a movie with a projector and screen to play for the children. After the movie, they had games and prizes divided into different age groups, and they brought a bag of fruit and candy for each child to take home. This was the last official function held at the Boy's Club of Lawton, Inc facility. In early January we received a very nice thank you letter from the Club's Executive Secretary expressing the Club's appreciation for the sponsorship. The party was written up in the second weekly article of the *Recruiting Journal* article.

CPT Whitehead was reassigned to Vietnam in late December, and I was proud to be selected to replace him as the battery commander on January 8, 1966. In early January our battery was redesignated as Battery D, 2nd Battalion under the OCS expansion program. OCS first expanded to eight batteries using the eight rows of WWII wooden building in the OCS area while construction was undertaken to build new barracks and dining facilities. On January 17, 1966, the Salvation Army held its Annual Civic Luncheon and a Dedication Service for the newly constructed Boy's Club building in the club's gymnasium at 1315 F Avenue. The 24 members of the new Salvation Army Boy's Club Advisory Board were inducted, and a plaque was unveiled in the lobby of the building.

During the first half of 1966, the new combined board of the Salvation Army Boy's Club met monthly, and we were pleased with both the new facilities and the financial support that was now available through the Lawton-Fort Sill United Fund. I enjoyed belonging to the local Downtown Kiwanis Club, one of three in the city of Lawton. I joined Al Hennessee on the Club's Boys and Girls Committee.

We lived about a 20-minute drive from the OCS area at Fort Sill. One Saturday evening I agreed to pick up supper from the McDonalds across the highway. A taxi was parking just as I arrived at McDonalds, and the driver was in line just in front of me. I overheard him order 40 hamburger meals all with Cokes. I wondered if this might be an order for an OCS unit, as they were not allowed to have food delivered. I immediately went back out to my car and pulled out of the parking lot, drove a few hundred feet down the road back to Fort Sill, pulled into a service station and waited.

Once the taxi was loaded with meals, the driver pulled out and headed for Fort Sill. I let one car get between me and the taxi and proceeded to follow him. This was long before the days of mobile phones, so I could not call my wife to let her know I would be delayed in bringing supper home. Sure enough, the taxi drove straight to Fort Sill and then straight to the OCS area and stopped right in front of one of the four platoons in my battery. I drove to the battalion headquarters about a block away and asked the duty officer to come help me round up hamburgers in one of my platoons. I called my wife to let her know it was going to be another hour before I brought supper home. We ran back to the barracks and arrived just as the taxi was pulling away.

I asked the duty officer to climb up to the second floor via the fire escape on the end of the building. We would both enter the building at the same time and shout "ATTENTION!" which would freeze everyone in place. I asked him to have everyone upstairs come downstairs with their food and drinks. Once everyone was assembled downstairs, I had the candidate platoon leader and candidate platoon sergeant get trash cans out of the latrine and gather up all the drinks in one and hamburgers and french fries in the other. I asked the lieutenant duty officer to supervise dumping out the drinks while we all waited. Once the drinks were disposed of, I had the duty officer take the trash can full of food back to his headquarters. I proceeded to lecture the assembled candidates on the importance of integrity and following rules if they expected to become officers. I let them know that they were lucky it was me who discovered their misdeed, because I knew one or two other battery commanders that might just consider terminating them all for a serious breach of discipline. I also let them know that as a former OCS candidate I knew the tricks of the trade and they should spread the word that it was very likely that I would know if anyone one else tried something as stupid as ordering meals for everyone in their platoon. I also informed them that I would *not* be very understanding of any additional attempts to break the rules. Finally, I told them that each and every one of them would jark twice the next weekend.

The last few weeks of Class 8-66 were filled with graduation planning and then parade and graduation rehearsals. It was an exciting time for the candidates as they welcomed their families to witness their graduation and the pinning on of their second lieutenant bars. I continue to maintain contact with many of the graduates and I was very proud to witness their graduation

OCS BATTERY COMMANDER
BATTERY I, OFFICER CANDIDATE SCHOOL,
(APRIL 18, 1966 TO JUNE 25, 1966)
CLASS 17-66, JANUARY 16, 1966 TO JUNE 25, 1966

Class 8-66 graduated on Friday, April 15, 1966, with 99 candidates pinning on 2nd lieutenant bars. I was informed that a new battery commander would take over Delta Battery and I would be assuming command of Battery I the next Monday. The commander of Battery I had been transferred and I would take over for the last ten weeks of OCS Class 17-66. It was a challenge to get to know the officers and NCOs in my new battery and to begin assisting in the decision making as to who might still be turned back to another class or be recommended for relief. Battery I was in buildings that had been taken over from another unit and our headquarters was in need of renovation. Fortunately, the facility engineers were willing to provide us paint, patching plaster, brushes, rollers, and sand paper. During my first two weeks we spent time repairing walls and repainting the building inside so that it was in much more presentable condition.

As I highlighted in describing my own OCS class, every class participated in an escape and evasion course a few weeks before graduation. While there were some differences in every planned course, there were some new things instituted for this Class. First, the class was divided up into teams of four individuals, and each team was given a radio and a map. Each team had a slightly different radio frequency so that teams could not communicate among themselves. This was not only an escape and evasion course, but was also a test of the student's map reading skills, as each team had to report their location to a control center each hour. Candidates were told they could not go outside the exercise area, which was clearly marked on each map. Soldiers tasked to act as the enemy would not have access to this information. As usual, the candidates were loaded into trucks to be driven to the exercise area. They were told the convoy would be ambushed somewhere along the route and they should do their best to escape into the woods and attempt to reach the daytime safe area. Then after supper, they should attempt to reach the nighttime safe area without being captured. If captured, they would be taken to an interrogation area, subjected to some interrogation techniques they had heard about during their escape and evasion training, and then released to rejoin the exercise.

As the battery commander, I knew where the ambush would take place and was able to position myself on a hill overlooking the ambush site. I watched the action through my binoculars. The aggressors captured about 20 candidates, and after lining them up and taking their names and serial numbers, put them on a truck to be taken to a fenced interrogation area.

I then drove to our operations center where we had prepared a large exercise area map with pins holding flags with each team number on them. One candidate who had been captured reminded me recently that candidates were being put in small wooden boxes covered by a lid in the interrogation area. He had refused to get in a box, so he was brought to the operations center where I questioned him about what happened. He explained that he was claustrophobic and was afraid to get in the box. I told him he would need to think about what would happen if he were really captured as he would *not* have the opportunity to refuse. I had him rejoin his team for the night phase of the exercise. He thanked me in his recent message for understanding and letting him return to the exercise.

After about 30 minutes of interrogation, the candidates were formed into new teams, given radios and maps, and were taken back to where they were captured. They were released and instructed to attempt to get to the safe area without being captured a second time. As locations were reported, the pins were stuck into the map. At the end of the first hour all the teams had reported in, and everyone seemed to be moving in the correct direction. At the end of the second hour, one team reported itself near the edge of the exercise area moving toward the artillery impact area. We radioed the team to find out why they were moving in the wrong direction, but we could not get in contact with them. When reports came in at the end of the third hour, this team reported that they were approaching the artillery impact area, well outside the exercise area.

We frantically tried to reach the team with no success, so we reported the information to range control. Immediately, four artillery firing units were directed to cease firing live artillery until we could locate and get the candidates out of the impact area. A helicopter was requested to search the reported grid coordinates, but it took nearly an hour before the helicopter could get on station. At the end of the fourth hour, this team reported themselves in the middle of the impact area. The helicopter flew at very low levels all around the reported coordinates and found no one. After consulting with range control and post operations, which were very unhappy folks, a second helicopter was sent and given the radio frequency of the team. Both helicopters had radio direction finding equipment on board, and the two helicopters were directed to orbit on opposite sides of the impact area and attempt to get a direction on the next radio transmission.

On hour five, the candidates reported approaching the far side of the impact area. We had plotted the general location of each orbiting helicopter, and both helicopters sent us an azimuth which we immediately plotted. Lo and behold, the azimuths crossed on a hill behind the nighttime safe area, just a short distance from the perimeter road of the installation. My XO and I jumped in my school support jeep, and our driver drove us up to the coordinates we were given. Sure enough, we found four candidates, a nearly empty case of beer, and the team radio.

They said they had gone straight to the perimeter road and climbed over the fence. They lost their map somewhere in this process, so they hitched a ride to a small convenience store where they bought a case of beer. One said he knew where the safe area was, so they climbed back over the fence and went to the hill behind the safe area. When asked about reporting being in the impact area, they said without a map they were just "making up coordinates" that they thought would disguise their location. We reported to range control and then base operations that there were no candidates in or near the impact area. We then took the beer and the radio and told the four candidates to follow us back down the hill to the safe area.

The next morning, I received a message to please come to the OCS headquarters, where the XO explained that their phone had been ringing off the hook asking whether the four candidates involved in the escape and evasion exercise prank were going to be kicked out of school for all the grief they had caused the entire post the day before. I asked the XO if he or the commandant was ordering me to relieve them. He said no, but he would be interested in hearing what I recommended if kicking them out was not my plan.

I asked the XO if he believed I was a good lieutenant. He said he thought I was one of the best he had ever met. I said, "Well sir, I am presented with something of a moral dilemma as my OCS battery commander was presented with a very similar decision the morning after our escape and evasion course." I gave him a short description of how four of us had broken the rules and could have been kicked out of school if the battery commander had decided one prank did not outweigh 20 plus weeks of outstanding performance. I told him that while these candidates were not going to be the distinguished graduates of their class like we were, they had performed well. I told him I thought the candidates had learned a valuable lesson, just as I had, and I planned to do the same thing my battery commander had done, chew their asses make them jark with weapons, and explain to them they were being given a chance to make up for their misdeeds by proving that I did not make a bad decision by becoming outstanding lieutenants. The XO said he would brief the commandant and if my plan was unacceptable, he would let me know. I never heard from him, and the four candidates graduated.

During the last few weeks of Class 17-66 residence, we were beginning to see the effects of the planned growth of OCS. OCS added three new batteries, G, H and I, taking over all the WW II buildings that had previously belonged to other units. We were briefed on plans to have one or two classes start per week. We had been witnessing the construction of a huge new complex next to us that would accommodate the rapid growth planned. We were informed that within two months both the existing batteries and new ones would be formed into battalions. Once again I was very proud to see the candidates graduate knowing that we had done our best to prepare them to be officers.

OCS BATTERY COMMANDER
REDESIGNATED TO BATTERY A, 2ND BATTALION,
(JUNE 26, 1966 TO SEPTEMBER 7, 1966)
REDESIGNATED TO BATTERY B, 3RD BATTALION,
(SEPTEMBER 8, 1966 TO MARCH 6, 1967)
CLASS 1-67, JULY 10, 1966 TO DECEMBER 17, 1966

I remained in command of Battery I and we prepared to welcome Class 1-67 to begin their 23-week course of instruction. This was a turbulent time in OCS as new batteries were being formed every week or two. And new battalions were as well. In late July, just two weeks after in processing Class 1-67, we were redesignated as Battery A, 2nd Battalion and then in September we were again redesignated as Battery B, 3rd Battalion.

Early in 1966, recruiting standards for OCS were lowered. Where in the past, most classes had a number of noncommissioned officers with some years of service, classes now had very few. Having a class with little or no previous military or leadership experience made it much harder to develop the leadership necessary to become a successful officer. Class 1-67 reported with close to 120 candidates. A few weeks into the life of the Class, everyone in OCS received a directive from the commanding general that the new standard for OCS classes was that commanders should keep attrition below 10%, as the need for lieutenants in Vietnam had grown by leaps and bounds. By the time we received the directive, we were approaching 10%. Soon word was spread that Class 1-67 was exceeding the attrition rate expected so that if it was determined a candidate needed to be turned back to another class for additional training, it would be a good idea to send turn-backs to our battery. This created many challenges for us over the next months, but we simply had to deal with it.

The attrition rate continued to grow for a couple of reasons. First, I refused to lower the standards that had been set for what it took to become an officer, even as most other units took the easy way out and graduated almost everyone. I later heard about at least two candidates that we recommended for relief that were turned back to another class and graduated. Both were relieved of duty in Vietnam. Second, we were being sent candidates others had recommended for relief. A quick review of their records usually showed that they had no business being officers. None the less, we put them in leadership positions, evaluated them, wrote detailed evaluation reports outlining their leadership failures and then recommended them for relief. As the class progressed, we passed 50% then 75%, and then a 100% attrition rate. That is, we had turned back or relieved more candidates than we had remaining in the class.

Sometime in the last two weeks of the residency of Class 1-67, I was summoned to the Commanding General's office. I was told that MG Harry Critz wanted to relieve me for my 100%+ attrition rate.

The commandant and I appeared at his office, and I brought with me two bulging boxes of the records of all those relieved or turned back. When I reported in, the first thing the general said was something to the effect of, "Sharpe, why do you think I should not fire you for failing to follow my guidance?" I simply said, "Sir, you are welcome to fire me, that is your prerogative. But before you do, I would ask that you take just a few minutes of your time to go over to that table and pick any three candidate records you wish. If you find one that you think did *not* deserve to be relieved or turned back, you should fire me." He got up from his desk, went over to the table, and looked at way more than three records. Finally, he said something to the effect of, "Sharpe, get your ass out of here and get back to work." I never heard another word about getting relieved of command. Class 1-67 graduated on December 16, 1966, with 60 candidates crossing the stage and receiving their lieutenants' bars. By graduation, more than 100 candidates had either been turned back to other classes or were released from the Officer Candidate School from Class 1-67.

Barbara Michele "Shelley" Morrison (Kennedy) was born in Iowa City, IA on November 27, 1966, to Anne Louise Kennedy, the younger sister of my wife Kathy, and Gordon Edward Morrison. We would later adopt Shelley and change her last name to Sharpe. At the time Anne, known as "Annie Lou" was traveling around the Midwest as a vocalist with a band led by Gordon. They would play six or seven dates on ten-day to two-week road trips. Soon after Shelley was born, Annie Lou asked us if we would keep Shelley while she went on her road trips. We agreed to do so, but what really happened was that Shelley basically lived with us and her mother would simply visit for a few hours each time she returned to her home base in Lawton.

On December 13, 1966, the Kiwanis Club of Lawton held its annual Installation of Officers at the Hotel Lawtonian. One of the new members inducted was Mr. Clinton Herring, who served with me on the Boy's Club Board of Directors. He would go on to purchase the local Lincoln/Mercury automobile dealership. I received a late Christmas present by being promoted to Captain on December 28, 1966.

On Wednesday evening, January 18, 1967, I was privileged to meet my first International President of a civic organization when the International President of Kiwanis International spoke at a joint meeting of the three Kiwanis clubs in Lawton; Downtown, Northwest, and Sheridan. Dr. R. Glen Reed Jr. was a dentist from Marietta, GA who served as the 50th International President. I had a picture taken with him but somehow it has disappeared over the years. On Monday, February 6, 1967, the Salvation Army once again held its Annual Civic Luncheon. A decision was made to have the 24 members of the Salvation Army Boy's Club Advisory Board serve for three-year terms, so some were inducted for a second year while others were inducted to serve through 1968. I was elected Board Secretary and was inducted to serve through 1968. This was cut short a few weeks later when I received orders for my next assignment.

By the time Class 1-67 graduated, OCS had grown to six battalions and 42 batteries. Thirteen new, large barracks buildings were built, and OCS was capable of housing and supporting 9,600 candidates. In February 1967 a US Army General Order redesignated the Office Candidate School to the Officer Candidate Brigade. The leader's title was changed from School Commandant to Brigade Commanding Officer. My final assignment in OCS was to command Battery B, 3rd Battalion for just three months before being reassigned to Turkey.

After receiving my orders for Turkey, I was informed that the plan was that I was to become the S2 /Intelligence officer of the headquarter detachment of the Special Ammunition Support Command (SASCOM) headquarters in Turkey. The detachment was in Chakmakli, 25 miles west of Istanbul on the European side of the Bosporus. In February I requested that I be sent to some type of preparatory training. Fortunately, there was a course at the Intelligence School at Fort Holabird, MD entitled "The Installation Intelligence Officers Course" that was scheduled from May 5 to 26, 1967. I was issued orders to attend on temporary duty enroute to Turkey. This would be the start of a secondary specialty in the army in which I would serve as an intelligence officer at virtually every level of the army over the next 14 years.

One postscript to this chapter. One of the graduates of OCS Class 1-67, 2nd Lieutenant Harold B. "Pinky" Durham Jr. gave his life for his fellow soldiers in Vietnam on October 17, 1967. He was posthumously awarded the Medal of Honor for his conspicuous gallantry while in close combat. His Medal of Honor Citation and a number of photos have been included at the end of this chapter.

Marriage to Mary K. Kennedy, April 3, 1965 in El Paso, Texas, near her parent's home.

Maid of Honor, Kathy's sister, Liz; Best Man, Willard.

L to R: Robert, Eileen, Dad, Kathy, Me, Mom, Charlie.

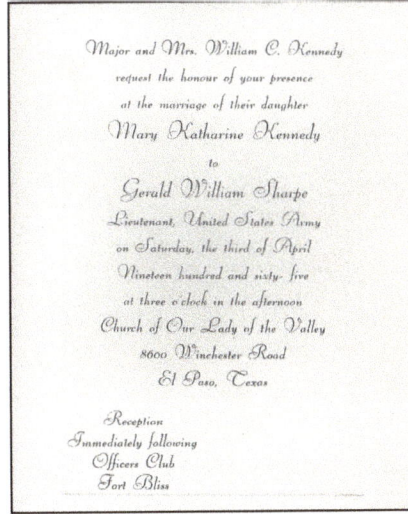

Major and Mrs. William C. Kennedy
request the honour of your presence
at the marriage of their daughter
Mary Katharine Kennedy
to
Gerald William Sharpe
Lieutenant, United States Army
on Saturday, the third of April
Nineteen hundred and sixty-five
at three o'clock in the afternoon
Church of Our Lady of the Valley
8600 Winchester Road
El Paso, Texas

Reception
Immediately following
Officers Club
Fort Bliss

Katharine and Gerald
April 3, 1965

My first portrait photo as a lieutenant.

L to R: Dad, Charlie, Mom, Robert, Milton, Eileen, Stanley, and Me.

Home we purchased at 1638 NW 50th Avenue in Lawton, Oklahoma. We retained this home for many years renting it out.

Partial views of the covered sand box I built in the back yard.

Horseback riding in El Paso on the visit Kathy and I made prior to our getting married. Kathy's sister, Ruth, is on the horse with me.

Kathy's father, Major (Ret) William "Bill" Kennedy is sitting on his couch and in front of our home in Lawton.

Caballero Motel where I stayed and where the family had dinner.

Me

I led Headquarters Battery to a win in the battalion run competition. Articles on the next page.

Fort Sill Troops Run For Fun

All work and no play makes Jack a dull boy, as the aphorism goes.

So the Army sets aside periods of "athletics and recreation." The troops call it "A and R" and look forward to the periods as a chance to play baseball, football or other sports.

Last week Lt. Col. Leo D. Johns, commander of the 3rd Target Acquisition Battalion, 26th Artillery, Fort Sill, decided that his men might enjoy something slightly more athletic than ping pong during their A and R period.

Col. Johns put his men in field gear, gave them rifles to carry and outlined rules for a 10-mile race through West Range.

To add enthusiasm to the project, Col. Johns put his four batteries in competition and offered the winners a half day off.

Slightly more than two hours after the race began, Lt. Gerald W. Sharpe, Headquarters Battery executive officer, trotted across the finish line with his battery behind him. Batteries B, A and C followed in that order.

"I had to hold them back," panted Lt. Sharpe after leading his battery to victory. "They wanted to run all the way." He noted that the six-month reservists in his battery seemed exceptionally fit.

No more races are planned for the near future, but Headquarters Battery is looking forward to the afternoon off on Friday.

"MY ACHING FEET"

Never before was a training holiday declared at a more opportune time for the 3d TAB, 26th Arty than last Saturday, October 24. Soaking of the feet was the agenda of the day for Hq Btry. The reason was the ten mile hike in which the 3d TAB, 26th Arty participated on Friday, October 23.

The hike was a competition among the batteries. Hq Btry emerged victorious with a total time of two hours, ten minutes and seventeen seconds. Coming in behind Hq Btry were B Btry with two hours, seventeen minutes and 45 seconds, A Btry with two hours, 21 minutes and 59 seconds and C Btry with two hours, 30 minutes and fifteen seconds. The total time was computed by adding to the actual time, which was two hours and 63 seconds, a penalty of ten seconds for every man not finishing. The actual time is deceiving though as the route taken had many time-consuming obstacles. Had it not been for the obstacles, the actual time of Hq Btry could have been in the one hour plus range. Having to cross water three times, at one time reaching almost knee depth, hiking the peak of Hoyles Hill and crossing fields with many cumbersome clumps of weeds and chuckholes were the biggest obstacles.

The whole battalion followed the same route with the starting order A, Hq, B and C Batteries each starting at ten minute intervals. Led by 2d Lt Gerald W. Sharpe, Hq Btry closed the ten minute gap and passed A Btry at the halfway mark. Lt Sharpe and MSgt Willie W. Garrett were large factors in the battery's success. Continually running between the front and rear of the formation, Lt Sharpe was a great booster of the men's morale. With Sgt Garrett at the rear lending morale support, many potential dropouts were discouraged.

Farewell plaque from the 1st brigade with the crest of each unit in the brigade. The crest from my previous unit, the 3rd Target Acquisition Battalion, 26th Field Artillery, is the red one at the end of the 2nd row.

Photos taken after I became the battery commander for OCS Class 8-66. It began on October 31, 1966 and I assumed command in January 1967. The full story of the class is in Appendix 4.

Class 8-66, Distinguished Graduates.

Tactical officers of Battery D.

A "Jark" up MB4 and a "Nuked" Cubicle.

In the dining hall and cleaning the barracks

PRESENT ARMS — Lt. Gerald W. Sharpe, platoon leader, snaps a weapon from the hands of Candidate Joseph W. Howze, Cleveland, Ohio. Lt. Sharpe inspected the personal appearance and weapons of the candidates in his platoon last Saturday, Nov. 6.

Second In A Series On OCS

First Training Week Means Rigid Schedule

Two inspections and one written exam highlighted Class 8's first academic week of OCS training. After a week of processing, the 113 candidates began a rigid schedule of classroom instruction.

Although subject to inspection 24 hours a day, a formal inspection of personal appearance and individual weapons was conducted on Saturday, Nov. 6 by the battery commander, Capt. Floyd D. Whitehead, and his platoon leaders.

Class 8 received its first written exam from the Artillery Transport Department of the Artillery and Missile School on Tuesday. The exam covered the classes the candidates had received on the Army's maintenance program.

ON WEDNESDAY, the candidates finally were given a chance to conduct their own inspections as they were taught how to inspect various army vehicles.

A junior officer is often assigned to the task of motor officer and is responsible for the condition of the equipment assigned to his battery. The instruction on Wednesday gave the class practice in knowing where to look to determine the readiness of a vehicle.

An integral part of OCS training is the close relationship each candidate has with his platoon leader. On Wednesday afternoon, each candidate was counseled by the officer responsible for his training. The candidates were questioned on their background and individual problems.

VETERANS DAY was a special day for Class 8 as they were allowed to leave the OCS area and use the recreational facilities of Fort Sill. Although the time was used by some to polish boots and do the hundreds of things that couldn't be accomplished on busier days, some candidates took the time to see a movie at one of Sill's three Post Theatres.

Friday afternoon the candidates received the beginning classes of what is one of the most important blocks of instruction in their 23 weeks. The class was taught the proper method of giving commands and teaching drill and physical training. They will spend many hours doing practical exercises in these subjects before their training is completed.

The usual inspection by the officers of Delta Battery took place on Saturday morning. The errors of the previous week were corrected but the well-trained eyes of the TAC staff were still able to find some flaws that the candidates had missed.

Next week the candidates will get a look of Sill's range area as they begin their instruction in map reading. They will learn how to locate their position on a map by observing the area around them. This last week they knew where they were — in OCS.

New Boys Club Officers To Be Installed

New officers will be installed during the annual meeting and awards banquet of the Lawton Boys Club Thursday evening at the Elks Lodge, Harold May, executive director, announced Saturday.

Bob Lotridge, assistant regional director of the Boys Club of America, will be principal speaker at the 7:30 p.m. banquet.

Joe Sotis, Dallas, regional director, will present special awards and Lew Johnson, president of the Lawton group, will make two awards to businessmen who have supported the club movement.

Entertainment will be by the Lawton Barbershop Quartet. Rev. Donald McDougal will give the invocation. Lt. Gerald Sharpe will serve as master of ceremonies and other officials will present reports. New officers will be installed by Toby Morris.

The meeting is open to the public. Tickets may be purchased at R&S Sporting Goods and Lew Johnson's Restaurant.

Above left, my first week inspection of the officer candidates in my platoon of Battery D. Top Right, Article from the Lawton Constitution newspaper on the candidates first week of training. Above, article on the new Boys Club,

BOYS CLUB OFFICIALS—Don Ronish (second from left) and Lt. Gerald Sharpe (third from left) post with Boys' Club of America officials Ronald Hale (left) and Al Killian after receiving distinguished service awards at the Boys' Club of Lawton, Inc. banquet last night. Ronish is treasurer of the Boys' Clubs of Lawton Council, and Lt. Sharpe is second Vice President of the organization. Hale is assistant Southwest Regional Director of the Boys' Clubs of America, and Killian is executive director of the local club.   (Staff Photo)

Merger Of Two Lawto Boys' Clubs Announce

Merger of the city's two boys' clubs was announced today by Salvation Army Maj. Harold Davis.

Organized as one club are the Lawton Boys' Club and the Salvation Army Boys' Club.

The club will be known as the Salvation Army Boys' Club and will be housed in a new structure being built by the Salvation Army at 14th and F with matching funds from the McMahan Foundation.

Harold May, director of the former Lawton Boys' Club, has been named executive director of the Salvation Army organization.

A Boys' Club Council, headed by Alford Hennessee, assistant police chief, has been organized to operate the merged program of the two clubs.

The council is the outgrowth of negotiations and meetings between the two clubs with the United Fund acting as arbitrator.

The Lawton Boys' Club since its founding, has sought United Fund support. The Salvation Army is a member agency of the United Fund program.

Hennessee said neither the community nor the United Fund can afford to support two boys' clubs and that the merger was essential."

"We believe a better progr for our boys will be the resu he added.

The newly-formed council ported the Lawton Boys' C has promoted an excellent p gram in the old Youth Cen building at 14th and Bell, has been handicapped "by l ited facilities and funds."

The Salvation Army Bo Club has been limited by l of facilities and a gymnasi and limited financial aid.

The new Boys' Club buildi will be dedicated and the n council installed Jan. 17 at Salvation Army Adviso Board's annual civic luncheon

The council will meet at 7 m. the third Tuesday of ea month in the new club buildi

Twelve members of the cou cil were each selected by t Lawton Boys' Club board a the Salvation Army Adviso Board.

Other members, ex-offici are Maj. Davis, May and M George Marshall, Salvati Army state commander.

Named to the council by t Lawton Boys' Club group we Johnnie Fletcher, Hennesse Clinton Herring, Mildred Hoo Lew Johnson, Arch Marc Rev. Donald McDougie, Tob Morris, Jack Munn, Dr. O. Parsons, Clarence Sadler ar Lt. Gerald Sharpe.

Named by the Salvatio Army Advisory Board wer Brig. Gen. (Ret.) John Bird, James Cottingham, W. Ford, Jr., Robert P. Hendric W. F. Hutson, James W. Joh ston, Carl Miller, Don Ronish Robert H. Scott, Charles Sn der, Rev. George Stauffer an Gene Walker.

Other officers elected wer Ford, vice chairman; Rev Donald McDougle, secretary and Don Ronish, treasurer.

Johnson Renamed Boys Club Head; Awards Presented

"Price Tags and Boys" was the title of a talk given at the annual meeting and awards banquet of Lawton Boy's Club Thursday night by Bob Lotridge, assistant regional director from Dallas.

Lotridge's talk preceded presentation of awards and election of officers for 1966. Master of ceremonies for the event, held in the Elks Lodge, was Lt. Gerald Sharpe.

"Boys Award" went to Bill Crawford. The award was for the person outside of the board of directors who did the most to advance Boy's Club work in the previous year.

Toby Morris received the 'Boy and Man Award." It was for the person within the board of directors who did the most to advance Boy's Club work during the year.

Presentation of both awards were by Lew Johnson, president of the organization.

Special awards were presented to Johnson and Porter Hood by Joe Sotis, regional director, Dallas.

Johnson was re-elected president; Arch March was named to replace Clinton Herring as vice president; Gerald Sharpe was appointed to the newly-created post of second vice president; Al Hennessee was reelect-

LEW JOHNSON

ed treasurer and Mildred Hood will again serve as secretary.

The Lawton Boy's Club was incorporated March 11, 1964 and began operations Sept. 5, 1964. Since its beginning the club has enrolled 1,372 boys with an average daily attendance of 150.

Lawton Boy's Club Awards Presented At Banquet Here

"Price Tags and Boys" was the title of a talk given at the annual meeting and awards banquet of Lawton Boy's Club Thursday night by Bob Lotridge, assistant regional director from Dallas.

Lotridge's talk preceded presentation of awards and election of officers for 1966. Master of ceremonies for the event, held in the Elks Lodge, was Lt. Gerald Sharpe.

"Boys Award" went to Bill Crawford. The award was for the person outside of the board of directors who did the most to advance Boy's Club work in the previous year.

Toby Morris received the "Boy and Man Award." It was for the person within the board of directors who did the most to advance Boy's Club work during the year.

Presentation of both awards were by Lew Johnson, president of the organization.

Special awards were presented to Johnson and Porter Hood by Jed Sotis, regional director, Dallas.

Johnson was re-elected president; Arch March was named to replace Clinton Herring as vice president; Gerald Sharpe was appointed to the newly-created post of second vice president; Al Hennessee was reelected treasurer and Mildred Hood will again serve as secretary.

The Lawton Boy's Club was incorporated March 11, 1964 and began operations Sept. 5, 1964. Since its beginning the club has enrolled 1,372 boys with an average daily attendance of 150.

The Salvation Army

FOUNDED IN 1865 BY WILLIAM BOOTH

CITADEL 1314 E AVENUE

POST OFFICE BOX 1280   PHONE EL 3-8222

WELFARE DEPARTMENT — 622 B AVE. — EL 3-3180

RED SHIELD BOYS' CLUB — 1314 E AVE. — EL 3-8222

RED SHIELD EMERGENCY LODGE — 508 S. 14TH — EL 3-8222

LAWTON, OKLAHOMA

MAJOR GEORGE MARSHALL
DIVISIONAL COMMANDER

MAJOR & MRS. HAROLD DAVIS
COMMANDING OFFICERS

LIEUTENANT STANLEY MELTON
ASSISTANT OFFICER

January 3, 1966

Lt. Gerald Sharpe
1309 Maple
Lawton, Oklahoma

Dear Lt. Sharpe:

It is a pleasure to welcome you to membership on The Salvation
Army Boys' Club Council. We understand you have agreed to
serve and we appreciate this obvious interest in Lawton youth.

The organization meeting to elect officers and handle other
business will be held this Wednesday January 5th at a noon
luncheon in the Citadel, 1314 "E" Ave. Cost of the meal will
be $1.25. It is very important that you be present for this
first meeting.

Attached is a copy of the official notice of this organizational
meeting, for your information.

We look forward to working with you.

Sincerely yours,

Robert Hooper

Robert Hooper, Chairman
Citizens' Advisory Board

Enclosure

The last day of May 1966 was a momentous occasion in the Sharpe household. Our Mother received her Bachelor of Arts Degree, Charlie graduated from high school, and Robert graduated from 8th grade. Above left: Charlie is on the right in the top row, Mother below him, and Robert below her.

Belleview College

Denver, Colorado

on the recommendation of the Faculty
has conferred upon

Thelma Ritchie Sharpe

the degree of **Bachelor of Arts** and by this
Diploma make known that she is entitled to all the **Honors**
Rights and **Privileges** to that degree appertaining
Given this thirtieth day of May in the year of our
Lord one thousand nine hundred and sixty-six at Denver,
State of Colorado.

In witness whereof the seal of the College and
signatures of its officers are hereunto affixed

Arthur K. White.
President

Donald J. Wolfram
Dean

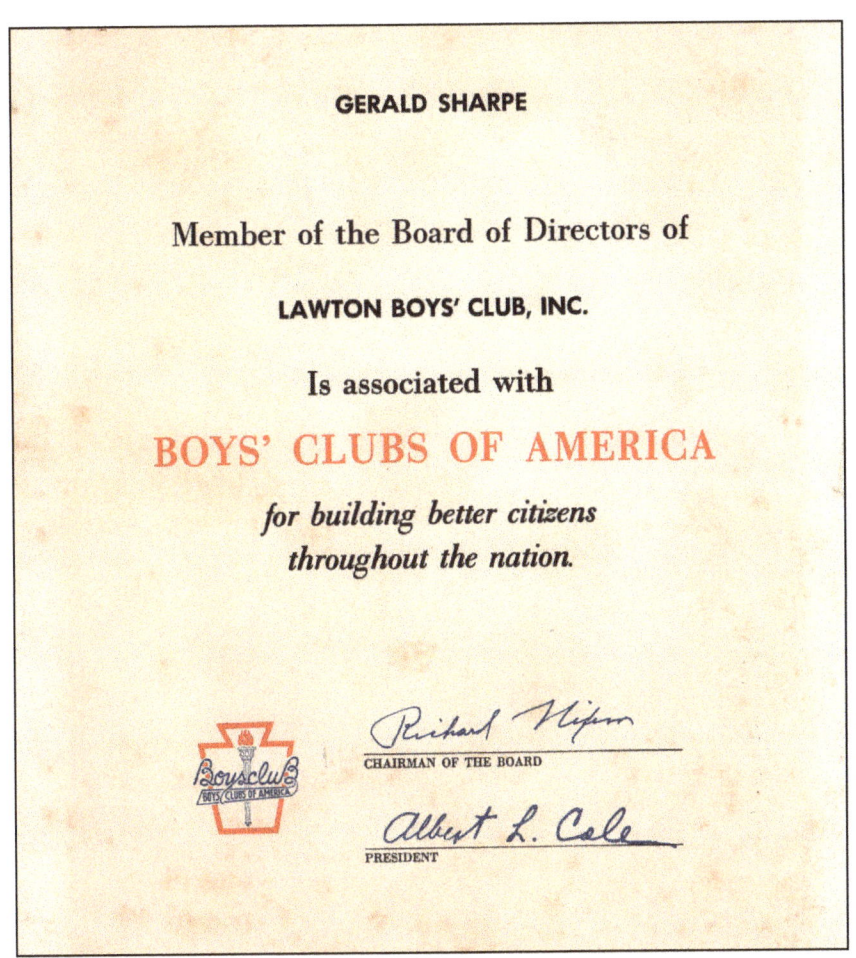

Signed by Richard Nixon before he became President

Boy's Club Award for Service

Command photo from OCS.

OCS Battalion Will Sponsor Kids' Party

As the Christmas season draws near, the men of the Officer Candidate School's 3rd Battalion have undertaken sponsorship of a Christmas party to be held at The Boys' Clubs of Lawton, 14th and F, on 10 Dec. at 1300 hours.

The party is to be open to all children in the Lawton-Fort Sill area and featured will be Santa Claus, cartoons and numerous toys and prizes to be given away.

Each child will receive a Christmas stocking filled with goodies.

Major R. S. Wheeler, the 3rd Battalion Commander, has received over four hundred dollars to be used in tomorrows affair, as the candidates and OC wives of the 3rd Battalion gave generously to insure that each child would have a merry Christmas.

This is the second year that OCS has sponsored a Boys' Club Christmas party and the second year that Lt. Gerald Sharpe has been the project officer. Highlight of the afternoon will be a drawing for at least 14 frozen turkeys and other foodstuffs. Every child attending will have a chance to win one of the many prizes to be given away.

The members of The Boys' Club have been working on decorations, such as home made Christmas Trees and Angels, as well as many other projects to insure a proper atmosphere for the children in attendance.

The wives of the Officer Candidates decorated a Christmas Tree and helped wrap many of the gifts.

Anyone who would care to donate time, money or effort to the Boys' Club program may call the director at EL 7-7541.

Salvation Army Board, Council To Be Installed

New officers and members of the Salvation Army Advisory Board and Boys Club Advisory Council will be installed at a Monday luncheon in the Salvation Army Citadel.

Lt. Col. C. William Jaynes will be principal speaker for the luncheon at 1314 F.

Advisory board officers to be installed are Ross Baker, chairman; Dr. Charles Andrus, vice chairman; Mrs. Lodell Goochey, secretary, and Robert H. Scott, treasurer.

Boys Club Advisory Council officers include W. D. Ford Jr., chairman; Don Ronish, vice chairman; Capt. Gerald Sharpe, secretary; and Jerry Bucklew, treasurer.

Annual Civic Luncheon

and

Service of Dedication

THE SALVATION ARMY BOYS' CLUB BUILDING

14th and "F" Streets — Lawton, Oklahoma

12:00 Noon

MONDAY, JANUARY 17, 1966

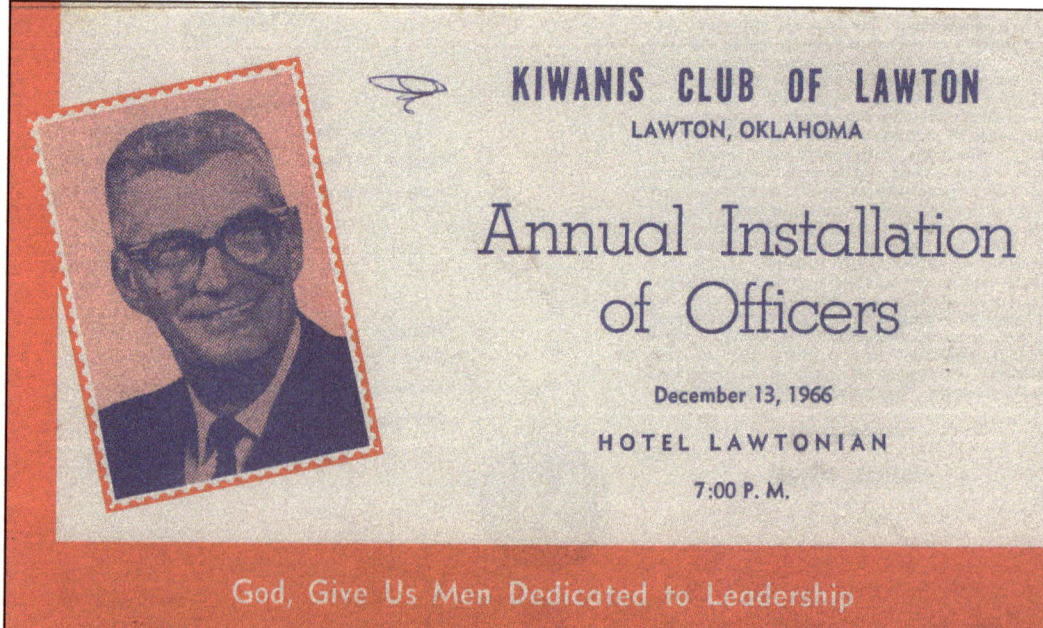

KIWANIS CLUB OF LAWTON
LAWTON, OKLAHOMA

Annual Installation of Officers

December 13, 1966

HOTEL LAWTONIAN

7:00 P. M.

God, Give Us Men Dedicated to Leadership

"A Call To Leadership"

Kiwanis Club of Lawton

(DOWNTOWN CLUB)

45 Years of Service

in

Lawton, Oklahoma

THE OBJECTS OF KIWANIS INTERNATIONAL

TO GIVE primacy to the human and spiritual, rather than to the material values of life.

TO ENCOURAGE the daily living of the Golden Rule in all human relationships.

TO PROMOTE the adoption and the application of higher social, business, and professional standards.

TO DEVELOP, by precept and example, a more intelligent, aggressive, and serviceable citizenship.

TO PROVIDE, through Kiwanis clubs, a practical means to form enduring friendships, to render altruistic service, and to build better communities.

TO COOPERATE in creating and maintaining that sound public opinion and high idealism which make possible the increase of righteousness, justice, patriotism, and good will.

WE BUILD

BY OUR LABORS WE MOLD—Man, who is created in the image of God, must in these times have faith, abundant hope, and most of all, charity. Charity is a virtue of the heart and in the concern of all mankind. Effective charity demands action, and the activity of yesterday is meaningless unless we perform new deeds of service today. We who have lofty ideals which inspire ideas, by our labors mold them into acts. The work we have done is as naught unless we go ever forward and redouble our efforts, and God would have it so. All men are not gifted with the same talents, but each in his own way and with earnest heart can work sincerely for his God, his community and his nation. Not every man can be a statesman, or a scientist, or a warrior, nor did God intend it so; but each as a child of God can be the best of whatever he is as he builds.

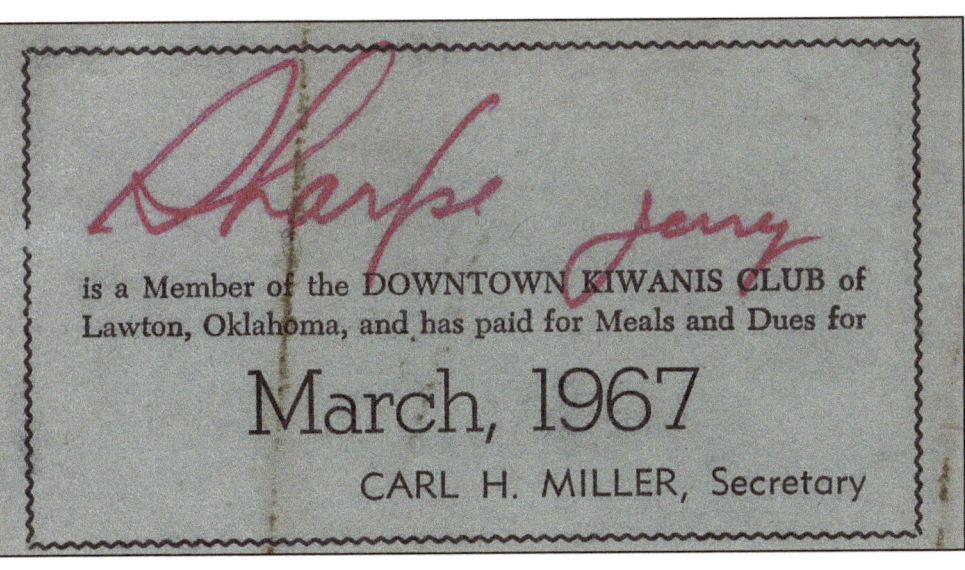

is a Member of the DOWNTOWN KIWANIS CLUB of Lawton, Oklahoma, and has paid for Meals and Dues for

March, 1967

CARL H. MILLER, Secretary

INDIA

The "Invincibles" have gone through three big changes within the past two weeks. Lt. Joel D. Boyd, our newly arrived Battery Commander received orders and departed to that "Happy Land Over There." We of India certainly will miss him. Lt. Gerald W. Sharpe is our new Battery Commander and we hope he stays with us. "Welcome to India, Lt. Sharpe."

Now that Precision Fire has been completed, the "Invincibles" can proudly walk down those "Dusty Trails" that all Upperclassmen walk. India has finally donned the "Red Tabs" and "Clickers" and plans to keep "Driving On."

BOYS' CLUB OF LAWTON, INC.

EXECUTIVE DIRECTOR
HAROLD MAY
3144 KINYON
LAWTON, OKLA.

January 9, 1966

Commandant
O.C.S. 3-66 Delta Battery
Fort Sill, Oklahoma

Dear Sir,

I would like to convey my appreciation to the men of O.C.S. 3-66 Delta Battery for their sponsorship of the Lawton Boys' Club, Inc. Christmas Party.

It is gratifying when young men put forth so much effort for the enjoyment of the young boys of the Lawton - Fort Sill Community.

On behalf of all the boys and myself, I again extend my sincere thanks.

Sincerely yours,

Harold May

Harold May
Executive Director

1966 FINANCIAL REPORT

Income
United Fund   13,002.50
Dues, donations, etc.   3,312.04
    Total $16,314.54
Expense
Salaries, ins., etc. $10,323.48
Maintenance and program 5,982.43
Bal. 12/31/66   8.63
    Total $16,314.54

A UNITED FUND AGENCY

WE PROUDLY PRESENT
OUR BOY'S CLUB ADVISORY COUNCIL

1966 Officers
Chairman Alford Hennessee, Vice Chairman W. D. Ford
Secretary Rev. Don McDougle, Treasurer Don Ronish

1967 Officers
Chairman W. D. Ford, Vice Chairman Don Ronish,
Secretary Captain Gerald Sharp, Treasurer Jerry Bucklew

'Elected' Members			Serving thru
Alford Hennessee	1966	Don Hall	1966
Lew Johnson	1966	Toby Morris	1966
Rev. Don McDougle	1967	Jack Munn	1967
Dr. O. L. Parsons	1967	Clarence Sadler	1967
Johnnie Fletcher	1968	Clinton Herring	1968
Arch March	1968	Captain Gerald Sharp	1968
Don Hall	1969	Don Ronish	1969
Richard Shaw	1969	Gene Walker	1969

'Designated' Members			
Robert P. Hendrick	1966	Brig. Gen. John Bird	1966
Don Ronish	1966	William F. Hutson	1966
Gene Walker	1966	Charles Snyder	1966
Jerry Bucklew	1967	Rev. George Stouffer	1966
David Carley	1967	LtCol Joe Burton (R)	1967
Jeptha Dalston	1967	James Cottingham	1967
Bill G. Frazar	1967	W. D. Ford, Jr.	1967
Carl Miller	1967	Carey Johnson	1967
Ted Stephens	1967	Robert H. Scott	1967
		John Winsted	1967

Lt. Col. William Jaynes
Divisional Commander

Major Harold Davis
Commanding Officer

Al Killian
Executive Director

BOYS' CLUB

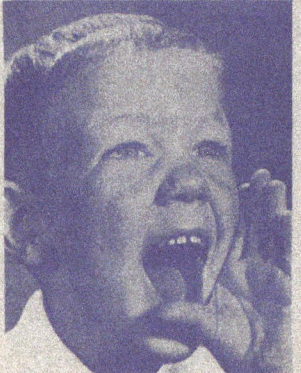

1967
ANNUAL REPORT
Lawton, Oklahoma

January 1967

THANKS, FELLOW LAWTONIANS

Lawton can be proud of its Boys' Club.

Thanks to your generous support thru the United Fund, the Club has made great strides during the past year.

It was just one year ago that the interests and know-how of two dedicated organizations, Lawton Boys' Club, Inc., and The Salvation Army Boys' Club, were pooled in an effort to provide Lawton with the finest Boys' Club possible.

Members of The Salvation Army Boys' Club Advisory Council join me in expressing our very deep appreciation to the United Fund, the news media, The Salvation Army Advisory Board, and a host of friends who have made possible our successful year of service to boys.

Yours for Youth,

Alford Hennessee, Chairman

The Salvation Army
Boys' Club Advisory Council

TODAY'S BOYS — TOMORROW'S LEADERS

HIS CLUB IS - - -

- an open door to fun and fellowship.
- 'HIS' Club - for boys only - he 'belongs'.
- the answer to HIS need for 'gang' identity
- place for varied recreational opportunities.
- offering training activities to develop his know-how.
- providing character-building activities in wholesome Christian environment.
- 'fun with a purpose'

SUBJECT: Personality of the Week

TO: Commandant
         USAAMOCS
         ATTN: Asst S3
         Fort Sill, Oklahoma 73503

1st Lt. Gerald W. Sharpe, commanding officer of Bravo Battery, Third Battalion is the personality of the week. Lt. Sharpe is quite familiar with the traditions of OCS as he was the honor graduate in Class 5-64. Prior to receiving a commission, Lt. Sharpe distinguished himself in the enlisted ranks. Numerous awards were presented to him: December 1962, Soldier of the Quarter, Third Armored Division Artillery, Friedberg, Germany; April 1963, Bad Tolz, Germany, Distinguished Graduate, 7th Army NCO Academy, and the General Doyle O. Hickey Award for the Outstanding Soldier in the Third Armored Division for the period January through June, 1963.

Lt. Sharpe is a native of Colorado and attended Belleview Prep School in Westminister, Colorado. Upon graduation, he furthered his education at Western State College in Gunnison, Colorado.

During the two years Lt. Sharpe spent at Western State College he engaged himself in a number of varied activities. Athletically he participated in intramural wrestling, softball, and flag football. Intellectually, he was sports editor for the school newspaper, a member of the student council, and vice-president of the chess club. Politically, he was a member of the Young Democrats, The International Relations Club and the student traffic court.

Lt. Sharpe is married and has a one year old daughter.

Lt. Sharpe is continuing an outstanding career as Battery Commander of Bravo Battery, Third Battalion. Hopefully, his fine leadership will produce officers of his caliber from class 1-67.

CHARLES C. SPARKS
O/C   Artillery
O/C   CINCO

INTERNATIONAL
PRESIDENT'S DAY

honoring

Dr. R. Glenn Reed, Jr.

President

KIWANIS INTERNATIONAL

TEXAS-OKLAHOMA DISTRICT

KIWANIS INTERNATIONAL

WEDNESDAY, JANUARY 18, 1967

7:00 P.M.

HOTEL LAWTONIAN — LAWTON, OKLAHOMA

Members Of

The Salvation Army Advisory Board

and

The Salvation Army Boys' Club Advisory Council

request the honor of your presence

at their

ANNUAL CIVIC LUNCHEON

Monday, February 6, 1967 at 12:00 Noon

Lt. Colonel William Jaynes, Divisional Commander

Oklahoma - Arkansas Division

THE SALVATION ARMY CITADEL

1314 "E" Avenue
Lawton, Oklahoma

Lunch $1.50
Reply Requested

Monday, February 6, 1967

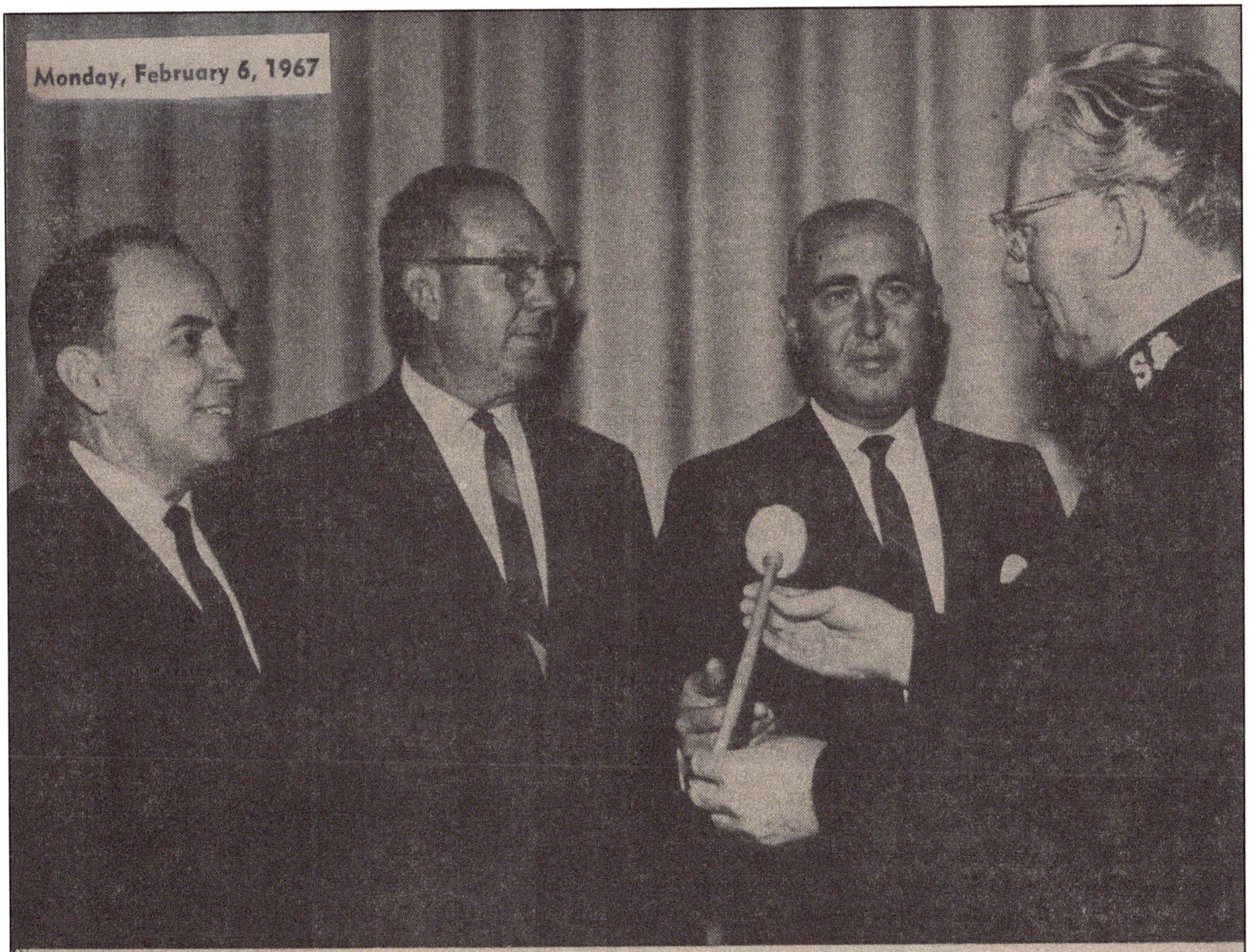

NEW OFFICERS—The 1967 officers for the Salvation Army Advisory Board and Boy's Club Advisory Council were installed at a noon luncheon today by Lt. Col. C. William Jaynes, divisional commander, in the Salvation Army Citadel. Pictured from left are, Dr. Charles Andrus, vice-chairman of the Advisory Board; W. D. Ford Jr., chairman for the Boy's Club Council; Ross Baker, chairman of the Advisory Board, and Lt. Col. Jaynes. Baker received the Salvation Army base drum stick, a traditional symbol of leadership through service and work.

(Staff Photo)

U. S. ARMY

ARTILLERY AND MISSILE SCHOOL

FORT SILL, OKLAHOMA

CEDAT FORTUNA PERITIS

OFFICER CANDIDATE CLASS 8-66

GRADUATION EXERCISES

14 APRIL 1966 - 15 APRIL 1966

GRADUATION REVIEW

Thursday, 1710 Hours, 14 April 1966

OCS Parade Field

* * *

GRADUATION RECEPTION AND DANCE

Thursday, 14 April 1966

Ball Room

Fort Sill Officers' Open Mess

Reception: 2000-2030 Hours

Dance: 2030-2300 Hours

—2—

GRADUATION CEREMONY

Snow Hall Auditorium

1000 Hours, Friday, 15 April 1966

Opening Selection . . 77th & 97th Army Bands

Invocation . . . . . Chaplain John D. Quick
(Audience stand)

Introduction . Brigadier General John S. Hughes

Address . . . . . . Colonel Paul S. Cullen

Oath of Office . . . Colonel K. W. Washbourne

Presentation of Honor Graduate
Award . . . . Colonel Paul S. Cullen

Presentation of Association of US Army
Award . . Brig Gen James F. Brittingham (Ret)

Presentation of Diplomas and
Certificates . . . . . . Colonel Paul S. Cullen

National Anthem . . 77th & 97th Army Bands
(Audience stand)

Closing Selection . . 77th & 97th Army Bands

—3—

OFFICER CANDIDATE SCHOOL
TACTICAL STAFF

"A"

Capt Joseph R. Regelski
1st Lt Herbert G. Vaughn
2d Lt Allen W. Austin
2d Lt Ronald A. Tate
2d Lt Levi Lewis
1st Sgt Casimer F. Konopacki

"E"

1st Lt Kenneth P. Legum
2d Lt Keith W. Vradenburg
2d Lt Richard W. Hensel
2d Lt Frank S. Gentzke
2d Lt David C. Robinson
1st Sgt Lloyd C. Cullier

"B"

Capt Pat C. Hoy, II
1st Lt Thomas A. Duke
2d Lt James M. Taylor
2d Lt Richard J. Bednarczyk
1st Sgt Walden E. Powell

"F"

Capt Thomas A. Herre
1st Lt Stanley W. Brown, Jr.
2d Lt Douglas R. Logan
2d Lt Terrance B. Harris
2d Lt Robert N. Figeria
2d Lt Raymond R. Ragauskas
1st Sgt Wilburn Smith

"C"

Capt Lary W. Padgett
1st Lt John V. Kost
2d Lt Leo C. Pennington
2d Lt Rex C. Burns
2d Lt Steven D. Oulman
2d Lt Andrew M. Pogen
1st Sgt Richard L. Heston

"G"

Capt Ambrose D. Scroggins
2d Lt Allen B. Green
2d Lt Joseph A. Fleming
2d Lt Daniel J. Wargo
2d Lt Asa J. Razook
2d Lt Jerry W. Shelton
1st Sgt George R. Critchley

"D"

1st Lt Gerald W. Sharp
1st Lt John C. Gambaccini
2d Lt David Boyd
2d Lt John R. Treanor
2d Lt Richard C. Nowakowski
1st Sgt Ralph Cheatwood

"I"

Capt Carl H. Moser
2d Lt Edward E. Rutledge
2d Lt John W. Banks
2d Lt Stonnie S. Sullivan
2d Lt Robert D. Kushman
1st Sgt Edward C. Bryant

—7—

Graduation photo from OCS Class 8-66.

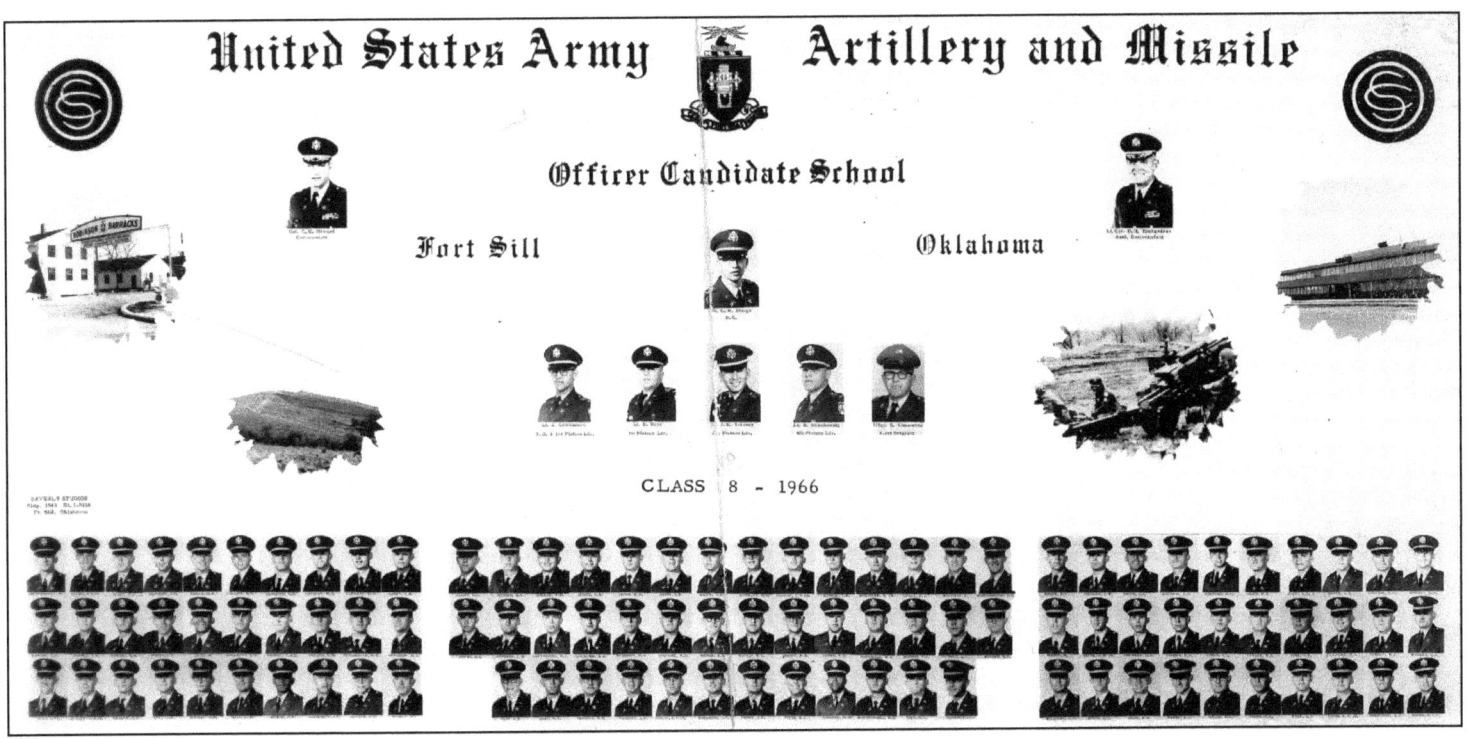

CLASS 8 - 1966

Graduation booklet from OCS Class 17-66.

U. S. ARMY
ARTILLERY AND MISSILE SCHOOL
FORT SILL, OKLAHOMA

CEDAT FORTUNA PERITIS

OFFICER CANDIDATE CLASS 17-66
GRADUATION EXERCISES
24 June 1966 - 25 June 1966

GRADUATION REVIEW
Friday, 1710 Hours, 24 June 1966
OCS Parade Field

* * *

GRADUATION RECEPTION AND DANCE
Friday, 24 June 1966
Ball Room
Fort Sill Officers' Open Mess
Reception: 2000-2030 Hours
Dance: 2030-2300 Hours

—2—

GRADUATION CEREMONY

Snow Hall Auditorium

1000 Hours, Saturday, 25 June 1966

Opening Selection . . . 77th & 97th Army Bands

Invocation . . . . . Chaplain John D. Quick
(Audience stand)

Introduction . . . Major General Harry H. Critz

Address . . . . Colonel Edmund Wendel, Jr.

Oath of Office . . Colonel K. W. Washbourne

Presentation of Honor Graduate
Award . . . . . Colonel Edmund Wendel, Jr.

Presentation of Association of US Army
Award . . Brig Gen James F. Brittingham (Ret)

Presentation of Diplomas and
Certificates . . . Colonel Edmund Wendel, Jr.

National Anthem . . 77th & 97th Army Bands
(Audience stand)

Closing Selection . . . 77th & 97th Army Bands

—3—

OFFICER CANDIDATE SCHOOL
TACTICAL STAFF

"A"
Capt Joseph R. Regelski
1st Lt Herbert G. Vaughn
1st Lt Levi Lewis
1st Lt Richard M. Pierson
2d Lt Ronald A. Tate
1st Sgt Casimer F. Konopacki

"E"
1st Lt Kenneth P. Legum
1st Lt Keith W. Vradenburg
2d Lt Richard W. Hensel
2d Lt Frank S. Gentzke
2d Lt David C. Robinson
2d Lt Emory B. Elliott
1st Sgt Lloyd C. Cullier

"B"
Capt Pat C. Hoy, II
1st Lt Richard J. Bednarczyk
1st Lt Coy R. Goodwin
1st Lt Richard J. Hladysh
2d Lt James M. Taylor
1st Sgt Walden E. Powell

"F"
Capt Thomas A. Herre
1st Lt Stanley W. Brown, Jr.
1st Lt Terrance B. Harris
1st Lt Robert N. Figeria
1st Lt Raymond R. Ragauskas
1st Sgt Wilburn Smith

"C"
Capt John V. Kost
1st Lt Leo C. Pennington
2d Lt Rex C. Burns
2d Lt Robert R. Thompson
1st Sgt J. H. Byram

"G"
Capt Ambrose D. Scroggins
1st Lt Gerald Watson
1st Lt William L. Ford
2d Lt Daniel J. Wargo
2d Lt Jerry W. Shelton
1st Sgt George R. Critchley

"D"
Capt James R. Heldman
1st Lt David Boyd
1st Lt Richard C. Nowakowski
2d Lt Robert L. Barber
2d Lt William C. Bilo
1st Sgt Ralph Cheatwood

"I"
1st Lt Gerald W. Sharpe
2d Lt Howard E. C. Brown
2d Lt John W. Banks
2d Lt Stonnie S. Sullivan
2d Lt Robert D. Kushman
1st Sgt Edward C. Bryant

—7—

Graduation photo from OCS Class 17–66.

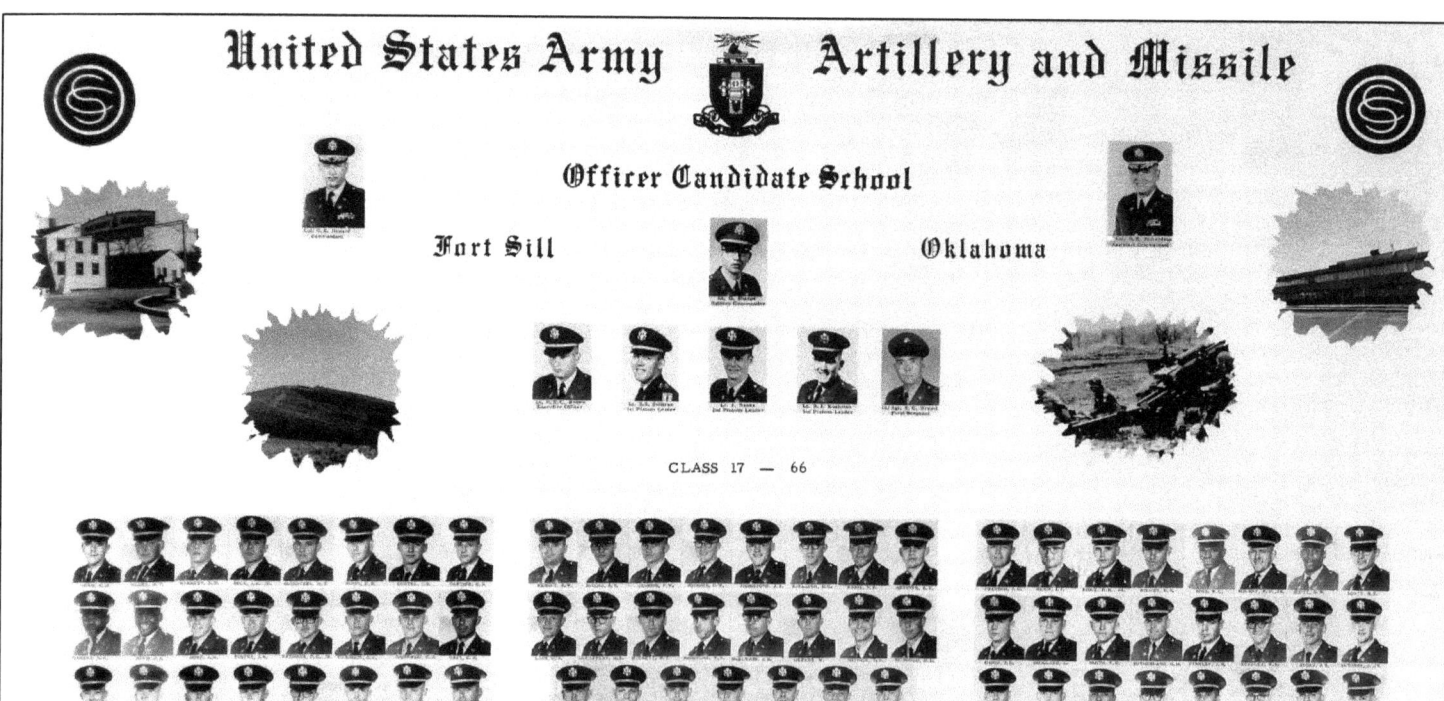

U. S. ARMY
ARTILLERY AND MISSILE SCHOOL
FORT SILL, OKLAHOMA

OFFICER CANDIDATE CLASS 1-67
GRADUATION EXERCISES
15 December - 17 December

GRADUATION RECEPTION AND DANCE
Thursday, 15 December 1966

Ball Room

Fort Sill Officers' Open Mess

Reception: 2000-2030 Hours

Dance: 2030-2300 Hours

* * *

GRADUATION REVIEW
Friday, 1650 hours, 16 December 1966

OCS Parade Field

—2—

GRADUATION CEREMONY
SNOW HALL AUDITORIUM

1000 Hours, Saturday, 17 December 1966

Opening Selection   .   .   77th & 97th Army Bands

Invocation   .   .   .   .   Ch (Maj) Eleson M. Herrick
(Audience stand)

Introduction   .   .   .   Major General Harry H. Critz

Address   .   .   .   .   .   Colonel William J. Lanen

Oath of Office   .   .   .   Colonel K. W. Washbourne

Presentation of Honor Graduate
Award   .   .   .   .   .   .   Colonel William J. Lanen

Presentation of Association of US Army
Award   .   .   .   .   Colonel Herman J. Crigger (Ret)

Presentation of Diplomas and
Certificates   .   .   .   Colonel William J. Lanen

National Anthem   .   .   77th & 97th Army Bands
(Audience stand)

Closing Selection   .   .   .   77th & 97th Army Bands

—3—

3d BATTALION COMMANDER
MAJ Richard S. Wheeler

EXECUTIVE OFFICER
MAJ Warren E. Boisselle

SERGEANT MAJOR
SGM C. J. Posey

GRADUATING BATTERY
"Bravo Battery"

BATTERY COMMANDER
1LT Gerald W. Sharpe

BATTERY EXECUTIVE OFFICER
1LT Roy G. Lake

PLATOON LEADERS
2LT Arthur E. Retzlaff
2LT Thomas J. Dunn
2LT Simon A. Haines, Jr.
2LT William L. Forkner

FIRST SERGEANT
1SG Cletus W. Warren

—7—

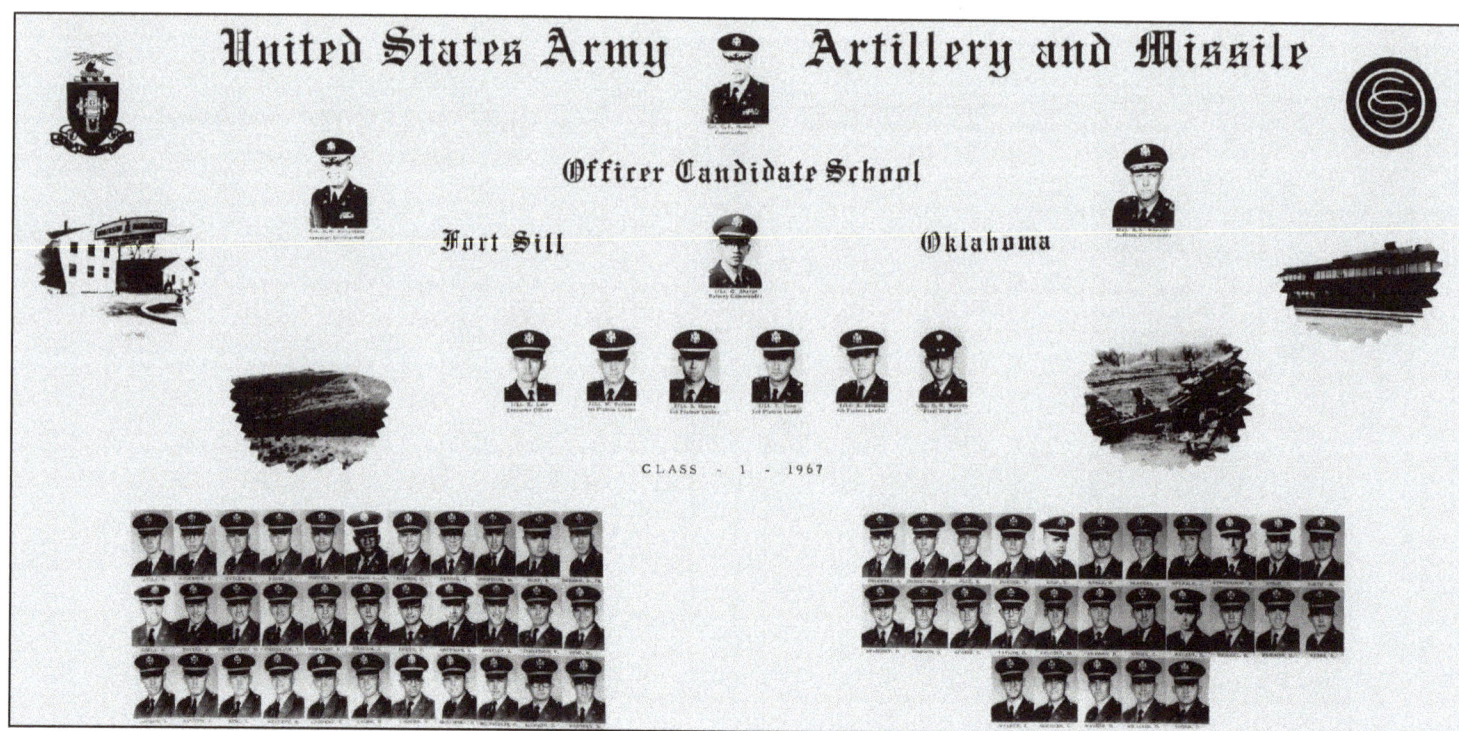

Graduation photo from OCS Class 1-67.

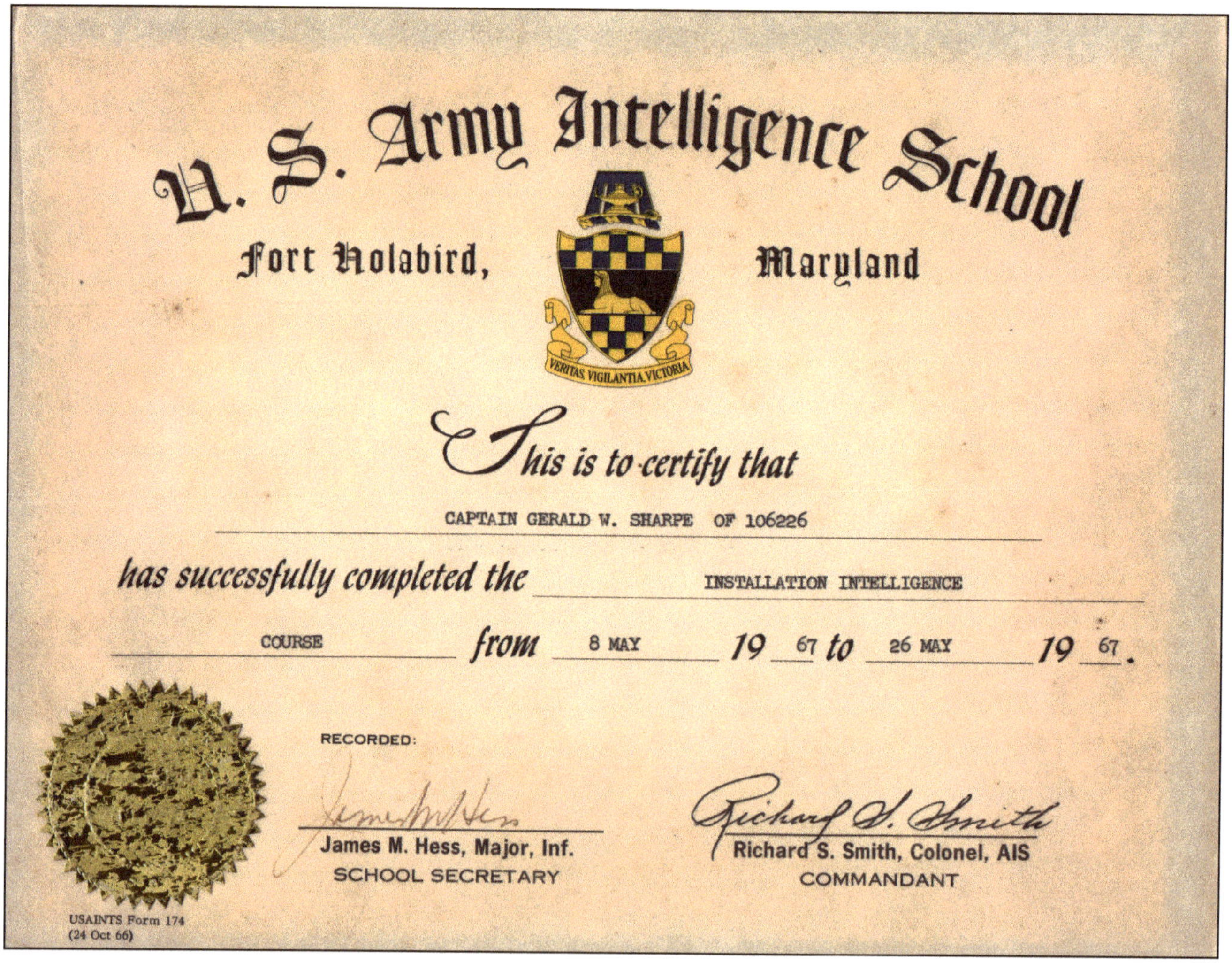

U. S. Army Intelligence School

Fort Holabird, Maryland

VERITAS VIGILANTIA VICTORIA

This is to certify that

CAPTAIN GERALD W. SHARPE   OF 106226

has successfully completed the   INSTALLATION INTELLIGENCE

COURSE   *from*   8 MAY   19 67   *to*   26 MAY   19 67.

RECORDED:

James M. Hess, Major, Inf.
SCHOOL SECRETARY

Richard S. Smith, Colonel, AIS
COMMANDANT

USAINTS Form 174
(24 Oct 66)

Graduation certificate from the Installation Intelligence Officers Course I attended on my way to my assignment in Turkey.

April 1, 1967

Major General Harry H. Critz
Commanding General
USAAMC
Fort Sill, Oklahoma   73503

Dear General Critz:

We would like to express our appreciation to you and to
Captain Gerald W. Sharpe, Hqs., 3rd Bn., Officer Candidate
Brigade, USAAMC of Fort Sill.

We wish to bring to your attention the outstanding
service and contributions Captain Sharpe has made to the
youth of Lawton and Fort Sill through the trying and
difficult period of the institution of a Boys Club program
in this community.

We recognize his work for youth was accomplished at
great personal sacrifice due to the demanding nature of
his military assignment as an OCS Battery Commander.

Once again, please accept our thanks and appreciation
for this service to the youth of Lawton. It is work such
as this and men as you and Captain Sharpe that have entwined
the communities of Fort Sill and Lawton.

Sincerely,

JACK MUNN
Vice President
Lawton Boys Club

CC:
Captain Gerald W. Sharpe

Thank you letter to the Fort Sill CG for my service to the Boy's Club.

OCS AREA DURING THE 1967 EXPANSION - LOOKING NORTH

All of my time as a candidate and as a battery commander was spent in the WWII barracks on the right. The expanded OCS area is on the left.

The President of the United States of America, authorized by Act of Congress, March 3, 1863, has awarded in the name of The Congress the Medal of Honor posthumously to

SECOND LIEUTENANT HAROLD B. DURHAM, JR.
UNITED STATES ARMY

for conspicuous gallantry and intrepidity in action at the risk of his life above and beyond the call of duty:

Second Lieutenant Harold B. Durham, Jr., Artillery, distinguished himself by conspicuous gallantry and intrepidity at the cost of his life above and beyond the call of duty on 17 October 1967 while assigned to Battery C, 6th Battalion, 15th Artillery, 1st Infantry Division, in the Republic of Vietnam. On this date, Lieutenant Durham was serving as a forward observer with Company D, 2d Battalion, 28th Infantry, during a battalion reconnaissance in force mission. At approximately 1015 hours contact was made with an enemy force concealed in well-camouflaged positions and fortified bunkers. Lieutenant Durham immediately moved into an exposed position to adjust the supporting artillery fire onto the insurgents. During a brief lull in the battle he administered emergency first aid to the wounded in spite of heavy enemy sniper fire directed toward him. Moments later, as enemy units assaulted friendly positions, he learned that Company A, bearing the brunt of the attack, had lost its forward observer. While he was moving to replace the wounded observer, the enemy detonated a claymore mine, severely wounding him in the head and impairing his vision. In spite of the intense pain, he continued to direct the supporting artillery fire and to employ his individual weapon in support of the hard pressed infantrymen. As the enemy pressed their attack, Lieutenant Durham called for supporting fire to be placed almost directly on his position. Twice the insurgents were driven back, leaving many dead and wounded behind. Lieutenant Durham was then taken to a secondary defensive position. Even in his extremely weakened condition, he continued to call artillery fire onto the enemy. He refused to seek cover and instead positioned himself in a small clearing which afforded a better vantage point from which to adjust the fire. Suddenly, he was severely wounded a second time by enemy machine gun fire. As he lay on the ground near death, he saw two Viet Cong approaching, shooting the defenseless wounded men. With his last effort, Lieutenant Durham shouted a warning to a nearby soldier who immediately killed the insurgents. Lieutenant Durham died moments later, still grasping the radio handset. Lieutenant Durham's gallant actions in close combat with an enemy force are in keeping with the highest traditions of the military service and reflect great credit upon himself, his unit, and the United States Army.

Photo of "Pinky" Durham

Medal of Honor presented

Gravestone, Tifton, GA

Vice President Spiro T. Agnew presented the nation's highest award for valor, The Medal of Honor, to Grace Jolley, the mother of Army 2nd Lt. Harold "Pinky" Durham Jr., on October 31, 1969. The presentation was witnessed by the Chief of Staff of the Army, General William Westmoreland and Senator Edward Gurney of Florida, who was the designated escort for "Pinky's" mother. Durham earned the medal posthumously for his bravery at the battle of Ong Thanh in Vietnam on October 17, 1967. Lt Durham gave his life just five days after his 25th birthday. In the left photo, to the back, left to right, are John and Marilyn Durham, the honoree's brother and sister-in-law, LTC Frank Serio who was at the battle, and Genie Durham, the honoree's sister. Unfortunately, Durham's father died the same year as Harold did, and never knew of his son's honor. Durham is buried in the Oakridge Cemetery in his home town of Tifton, Georgia. Pinky has been honored with the naming of buildings and streets in his honor. His classmates in OCS Class 1-67 have honored him in various ways and attend ceremonies on his behalf. (Many thanks to John Durham, "Pinky's" brother, for his assistance in putting together this tribute.)

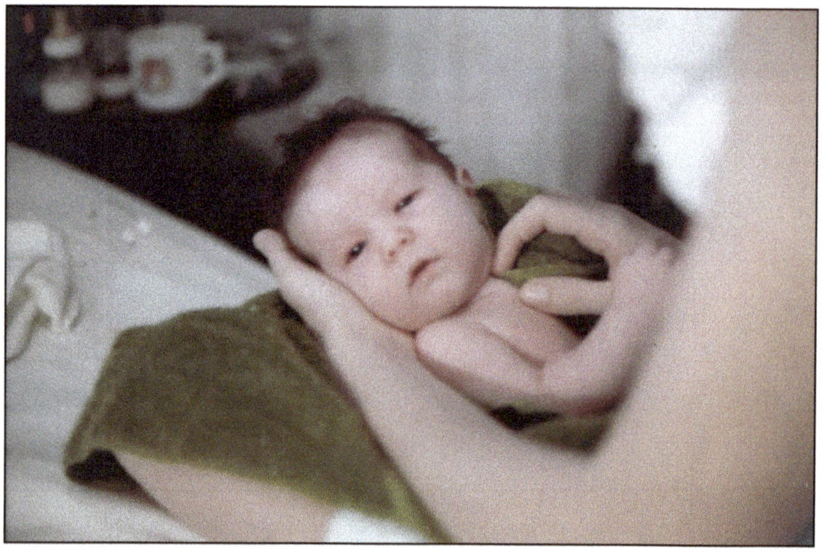

Sherrin Louise born Oct 12, 1965.

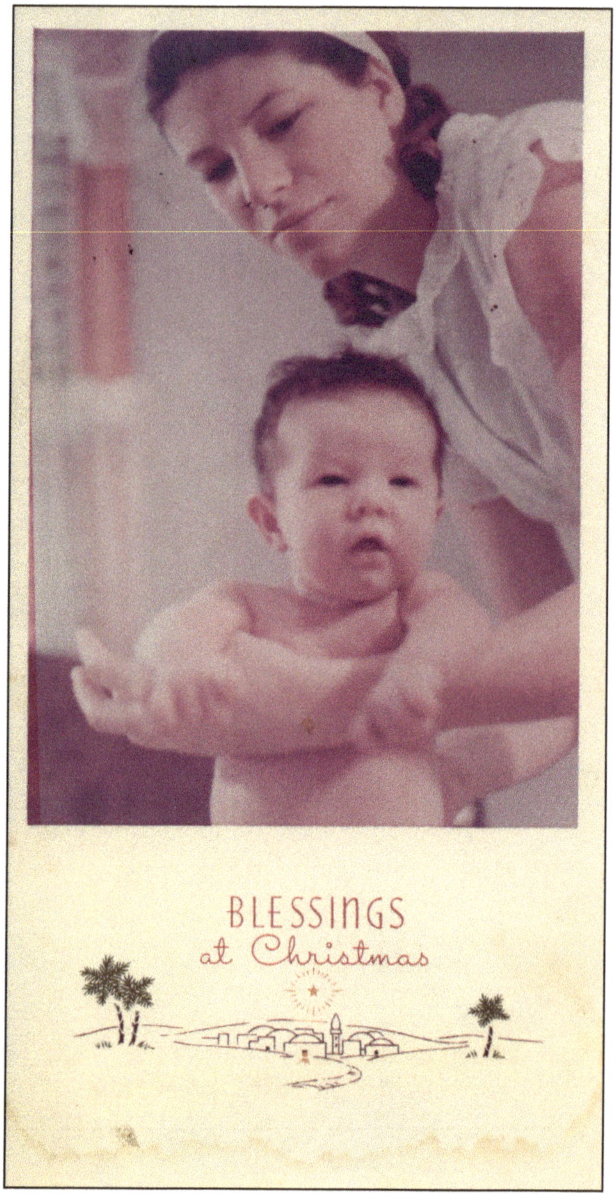

Christmas card, 1965.

BLESSINGS
at Christmas

Sherrin, age four months in February 1966.

Sherrin showing how big she is and on front lawn, Mar 1965

Helping daddy rake the front yard, Apr 1965.   *Candid photos.*

 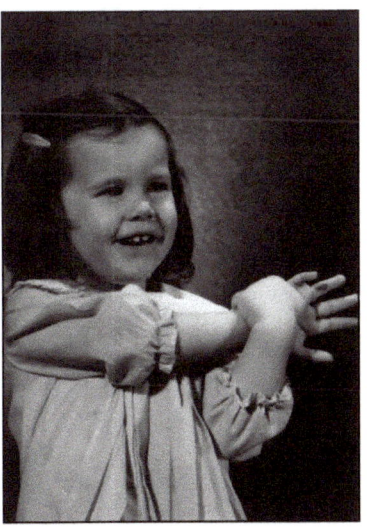

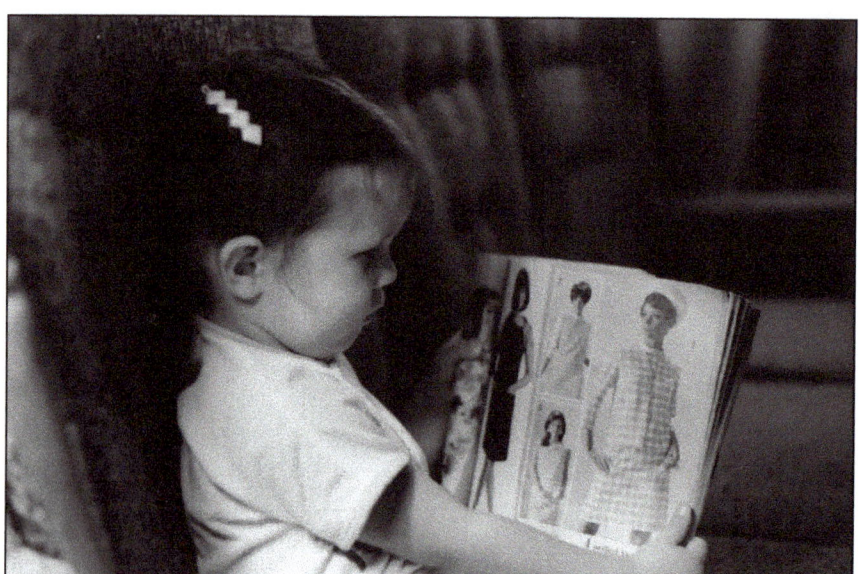

Candid photos. Sherrin checking out clothes for mom. Kathy with Sherrin at the mall.

Professional Photo 1966.   *Sherrin, Easter 1966.*

Sherrin in backyard with the sandbox I built.

Sherrin's 1st birthday, Oct 1966.

Merry Christmas!

Gerald, Kathy & Sherrin Louise

Christmas card 1966, printed before Shelley was born.

Kiwanis ladies' night, fashion show

Kiwanis ladies' night.

Sherrin's 2nd Birthday.

Sherrin's 2nd Christmas.

Visiting grandparents, El Paso, Jan 1967

7. TURKEY, CHAKMAKLI AND CORLU

UNACCOMPANIED TOUR
MARCH 7, 1967 TO MAY 26 1968

S2/INTELLIGENCE OFFICER
SASCOM DETACHMENT 67 (528TH FA GROUP)
(MARCH 7, 1967 TO AUGUST 31, 1967)

I ARRIVED IN TURKEY ON MAY 30,1967, FOLLOWING GRADUATION FROM THE COURSE AT FORT Holabird on May 26 and immediately assumed the duties of the group intelligence officer. It was an unaccompanied tour (no dependents allowed), so I left my family in our home on 50th Avenue in Lawton, OK. Sherrin was a year and a half old, and Shelley was a little more than three months old. We had not yet adopted her, but she had virtually lived with us since birth. Her mother was Lou Kennedy, my wife's sister. She was a singer with a band.

Before I departed Lawton, we decided that we would communicate both by letters and by sending three-inch reel to reel tapes back and forth. We purchased two tape recorders and about 50 tapes. I brought one recorder with me and half the tapes. We sent them back and forth about once a week. I am very fortunate to have kept the tapes and have had them transcribed into digital format. They have been very helpful in reminding me what happened when.

Special Ammunition Support Command in Turkey maintained custody of both 8-inch and Honest John rocket nuclear warheads. The four subordinate warhead support detachments supported Turkish artillery units by maintaining custody of nuclear warheads and conducting joint security training with them on the assembly and disassembly of these weapons using training rounds rather than the actual warheads. The largest of the four detachments, Detachment 74 located in Corlu, which I would later command, was a 50-minute drive west of the headquarters and supported one Honest John battalion and five 8-inch battalions. There was a ten-man explosive ordnance disposal detachment (Detachment 201) in Chorlu also commanded by a captain as well as a small Air Force Communication Sec-

COL(RET) GERALD W. "JERRY" SHARPE

tion. The next largest was Detachment 98 located about 700 miles across the country, just outside the town of Erzurum. They supported one Honest John battalion and two 8-inch battalions. According to the Turkish guidebook we were issued, the Turkish army considered Erzurum so remote that they only assigned officers there once in their careers.

Detachment 155 was located in Ortakoy, on the northern edge of Istanbul and supported one Honest John battalion and one 8-inch battalion. They also had a ten-man explosive ordnance disposal detachment (Detachment 101), also commanded by a captain. Finally, Detachment 97 was located just outside Izmit, about 75 miles east of the headquarters on the far eastern edge of the Sea of Marmara. It supported one Honest John battalion.

As had happened previously, my predecessor had already departed for the USA, so I had to quickly figure out what my responsibilities included. The course at Fort Holabird had been a blessing as it came with a complete checklist of responsibilities. My first task was to do a 100% inventory of and then sign for all classified documents. It had been suggested that in addition to checking current documents it would be a good idea to review the status of all documents that had had some disposition in the previous six months (destroyed, sent out of the headquarters, downgraded, or lost.) Army regulations at the time required that a 100% inventory be made every six months or at a changeover of custodians. A 10% inventory of all classified documents was required monthly. I found that we had more than 1,000 documents classified as secret or top secret.

The inventory took about a week, and I quickly discovered that we had secret and top-secret documents that were many years old. According to the commander, there was no known guidance about what should be kept for the historical record. Long story short, I prepared 400 destruction certificates, got witnesses to observe the burning of the 400 classified documents, and by the end of my second week we had a little more than 600 classified documents. No one ever seemed to notice the documents were gone. The good news was that I found no discrepancies in the accounting of the documents on hand or those disposed of in the previous six months.

One of the documents we were given at Fort Holabird was a sample SOP for S2 operations. It turned out no one had ever seen fit to prepare any kind of standard operating procedures for the section, and more importantly there did not appear to be any SOPs on file from the four warhead support detachments under our command. I drafted a Detachment Security SOP, had it reviewed by the headquarters staff, and then sent it out to the detachment commanders. Each had good suggestions for improvement, but it became clear that there were different procedures being used in each detachment. Over the next two months, I visited each detachment, conducted a security inspection based on checklists I had drafted in the security SOP, and one by one got each detachment commander to agree to modify his procedures to conform to the new SOP.

My first visit to Erzurum was an eye-opening experience. We flew aboard a twin-engine DC-3 that shook and rattled to the point that I wondered if it would survive the trip. The last 100 miles was over a mountain pass that was virtually the only road leading to Erzurum. We could feel the engines straining as the pilots attempted to get higher and higher. We could see mountains on both sides of us and could look down and see vehicles just below the airplane. We finally reached the top of the pass, and it appeared the pilots cleared it by no more than a hundred feet. We were very relieved to land on the other side of the pass a short time later.

One pleasant aspect of my duties was representing SASCOM headquarters at the weekly staff meeting of the US Ambassador to Turkey, the Honorable Parker T. Hart. The meetings normally lasted about two hours, and I got to know many of the staff personnel at the US Embassy. A young lady about my age made a habit of sitting next to me. I enjoyed talking to her, but having only been married for about two years, I had no interest in having more than a working friendship. After perhaps two months on the job, everyone at the meeting received an invitation to a party at the ambassador's home. Once again, the young lady seemed to attach herself to me, and we enjoyed some pleasant conversation. She invited me out to the ambassador's garden. She said it was fabulous and had quite a few bronze statues. I agreed to look and we strolled out into the garden. As we got further from the house the young lady turned to me and basically said she would really like to sleep with me as she was lonely and did not have a boyfriend. I told her that while I liked her and enjoyed our visits at the embassy meetings, I was newly married, and had zero interest in sleeping with her. She seemed to accept my explanation and we returned to the party. As I recall, she shifted her attention to someone else at the meetings and while we were pleasant to each other, we never sat together again.

My trips to the embassy meetings were always in the morning. Several times, my Turkish driver suggested we visit the Grand Bazaar. It had perhaps a thousand shops on three underground levels. Jewelry and brass items were unbelievably cheap. I would bargain with shop keepers, but I would not buy anything. We received many inspectors from the SASCOM headquarters in Europe, and they always wanted to spend at least one afternoon and evening in the Grand Bazaar. I became their escort, and over time developed friendships with several jewelry, brass, carpet, and souvenir shop keepers. Carriage lamps, samovars (fancy brass items used to brew tea), brass trays, brass mirrors, and jewelry were the most sought-after items. I began keeping a notebook on what things cost at the various shops so I could make sure the visitors were not paying more than they should.

Bargaining was simply a way of life for shopkeepers and prices always started at about twice what the final price might be. In one jewelry shop I picked out a very nice pair of opal earrings and a matching necklace for my wife. I made an agreement with the shopkeeper that if I brought him customers, he would reduce the price of the set I wished to buy. Again, I kept the price in my notebook and after each visit he would tell me the new price. I would cross out the old one and write down the new one. By the time I was ready to go home I filled a suitcase with items that cost me about one fourth of what I would have expected to pay.

One rather unpleasant duty I had was being appointed to investigate a complaint by a soldier in Erzurum that his commander was giving and receiving oral sex from a select group of men who were in turn being treated more favorably than those who were not so fortunate, or unfortunate, as the case might be. The commander was a major, and I was a captain, but my commander told me he thought it was my job as the security officer to do the investigation, and he assured me that I should not consider the difference in rank to be a problem.

Once again, one of my two security clerks and I rode the rickety DC-3 to Erzurum, making it over the pass in both directions by a very small margin. I informed the commander that I was there to conduct a security inspection as he had been in command less than 60 days. I enlisted one of the lieutenant team leaders to act as a witness, informing him that this was an official investigation, that he was being compelled to cooperate, and that he could speak to no one about what we were doing.

My first task was to take a sworn statement from the soldier who had complained. He named three additional witnesses and named four soldiers he stated were participants. I next took sworn statements from the three additional witnesses. All four swore they had observed the commander entering a stall in the bathroom of the barracks and heard what they believed were sexual sounds. I asked each witness if they believed there was one or more participants who had done so less than willingly. Three of the four witnesses named the same soldier, stating they believed he had been coerced into participation.

I had the soldier they had identified brought to the office I was using for my investigation, with the lieutenant as my witness. I informed the young man that he was suspected of participating in what was viewed as a violation of the Uniform Code of Military Justice. I informed him that I had the power to grant him immunity from prosecution if he voluntarily gave a statement as to what had happened, when it happened, how many times it happened, and whether he was witness to others participating with the commander.

At first, he denied any involvement so I told him I was sorry, but witnesses had seen his participation, and I would be forced to have him arrested and charged with others as a willing participant. The soldier broke down crying and stated he had participated, but not willingly. He alleged two other soldiers had threatened him if he did not participate in "having a little bit of fun." He gave a two-page statement about what had happened, and he named six other participants.

My security clerk typed each statement in three copies and I had them signed and witnessed before anyone left the office. Upon completion of my statement taking, I called Colonel Farrell, the Group Commander and relayed to him what I had found. He stated he was sending a captain to take over command of the detachment the next morning. The colonel stated he would inform the major the following day that he was relieved of his duties, that he was to sign over classified documents to the new commander, and he was to return with me to Istanbul a day later. The next day, after the major got off the phone with the colonel, he demanded that I come to his office, which I did.

I brought the lieutenant with me as a witness and told him that before he said one word, he should understand first that I was there as what was called an Article 32 Investigating officer appointed by the group commander. I handed the major a copy of my orders. I next informed him that he was accused of violating the Uniform Code of Military Justice, and that before he made any statements, he should understand that he had the right to have a lawyer, but that anything he said would be transcribed by me with a witness and anything he said could be used in a possible court martial. He decided to say nothing.

The lieutenant and I left his office, and I went to the airport to pick up the new commander. The next day when we arrived at the Istanbul airport, there were two CID Agents from Germany waiting for us. They informed the major of his rights, placed him under arrest, and later in the day they departed on a plane for Germany. Of the six soldiers that had been named, four were give nonjudicial punishment, and two, who had participated in coercing the soldier who broke the case open, were both sent to Germany and court martialed. All three were convicted and kicked out of the army. The five witnesses were transferred to different detachments to serve out their one-year tours.

I had the opportunity to make one trip to Germany during my time in the Group Headquarters. SASCOM held a two-day S2/Security Conference in Frankfurt on 22-23 June 1967. I went a day early and left the day following. I gave a presentation of the security situation in Turkey and briefly outlined how we were organized. SASCOM supported host nation forces in seven countries and ten S2s gave presentations. I also had a long PX shopping list for the main post exchange (PX) in Frankfurt from fellow officers in the headquarters. I returned with a suitcase full of "goodies."

Whenever we had an officer transfer out of the headquarters, we would have dinner at a nightclub in Istanbul called the Caravanserai. The nightclub, which still operates today, features two nightly belly dancing shows, Turkish folk dancing, singers, and a DJ. It offers both a meal and unlimited drinks for one admission price. I do not remember exactly when, but during one of these farewells, one of the very well-endowed belly dancers stood behind and between two of us and a picture was taken. Not sure what I was thinking, but I mailed a copy of the picture home. A few days later I got a very angry letter back from my wife, Kathy, accusing me of "cheating on her!" I wrote back and assured her that I was not so stupid as to send her a picture if I really was "cheating!"

I explained that there were 20+ officers at the farewell and the belly dancer kept all their clothes on and they simply came down into the audience to take pictures for tips. It did not seem to satisfy her, and I heard about it several more times. She told me in no uncertain terms that I should *not* attend such events. Needless to say, no more pictures were ever sent home. See the offending photo at the end of the chapter.

DETACHMENT COMMANDER
SASCOM DETACHMENT 74 (21ST FA MISSLE DETACHMENT)
(SEPTEMBER 1, 1967 TO MAY 26, 1968)

I assumed command of the TUSLOG Warhead Support Detachment 74 in Corlu from CPT William Forster on September 1, 1967. In US-only documents, we were the 21st Field Artillery Detachment, but that designation was not used in-country. I was authorized 57 personnel but had only 47, including six officers, one for each assembly team. We had a headquarters section, six maintenance and assembly teams (five eight-inch and one Honest John), and had in support a small US Air Force Communications Section and an Explosive Ordnance Detachment.

Unlike my time in Chakmakli, we were faced with considerably greater host nation support problems. The 1958 agreement between Turkey and the United States declared that when the warhead support detachments were established, the Turkish military would provide all support in the way of barracks, offices, transportation, food, fuel, water, and electricity. In 1967/68, Turkey was the third largest recipient of military assistance in the world, so it was not like the USA was not providing more than adequate funds to the Turkish Military. As each year passed, the Turkish forces provided less and less support, and we were required to divert US funds allocated for fresh fruits and vegetables and morale welfare to fixing our generators, water pumps, water heaters, and barracks heaters. The Turkish one-star corps artillery commander responsible for our support had little sympathy. His soldiers lived spartan lives, with no electricity or even glass in their barracks windows.

In a letter to my parents I wrote,

October 30, 1967

"I am now the commander of the largest detachment in Turkey and the second largest in all the European Command. We work with the Turkish Army, and I have been truly impressed with the hospitality and friendship I have been shown. I spent my first two weeks here out visiting the six units we support, meeting the lieutenant colonel commanders and their staffs. At each unit I was received like a long-lost son, and I was treated with the same warmth you would expect to find in your home. The Turkish officers I work with are all considerably older than I am, and they marvel at my age. A Turkish captain is likely to be in his mid-30s. I also work with many individuals who outrank me by a considerable amount such as full colonels and generals. In each instance, however, I have been treated as an equal and it is truly gratifying to have this type of relationship with people who could make life in Turkey uncomfortable to say the least."

As you will read, this was more than a little prophetic.

Shortly after taking command, my first sergeant told me there was another detachment we should go visit. It was on the Sea of Marmara at Tekirdag, about a 35-minute drive from Corlu. It was manned by the US Coast Guard and housed one of two large LORAN (long range navigation) towers around the Mediterranean that allowed ships and planes to get accurate location information. We were welcomed by the commander and given a short tour of the detachment. They were only about half the size of our detachment (two officers and 20 enlisted men), but 100% of their support came from the US Coast Guard. They had a speed boat, water skis, a gym, a theater, a club, and a basketball court. I was impressed and envious. They were resupplied monthly by sea plane. Their resupply included fresh milk—which I had not seen since arriving in Turkey. I am pretty sure I drank six or seven glasses of milk while visiting with the CO in the dining facility. One dissatisfaction they had was that they received only four movies each month. My detachment received six movies a month, so I offered to make a deal. In return for providing us with some fresh milk, we would trade our movies with them so that together we each had ten a month. This agreement worked out well.

One of the challenges I had as a detachment commander was how to spend our morale and welfare money. We received some number of dollars for each soldier each month, and when I assumed command there was about $12,000 in the fund. Visiting the Coast Guard station gave me the idea that perhaps we should get our own boat and water skis and keep it at the LORAN station. I wrote my father and asked him if he could find a company that would sell us a 24 to 26-foot outboard motor speed boat and ship it to Turkey. I suggested that he should use his talents to try to get a company to give us a much-reduced price as a public service for soldiers serving far from home. I told him I had $13,000 available to spend. About two weeks later I received a letter telling me that the Glastron Company out of New Orleans would sell us a 26-foot boat and would include life preservers, tow ropes, and water skis for $10,000 and they would ship it to Istanbul at their expense. The retail price was about $18,000. Glastron has been producing boats since 1956 and as of 2019 had sold more than a half million boats around the world.

The shipment arrived, it cleared customs, and Glaston even paid to have it sent to Tekirdag. Our soldiers began a rotation of taking a day off and spending the day at the LORAN station. Soldiers and coastguardsmen shared the two boats and had a great time. In addition to renovating our small club and theater with morale and welfare funds, I had a patio built with a small water fountain in the middle and was able to get chairs, tables, and umbrellas from the PX in Istanbul. I later found out my replacement filled in the pond and cemented it over to enlarge the patio.

About my third or fourth week in command, I went across the road from my trailer to check on the generator, perhaps two hours after dark, and noticed that the door to my small PX appeared to be open. I rushed over to the PX and was on about the second step when a Turkish soldier came running out with an armload of items. It was dark and he apparently did not see me as he had stuff piled high in front of him. He ran into me head-on and knocked me on my butt. He fell as well, dropping everything he had in his hands. We both got back on our feet, and he took off running with me yelling in hot pursuit. He knocked over a trash can in front of me at one point, and once again I went down. By the time I got back up, he had disappeared, but there were soldiers looking for him everywhere — to no avail.

I went back to the PX, and the PX manager (a sergeant) and my officers were gathering up what had been taken. The soldier had broken the single hasp holding the padlock on the door to gain entry. I had my interpreter call the post duty officer, a lieutenant colonel, and he arrived a short time later. I explained, through the interpreter, what had happened.

About 9:00 the next morning, I received a message to report to the Corp Artillery Commander's office right away. As we drove to his office, there were soldiers marching onto the post parade field. It appeared the entire garrison was assembling. When I arrived, the commander was mad as hell. He explained that thieves were not tolerated in the Turkish Army, and that if we found the thief this morning, he would personally execute him on the spot.

I explained that while I appreciated the opportunity to search for the thief, I never saw his face, and it was dark when I was chasing him. He said, "We are going to look every soldier in the eye if it takes us all day." Honestly, I am not sure even if I had seen the young man that I would have had the nerve to point him out. We spent the next five hours walking up and down ranks of soldiers stopping in front of each one, many of whom looked scared as hell!

Detachment 74 existed to store the Honest John and 8-inch nuclear warheads that would be allocated to the Turkish army in case of war with the Soviet Union. We also conducted training on the assembly and disassembly of nuclear weapons. The bunkers where the warheads were stored were surrounded by two rows of chain-link fencing. US soldiers patrolled inside the inner wire, and Turkish soldiers patrolled between the two fences. These areas were required to be lit with floodlights during all night-time hours and should the generator providing power fail, the number of both US and Turkish guards had to be doubled.

There was a very complicated form of messaging from the European Command in Europe for the release of nuclear weapons. One half of the messages were sent down through Turkish channels, and the other half came through a US channel. Exercises of this system were carried out at all hours of the day and night, and a Turkish officer and an American officer would have to open their respective safes to get out sealed authenticators. Each officer had information in their portion of the message that allowed them to verify the authenticity of the procedure. The entire process was timed, and failure to meet established standards had serious consequences.

I quickly learned that we had four generators but often only one 88 kw one was operational. A second 88kw generator and a 27kw stand-by generator in the exclusion area had not been operational for two years. Our 62kw backup generator was broken far more often than it was operational. When the primary 88kw generator needed repair, I would request repair from the host nation, but virtually every time I had to send someone to Corlu to buy the needed parts, listing the purchase as "fresh vegetables," as we were not allowed to spend US money for repairs.

I lived in a small 20-foot travel trailer, and it was parked right across the street from the generator building. I had two light bulbs in my trailer, and I could tell how the generator was running by whether or not the light bulbs burned brightly or faded in and out. We had Turkish soldiers manning the generator in three shifts, 24 hours a day. Their job was to adjust the RPM to keep it running smoothly. The soldier sat in a chair in front of the generator building in plain sight of the window in my trailer. Almost nightly, I would see my light bulbs start to dim but not come back bright. I would look out my window and would see the Turkish soldier leaned back against the building, apparently asleep. I would slip on my robe, walk across the street, kick the chair, causing the soldier to fall, and he would immediately jump up and tune the generator. I never had to say a word, but the exercise really got tiresome.

By early October I began seeing more and more lights across the Turkish garrison at night. I had the only generators on base, and commercial power was still a few years away from arriving from Istanbul, so it was clear that electricity was coming from our power source. It was about two hundred meters (656 feet) from the generator building to the exclusion area, and there were perhaps 15 poles that carried the power up to the exclusion area to operate the required lights. On my first night in command, my first sergeant and I walked the line of power poles and I noted that virtually every pole had pieces of WD 1 communications wire running up the pole and out across the garrison. At the end of each line was a single light bulb. On my first night I recorded that the 88kw generator was running at a comfortable 66kw. As the weeks progressed, I noted the generator was increasing in usage each night, creeping up one or two kilowatts a week. When we passed 75 kw, I knew we had to do something, as the generator was breaking down once or twice a week. I got my officers and NCOs together and told them we needed to conduct a test of how much electricity we were using. Peak usage seemed to be about 11:00 pm, so one night everyone in the detachment was told that at exactly 11:00 pm we were going to shut down everything that used electricity—including all the lighting in the exclusion area and all of our refrigeration units—for five minutes and see how much change took place in the kw usage on the generator. Usage dropped from 75 to 25 kw, meaning that the Turkish garrison was illegally diverting twice as much electricity as our detachment was using.

The next morning, I went to see the Corps Artillery Commander and explained to him that in my two months in command, usage of my generator had gone from 66kw to 75kw without anything changing in my detachment. I told him about the test we had conducted, and it appeared his garrison was using about 50kw to my detachment's 25kw. I requested that he take some action to reduce nonessential power usage. He said he would investigate it. We were pleased to note the next night that power usage dropped to under 60kw for the first time in months. I sent the commander a message thanking him for his cooperation, but the good news did not last long as within two weeks we were back to 70kw.

I again went to see the commander and this time he did not promise to investigate it but said much of what was being used was essential. I explained that I was required to have lights in the exclusion area and on four nights out of the previous nine we did *not* have lights because the generator was broken down. I suggested to him that if he could find a way to get the second 88kw generator working, everyone would have adequate power. Regrettably, nothing changed.

Over the next week I visited the deputy commander a couple of times as the commander was "too busy" to see me. I told the deputy that I had consulted with my higher headquarters and if necessary, we would remove all the wires from the poles. He suggested that I might wish to find some other solution as that would make the commander very angry. I told him we had to do something, as the only operational generator we had was breaking down from overuse. I suggested it would be in everyone's best interest to cut usage by at least 10kw. Again, nothing happened.

Two days later, during the last week of October, I assembled all my officers and NCOs and told them that at 7:00 pm that evening I wanted every wire cutter, ladder, and soldier available to remove every wire from the poles. I said they were to cut the wires back away from the poles so they could not be hooked up again and then to remove wires from every pole and lock up the wire. I also said I wanted guards posted to turn away anyone trying to reconnect the wires. It took us nearly three hours to remove all the wire, and one by one every light on the garrison blinked off. We were amazed no one came to complain.

I finally went to bed about 1:00 am without a single person having shown up. At about 3:00 am I was awakened by horns, lights, and much yelling outside my trailer. I quickly slipped on my uniform and stepped outside to find a swarm of colonels and lieutenant colonels all exceptionally angry. My interpreter informed me that the entire garrison had been out on a training exercise, had returned to find no one had electricity or lights, and the new three-star 5th corps commander was coming to inspect in the morning. They said they would need to work the rest of the night to get ready and could not do so without electricity. Through my interpreter I told them I was very sorry, but no one was hooking them back up. After much screaming, threatening to have the detachment sent packing, and telling me they would make sure the three-star came to the detachment in the morning and set me straight, they finally all went away—with no electricity.

The next morning I had our mess sergeant get four sheets of plywood and make a very large table in the middle of the dining room with table clothes, chairs, drinks, snacks, and a three-star, two-star, and one-star place set at one side of the table. I had no idea if anyone would come, but I wanted to be prepared. About 8:00 am I called the Group Commander and filled him in on the events of the last evening. His only comment was, "Son, that took a lot of balls, hope you survive it!"

A little after 10:00 am, a long convoy came down the road with flags flying. We stood at attention in front of the dining facility as the three-star car stopped next to us. The general got out, along with his two-star deputy, and immediately started speaking English (which was very unusual). I had met him before, and he shook each of my officer's hands and we went into the dining facility. The table rapidly filled up and I could see many of the officers from the night before with big smiles on their faces as they, I think, thought my world was about to end. The Corps Artillery Commander just sat quietly. The three-star first spoke to my officers and NCOs in the room telling us how much he appreciated our support and how our support of his Corps was vital to his accomplishing his mission. He then asked about my family and how my daughter was doing. I said Sherrin had turned two years old, and I was looking forward to getting back home.

Next came the big question. He said, "Captain Sharpe, I understand there was some trouble in your detachment last evening. Will you please tell me about it?" I sort of sucked it up and in about two or three minutes explained how many times my generator had broken down since I took command, how many nights the exclusion area had been without lights, how my detachment was using about 25kw of power, and how I had tried every way I knew to cut usage, get a second generator working, or in some way keep the power on. I explained that the previous night at 7:00 pm, the generator was running at 72kw and by the time we finished cutting wires off the poles, it was running at just 26kw. I said I would be pleased to share power, but General, I must have enough to secure your nuclear weapons in accordance with our host nation agreement.

He looked at me without saying a word for a minute or so, and inside I could feel my career slipping away. He then turned and in a very loud voice started reprimanding everyone at the table. My interpreter later told me he threatened to end every one of their careers if I had any more such problems. Smiles disappeared and real fear came across their faces as he talked loudly for two or three minutes. He paused, stood up, turned to me and said, "Captain Sharpe, keep up the good work, I do not think you will have any more power problems." Everyone filed out, got in their vehicles, and drove away. I immediately called my commander back and out briefed him. I think he breathed a big sigh of relief that I had not created an international incident.

I wish that was the end of the story, but it was not! The Corps Artillery Commander sent me a short message later in the day that basically said, "You may have won that battle, Sharpe, but nothing is going to be easy for the rest of your tour." He did try to make our life difficult for a while by cutting off our water daily at 7:00 pm, requiring everyone to wait at the gate for an hour or two to get on or off the base, by not picking up our garbage and trash, and by not letting our Turkish barber on base. We did agree to allow wires to be hooked up restoring some power to units across the base, but it was kept at a reasonable level. The good news was that our generator did not break down again even once before I left.

October 29 is Independence Day in Turkey. It celebrates the activities of October 29, 1923, the date Turkey declared itself to be the "Republic of Turkey." After that, a vote was held in the Grand National Assembly, and Kemal Atatürk was elected as the first President of the Republic of Turkey. Ataturk was, and I suspect continues to be, revered by everyone in the country. I was told that virtually every home has a picture of him hanging on the wall. I had noted that most of the offices I visited had both a framed picture of Ataturk and beside it a picture of President John F. Kennedy who was held in high regard by Turkish officers. In the same letter of October 30, 1967, to my parents that I quoted earlier, I wrote,

Yesterday was Turkish Independence Day, just as our 4th of July is, and my day was long but truly inspiring. The new 5th Corp Commander, that I had met a little more than a week ago, invited me to attend their Independence Day celebrations as his guest. We started the day with a ceremony at the Ataturk statue in the middle of Chorlu, the town outside our garrison. Ataturk is like our George Washington, Abraham Lincoln, and John F. Kennedy all in one. There were wreaths laid at the foot of the statue and my interpreter and I were the only ones present below the rank of full colonel. After the wreath laying, we went to the office of the governor, filed in in single file, and shook his hand to pay respect to him. Next, we went back to the town square for a big parade. I tried to stay off the reviewing stand, but the general would have none of it. He had me brought up and placed in the front row of dignitaries. As a result, I had probably the finest view of the parade of anyone in the city. It was a truly impressive parade as it appeared everyone in the city participated. Grade school children marched along with boy and girl scouts, middle and high school students, veteran soldiers who had fought in the War of Independence, merchants, taxicab drivers, and an array of military units. All in all, the parade lasted about two hours. The day concluded with a formal dinner at the Officer's Club. Once again, I was the guest of the Corps Commander and had a private table right on the dance floor just a few feet from him. I felt flattered as the Corp Commander stopped by my table on the way in, introduced his wife, and exchanged a few words while the remainder of the 150 people in the room were standing. During the dinner he sent a bottle of champagne to the table and on his way out he again stopped at the table and said he was glad I could attend, and please make an appointment to visit him in his office.

I had to attend meetings at the Chakmakli headquarters every few weeks, and the drive from Chorlu was fraught with danger. The Turkish countryside was still very undeveloped, and commercial power was very slowly making its way east. It would not arrive in Chorlu until more than a year after my departure. The problem was that poor people who drove ox carts or simply walked had no concept of automobiles. There were no automobiles around where they lived, so when they reached a paved highway, like the only one from Chorlu to Istanbul, they would simply drive or walk out across the road without realizing that cars were driving very fast and would be unable to stop. I must have seen two or three terrible accidents with several people killed in each as an ox cart full of people simply pulled out right in front of a speeding car. The happened more at night, but I observed a few in the daytime. As a result, my driver and I paid very close attention to anyone or anything on the side of the road, always slowing down until we were past the danger. See a picture of an accident at the end of the chapter.

After a few weeks of harassment at the hands of the corps artillery commander, I received an invitation to a birthday celebration dinner and dance from the three-star corps commander. There was a note attached that said please bring four of your officers. It was for the next Saturday. I took the invitation and went to see the one-star corps artillery commander who had refused to see me since the power cutting incident. I told his deputy I wanted to see him, and he told me he was sorry, but the general had issued instructions that he did not want to see me. I should add that in the Turkish Army, officers have life and death authority over subordinates, and while the deputy was a full colonel, when he went into the general's office he would be at rigid attention and quivering with fear that the general might end his career. With that in mind, you can imagine the deputy's consternation when I simply walked past his desk and went into the general's office.

About the time I stepped inside the general's office, the deputy grabbed me and tried to drag me out of the room. I told the general, "We need to talk now!" The general told the colonel to let me go, and I walked up to the general's desk and said something to the effect of, "General, it is time to end the harassment. You have had weeks now to take your revenge on us, but I want to remind you that the corps commander said I should let him know if we had any more trouble."

I showed the general the invitation to the birthday party and then told the general, "I am sure the corps commander is going to ask me how things are going in my detachment. I do not want to have to tell him any lies." The general did not say anything for a minute or two, he only stared at me with his well-known frown, then burst out laughing.

He basically said, "OK, Sharpe, I hope you have learned that I do not like being embarrassed by a captain. Now get out of here and go back to work. I will see you and your officers at the ball!"

It was a great event, much like the Independence Day ball some weeks earlier and not only did the three-star general again put us at a table right next to his, but also, much to our shock and amazement, when the dancing started, he would go out and take the hand of an officer's wife, bring them to the table and tell me or one of my officers to dance. This was a very scary proposition, because in Turkey even looking at a married woman was not considered proper. No one seemed to get upset, and we each danced perhaps two or three times during the evening. It was an enjoyable affair. Even the corps artillery commander brought his wife to the table and introduced us.

In November 1967, there was a large flare-up of controversy between Greece and Turkey relating to Cyprus. 15 Turkish Cypriot citizens were killed in Cyprus on November 15, and Turkey prepared to invade Cyprus. The United Nations, Great Britain, and the United States all became involved when President Johnson sent former Secretary of the Army and former Deputy Secretary of Defense, Cyrus Vance, to the Region on November 22. Mr. Vance would later become President Carter's Secretary of State and his daughter has recently married one of my US Army War College Classmates.

While everyone was conducting diplomacy, sometime during the last week of November, the Greek government was conducting a missile test, and it went astray with a missile landing on Turkish soil. As I recall the events, the Turkish government demanded an apology, and the Greek government refused to provide one. I should say in passing that I had learned in my time in Turkey that the very worst thing you could call a Turkish person was a "Greek." While Cyrus Vance had gotten Turkey to agree to hold off on invading Cyprus, this incident caused the Turkish Army to mobilize and march to the Greek/Turkish border with the threat of invading Greece itself. The main road from Istanbul to the Greek border went right by our detachment, and for the first 24 hours Turkish army units moved past us, one unit after another. Since we were storing nuclear warheads there was great concern at our headquarters and in Europe that the Turkish army would attempt to take our warheads—by force if necessary.

COL William J. Ferrell, the Group Commander, came to my detachment early the second morning and had only been in my office a few minutes when the Turkish Corps Artillery Commander came to my office and demanded I give him "his nuclear weapons." I got up from behind my desk and told the general, "Sir, I would be pleased to give you your nuclear weapons." COL Farrell jumped up out of his chair and started to say something and I simply turned and said, "Sir, let me finish." I continued, "General we have practiced how to make this turnover a number of times. Please go get your Special Ammunition Officer and come back so we can follow the two-man process for releasing your weapons. I will wait for your return." The general looked at me for a minute, turned, and left my office. I never saw him again during the crisis. COL Farrell breathed a big sigh of relief and said something to the effect of, "Sharpe, you really had me worried there for a minute. Well done." He stayed the rest of the day and by nightfall units started marching back from where they came. Our understanding was that President Johnson told Turkey if they did not turn around, every single dollar in US aid would cease.

Also, in November I received a flyer stating that a 1967 Christmas charter flight to JFK Airport in the USA was being arranged through the American Embassy on Pan American Airlines. The flyer said it would depart on 12 December and return on 6 January. I called the commander in Chakmakli and asked if I could take leave over Christmas. He said yes, simply designate your senior lieutenant to be the acting commander. I purchased a ticket and very much enjoyed my three weeks at home.

While there, Kathy and I met with Lou, Shelley's mother, and gave her an ultimatum: either take Shelley back and care for her, or let us adopt her as she already calls us Mommy and Daddy. We also insisted that she would have to agree that she would *not* let Shelley know she was her mother, and she would be "Aunt Lou" from that day forward. She agreed and I told her that when I returned from Turkey, we would get a lawyer to draw up the paperwork, present it to a judge, and finalize the adoption. Also, during my trip home, we drove up to Westminster, CO and spent a few days with my family.

Our barber, which the post commander would not let on base for a few weeks, would come two days a week, Tuesdays and Fridays. He set up shop in a small room with an old barber chair and a big mirror. The first time he cut my hair; he asked me in very broken English how I wanted my hair cut. I told him, "Exactly like it is now, very short on the sides and flat on top." He started on the sides and got it very short, but he seemed unsure as to how to cut the top. After walking around me once or twice looking at the top, he then took the clippers, and started to do the same thing on top as he had on the sides. I was watching him closely in the mirror and ducked my head and yelled stop! I got up out of the chair, told the barber to hand me the clippers, and I proceeded to cut the top a little shorter but still perfectly flat. When I was finished, I sat back down and from that day forward he would say, "Me side barber, you top barber!"

One of my memorable experiences while in Chorlu was being invited to a dinner hosted by the commander of the division supported by one of the four nuclear-capable 8-inch artillery battalions. The artillery battalion commander delivered the invitation and informed me that the division would be on a week-long training exercise up near the Bulgarian border. He said there would be a dinner on the last night of the exercise for all the commanders and senior personnel in the division. He gave me a map and the grid coordinates and asked if I needed someone to guide me to the location, which appeared to be about 30 miles off the main road in some very hilly country.

I declined, saying I was a good map reader. I thought I could find it with no problem. He said the dinner would start just after dark, but I should try to arrive earlier. I was not sure how long it would take to get there, but my guess was about four or five hours in my jeep. My driver and I left a little before lunch and proceeded without incident to the turnoff of the paved road that was very well identified by the battalion commander. From that point on, the route was very difficult to determine. We made a number of bad choices where trails split, and we had to backtrack several times. This was before the days of GPS, and I had no way to communicate with the unit, so we just kept trying our best to get to the place marked on the map. It was nearly dark when we could see lights off in the distance and used them to guide us the last 20 minutes or so.

We arrived just as it got dark and we were surprised to see a very large tent, certainly larger than any army tent I had ever seen. I was escorted to the tent, and someone offered to take care of my driver. I was surprised, to say the least, to find the inside of the tent carpeted. There was a huge table set for about 30 people with tablecloths and fine china. The battalion commander welcomed me and took me around the table introducing me to the generals, colonels, and lieutenant colonels in attendance. I was the only junior office in attendance, but I was welcomed with open arms by all. There were waiters in uniform, and we proceeded to eat a five or six course meal long into the night. The battalion commander sat on one side of me and a full colonel, who had trained in America, sat on the other side. They were kind enough to translate the MG commander's comments who also welcomed me in English during his talk about how pleased he was with the unit's performance during the exercise.

There was lots of drinking of raki, the Turkish national drink, sometime called "Lion's Milk." I knew the drink as "ouzo" and while I did not drink, I had tried a taste or two. It tasted like licorice and had an alcohol content of 30% to 45%.

I was made to feel very welcome and our role in training was recognized by the Division Commander. The dinner broke up about midnight, and I wondered to myself how we were going to find our way back to the highway in the dark. My jeep and driver were waiting just outside the tent, and I thanked as many people as I could for inviting me. As I walked to my jeep, the battalion commander took my arm and told me to get into the back of his jeep. He said that there was no way I would ever find my way back, so he was going to take me. He had an officer ride with my driver and told them to simply follow his jeep. We carried on a very pleasant conversation and the commander's driver obviously knew the way as an hour later we arrived at the main road back to Chorlu. We got out of the jeep, and the commander gave me a big bear hug, thanked me for coming, got back in his jeep, and disappeared into the darkness back the way we had come. This was just one example of the warm hospitality given by host nation officers that really impressed upon me what a valuable service we were rendering to them and to our nation.

Every Turkish unit had to undergo an unannounced NATO nuclear inspection annually. The inspectors would arrive in the morning with their assigned interpreters, and they would conduct a briefing for the battalion being inspected. There were questions asked of the officers, NCOs, and enlisted men, and as their final exercise, they had to assemble and then disassemble a nuclear training round.

About halfway through my command, we had an inspection for the Honest John missile battalion we supported on the Corlu garrison. During the morning and early afternoon, the unit went through

COL(RET) GERALD W. "JERRY" SHARPE

the various steps of removing the practice warhead from its container. Using a wrecker, they lifted the warhead out of the container and placed it on a stand, then went through the procedures of opening the door on the side of the warhead and setting the firing data given to them by a practice "Emergency Action Message" that had to be opened by both a US and Turkish officer.

The final exercise was to mate (attach) the warhead to a missile. To do this, the Turkish wrecker operator had to lift the warhead, raise it to the level of the missile, then slowly swivel it from the carriage it was sitting on up to where bolts could be used to attach the warhead to the missile. This was a very delicate operation, and the Turkish wrecker operator was a vital link in the process. After some maneuvering, the warhead was in the proper place, and the inspectors told the Turkish officer in charge that they wished to question the mating crew about the procedures they would perform and then observe them attaching the practice warhead to the missile.

The Turkish sergeant supervising the wrecker operator apparently explained to him that he should sit and wait for the inspectors to finish their work. To understand what happened next, you need to know that in front of the wrecker operator were vertical handles that the wrecker operator pushed or pulled to make the boom go left or right or up or down. Throughout the operation, a Turkish sergeant stood on the wrecker just behind the operator and told the operator what to do. For some unexplained reason, the sergeant supervising the operator came down off the wrecker for a few minutes, leaving the wrecker operator sitting in his seat with his arms folded in front of him. Suddenly, the Turkish wrecker operator leaned forward, apparently to look at something. We were later told that his body had pushed the control levers forward. The warhead immediately came away from the missile, went down, and the boom swiveled. The warhead struck a group of soldiers and inspectors before the operator sat back up in horror over what he had done. The Turkish sergeant in charge leapt back up onto the wrecker, and using a heavy stick he carried, beat the operator to a bloody pulp. The inspectors declared the inspection over for the day, and an ambulance was called to take the operator away to the hospital. The inspection team returned to my detachment and held a meeting in the dining facility. After consulting with their headquarters in Frankfurt, Germany, they informed the Turkish battalion commander that they had been authorized to conduct an immediate re-inspection the next morning. Fortunately, the next day, with a new wrecker operator under very close supervision by a new sergeant (we never again saw the old one), the unit passed and was declared nuclear capable for another year.

One of my last acts as the Detachment 74 Commander was to write and send the following letter to Senator Fred R. Harris, who served as the senator who represented Oklahoma including Fort Sill and Lawton from 1964 to 1973. I regret to say I never received an answer, and nothing changed with host nation support according to my successor, CPT Thomas J. Moore, who assumed command on May 26, 1968.

14 April 1968

Dear Senator Harris:

It is with much hesitation and only after deep thought that I undertake the writing of this letter. I may well be placing a successful career in the balance by doing so, but I feel the importance of the subject cannot be measured by how few people write or what the consequences of doing so are. As a citizen and a fellow resident of Lawton, I feel that addressing this letter to you is appropriate.

I am the commander of an isolated American unit of about 60 men (TUSLOG Detachment 74, APO NY, 09380) located outside Corlu, Turkey about 90 miles west of Istanbul. Our mission is to provide technical training and special ammunition support to the Turkish Army. I have been in Turkey for nearly 11 months and have been in command at this detachment since September of last year. My tour of duty will end on the 29th of May, and I will be returning to Fort Sill for further schooling.

In the time I have been here, however, I have seen American soldiers living and working under conditions that are simply unsatisfactory, to say the least. I do not profess to be an authority on international politics, but I believe that in 1958 the Turkish and American governments entered into an agreement whereby US Army units were stationed on Turkish soil. This arrangement was later implemented by a service-to-service technical agreement signed by the Turkish Ministry of Defense and the Commander in Chief of the United States Army Europe which provided for the Turkish Government to supply necessary support, without cost to the United States, in such areas as quarters, maintenance, electricity, water, heat, transportation, and general day-to-day living essentials. I am sure this agreement was signed in good faith by both parties at the time, but today it is not worth the paper it is written on.

I have records and files and letters which show that the Turkish Army is not supporting the American Detachments as they agreed to do, and in the case of my unit, were it not for having money allocated for food and morale welfare at our disposal, we would be without heat, electricity, and hot water at this very moment. These problems are major ones, but just as important are the hundreds of minor expenses borne by my unit each month to maintain a minimum health and living standard in areas that are solely the responsibility of the Turkish authorities. Every possible attempt has been made to get support through Turkish channels, including writing letters and making visits to commanders up the chain of command, all to no avail.

Senator, these problems may seem to be local ones or ones which result from our particular situation, but I assure you from personal experience that the ramification of support at this site leads to the top of the Turkish Army and to a basic policy by the Turkish government. On the other side of the coin is the Turkish argument that it cost 11 times as much to support one American soldier as it does one of their own, and that the American soldier wants to live in luxury. My answer to these arguments is that a basic standard has been set on how an American soldier should live when in garrison which includes a roof over his head, a good bed to sleep in, hot water to shower with, electricity to operate his mess hall, and heat for his billet. These are not luxuries, and to the best of my knowledge are minimum requirements.

As a unit commander I have seen my unit without water for three weeks when the Turkish unit across the road had water, I have seen my men cold because the Turkish Army would not spend $25 to fix a heater, I have seen my men go without hot water for a month because the Turkish Army would not spend 5 dollars to fix a water heater, I have seen my detachment go without electricity for

12 hours a day because the Turkish Army would not spend 60 dollars to fix a generator, and, Senator, I have seen my men work double shifts because the Turkish Army would not spend 3 dollars for light bulbs used for protective lighting in our nuclear exclusion area.

I do not exaggerate the situation here Senator, nor do I give you all the instances where my men have suffered hardship because of a few dollars expense. As a commander I cannot permit my men to suffer needlessly and as a result have spent American funds from wherever I could obtain them to repair or replace items that should have been provided by Turkish authorities. I could easily accept the easy way out and say to myself: "you are going home in a few days, why worry about it?" But Senator Harris I stop and think of the hundreds of men who have gone before me and the hundreds who will have to face these same problems day after day when I leave.

Senator, I read in my Readers Digest Almanac that Turkey received 134 million dollars under the United States Foreign Military Assistance Program last year which ranks third in the world for this type of aid, and yet I ask you, does it make sense that American soldiers and citizens should suffer for the lack of a hundred dollars or a thousand dollars? If our investment in Turkey, for military assistance alone, is worth 134 million dollars a year, then surly the expenditure of $5,000 more a year would be justified by their government for a unit that supports nearly one fourth of their NATO committed army forces. Although this reasoning seems logical enough, it is not an accepted fact since, if it were, our day-to-day living problems would be virtually eliminated.

I propose that if the Turkish government does not desire to support the American army that you and Congress have a moral obligation to do one of two things. Either make it clear to the Turkish government that required and agreed upon support must be provided; or accept the fact that Turkey is not prepared to support American forces and turn the responsibility over to present military organizations with an appropriate cut in US assistance to cover the cost.

These problems are not going to go away as long as we close our eyes or bury our head in the sand and pretend they do not exist as we did in the past. It is going to take aggressive action on the part of many individuals to bring about a change in policy at such a level as to alter the unsatisfactory trend in support that has developed over the past nine years.

It is my hope that you can be of some assistance in this matter as I believe the problem lies on a Congressional level, most certainly not on mine or even my headquarters up to the Commander in Chief, United States Army Europe. Numerous messages on these problems have been sent to the Turkish General Staff but to date no results are forthcoming.

Early in my letter I said I had many misgivings about writing. My experience in over six years in the service has shown that the man who wrote his congressman is pegged as a complainer or one who goes out of the chain of command to solve his problems. If I thought I qualified for either of these categories I would never have written this letter. I firmly believe the Army has been sold down the river in Turkey and that soldiers have suffered in silence long enough. The morale of my men throughout all their hardships has been excellent, and I think it is a tribute to the young men of America that they will serve in a far-off land, undergo every conceivable inconvenience, and still keep a smile on their face and perform their jobs in an outstanding manner. To ask more is sheer nonsense, but to accept less is not in the tradition of the service.

My thanks to you in advance for consideration of these problems. I know that you are a busy man, but every day that passes compounds the bad situation here. If some solution could be found, every soldier in Turkey would benefit. My best wishes for your continued success and my thanks for your outstanding representation of Lawton and Oklahoma.

GERALD W. SHARPE
Cpt, Arty
Commanding

My first six months in Turkey was with TUSLOG Det 67 in Chakmakli, 10 miles west of Istanbul.

Chakmakli is known as the "Chock." The Club, PX, and Support were housed in Quonset huts

Barracks and street scenes from Chakmakli.

My second six months was spent on a Turkish military camp in Chorlu, 40 miles west of Istanbul.

Above photo is of my change of command and photos above and below are of the detachment area.

Story on site, Nov 8, 1969, Stars & Stripes.

TUSLOG Det 52 where we kept a boat and water skis.

Gypsies in wagons caused many accidents.

Many palaces were built along the Bosporus and multiple Ferrys crossed before bridges were built.

One of many accidents seen on the road to Istanbul.

The Topkapi Palace, built in the 1460s, served as the center of the Ottoman Empire and was the main residence of its Sultans for several hundred years. The jewelry collection is worth millions of dollars and offers a glimpse into the unparalleled wealth and extravagance of Ottoman sultans.

Above photos taken in my office. Left is my command photo that was framed and hung on the wall while I was in command. Below is the news article published in the monthly SASCOM newspaper. Bottom is the flyer for our Christmas vacation flight to America.

CPT SHARPE TO NEW COMMAND

A change of command ceremony was recently held at TUSLOG Det 74. CPT William Forster presented the command to CPT Gerald Sharpe. CPT Sharpe was formerly assigned to TUSLOG Det 67. CPT Forster is departing for a CONUS assignment.

1967 X-MAS CHARTER FLIGHT TO USA

1. Times (all local): Dep. Istanbul: 1430 hrs. 12 Dec. 1967
   Arr. JFK   : 1955 hrs. 12 Dec. 1967

   Dep. JFK   : 1730 hrs. 6 Jan. 1968
   Arr. Istanbul: 1130 hrs. 7 Jan. 1968

2. This is a charter flight no refund will be made for missing this flight. The ticket is not transferable and you cannot use it for any other flights. Missin the flight means losing the full value of the ticket.

3. Recommend that you will be at the airport at least 5 hours prior to the above departures.

MERRY X-MAS AND HAPPY LANDINGS

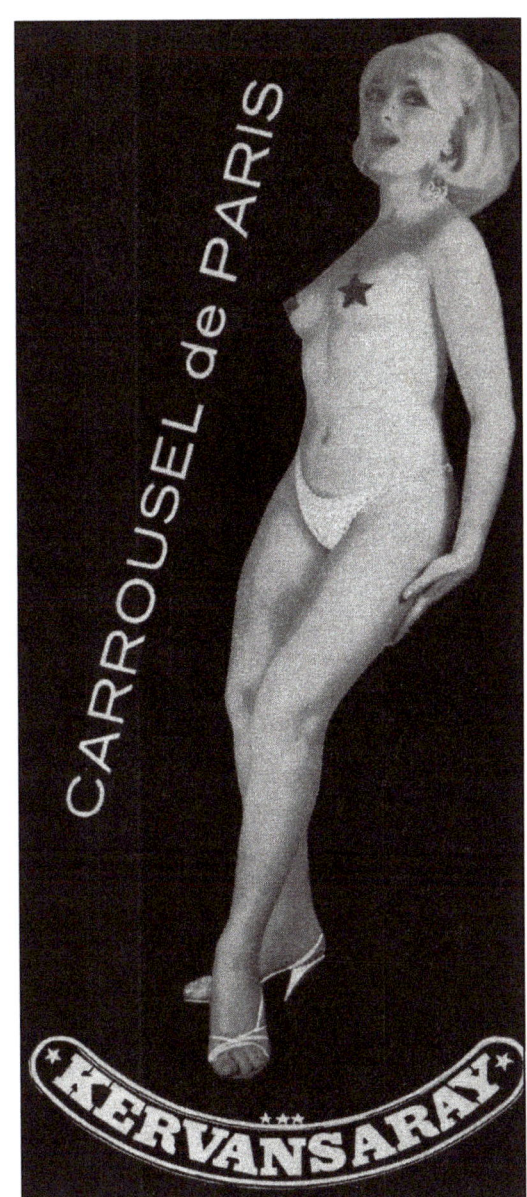

Photos of a farewell dinner at the Kervansaray. One caused my wife to be very unhappy.

Small concrete model of the patio, pool, and fountain I built next to our club. Model was given as a farewell gift at my going away party.

Grand Bazar, largest in the world, 4,000 shops — Dolmabahce Palace — replaced Topkapi Palace.

Interior and exterior views of the Blue Mosque, or the Sultan Ahmed Mosque, built in the early 1600s.

Egyptian obelisks brought in 320 AD, Sultanahmet mosque in background; Roman fortifications.

I have more than 100 3" tapes that my wife and I sent back and forth.

I was presented these unit badges from our supported corps, divisions and battalions.

Similar photos of Ataturk and Kennedy were displayed in offices and homes I visited in Turkey.

Shelley, May 1967, age 6 months; and July 1967, age 8 months.

Sherrin with a book in hand. She loved to read from an early age.

Sherrin, on the floor, October 1967.

Sherrin trying on daddy's shoes.

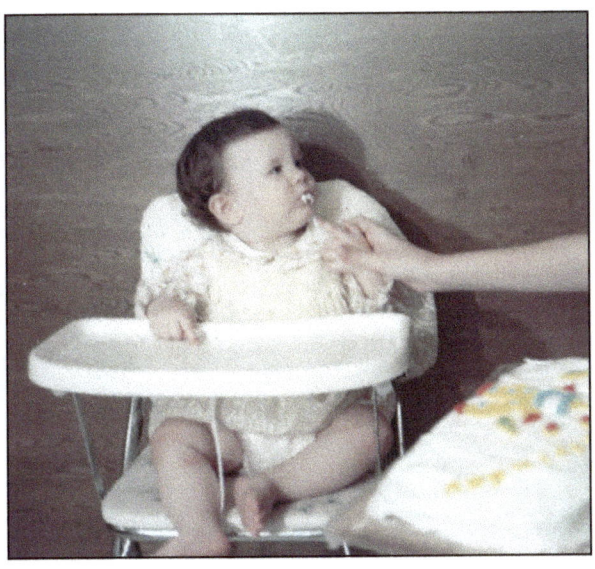

November 1967, Shelley's 1st birthday.

Sherrin "talking" to my photo. Shelley with me on Christmas vacation from Turkey.

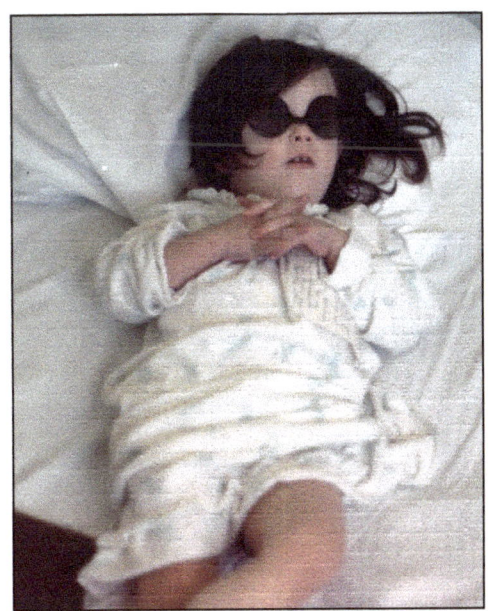

Shelley and Sherrin with "Aunt" Lou on our front lawn. Sherrin being "cool" in bed.

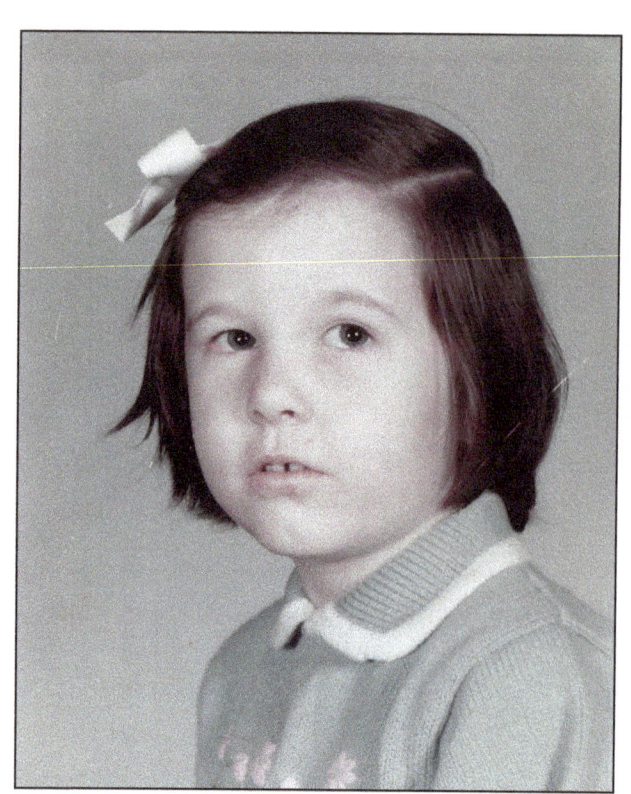 

Three photos of Sherrin. Below: Sherrin trying out a piano.

8. FORT SILL, OFFICER ADVANCED COURSE

MAY 27, 1968 TO JUNE 2, 1969

UPON RETURNING FROM TURKEY, I ENROLLED IN ARTILLERY OFFICER ADVANCE COURSE CLASS 1-69. There are four levels of schooling for US Army officers. OCS was my "Basic Course." The second level is called an "Advanced Course" for captains and is basically ten months of schooling to prepare an officer for company/battery command and battalion staff positions. The other two levels are the Command and General Staff College and the Army War College.

I was in a very unusual position in the class as there were only two of us out of 116 US Army students who had not been to Vietnam. We also had five marine captains and 13 international officers from five countries. This presented us with a number of challenges in that most of the instructors had also served in Vietnam. Every day, we heard comments like, "I do not need to cover this, because you have all experienced it in Vietnam." We would both raise our hands and ask for a better explanation, but more often than not, the instructor would blow us off with a comment like, "Just ask one of your classmates."

After having been gone since Christmas leave, I found Sherrin and Shelley had both grown like weeds. Shelley now saw Kathy and I as Mommy and Daddy. We had discussed in letters and tapes back and forth what to do about her mother spending less and less time with her. One of our first discussions now that I was home was how to implement the agreement we had made with Shelley's mother while I was home on Christmas leave. After some back and forth with Annie Lou, I went to the legal office at Fort Sill in early November and was told that adoptions were not handled by Army legal offices.

The next week, before my regular Thursday meeting of the Kiwanis Club, I asked Al Hennessee if he could recommend a lawyer. He reminded me that we had a retired Army full colonel lawyer in the club by the name of Al Ashton, and that he belonged to a firm that had several lawyers. After the meeting, I talked to Al, and he said he would be not only pleased to do it, but would not charge me their normal fee in appreciation for everything I was doing for the community.

Over the next few weeks Kathy and I met with Al and a young attorney he had drafting the paperwork. After the paperwork was drafted, Al suggested that he should meet with Annie Lou alone to explain the process and then have her appear with him before the judge to finalize the adoption. This is exactly what happened, but unfortunately the process drug on for months as getting Annie Lou to meet with the attorneys and then getting her to sign paperwork took much longer than expected. A court date was finally set for March 24, 1969, and the adoption was finalized. Both Gordon and Annie Lou were barred from raising any future issues with the adoption. Shelley was unaware of her adoption until much later in her life. We were told that our bill was about one fourth of the firm's normal price.

In late August my brother Robert came to live with us at the age of 15 for the 1968-1969 school year as a special favor to my father. He had been kicked out of Belleview High School near the end of his sophomore year for refusing to cut his hair and generally making trouble for his teachers. In return for coming to live with us he agreed to follow rules, cut his hair shorter, and try to get better grades. We enrolled him in Lawton High School for his junior year. In a letter to my Mom and Dad I wrote,

December 2, 1968,

"Robert has been doing poorly in school. He got his second six-week grades last week and he got an F in algebra and in American literature. I have laid it on the line to him that I will accept nothing less than a "C" in any subject these next six weeks. He seems to study quite a bit but has sometimes failed to hand in his required work. He received a "B" in American Literature, a "C" in algebra, and a "D" in chemistry the first six weeks and I talked to his chemistry teacher about it. The teacher said Robert needs to study more so I emphasized that to him. I have told him I will not continue to sacrifice my privacy and my time as well as my house and money if he is not going to apply himself and I mean it. I am willing to do almost anything to help him as long as he is willing to expend the minimum effort to get a "C" even though I believe he could get "A"s if he tried. I had thought about letting him get a part-time job, but I do not feel he has the time to spare from his studies. I have kept a tight rope on his activities but have been very disappointed with his grades."

One of our classmates was a senior captain—and likely a genius with a photographic memory —who had suffered medical issues about two thirds of the way through his previous class. He was now retaking the course with us. He was not a happy camper about spending more time listening to lieutenants and junior captain instructors teach, as he was sure he knew far more about whatever subject they were teaching than they did. He always sat in the back row and brought a book to read. He simply dared instructors to ask him questions. A few senior instructors would occasionally risk it and ask him a question. He would barely look away from his book and give the textbook answer, leaving the instructor speechless.

In one class, taught by a senior major, we took on a practical exercise in which the class was divided up into four teams. We were given a five-page handout that listed all the equipment that a division would have to ship to make a move to Europe. The requirement was to determine the number of tons the sealift command would have to move. It was a four-hour class with the first three and a half hours committed to each team trying to determine the combined weight of everything listed in the five-page handout, using only the reference material provided.

The teams then had 30 minutes to describe how they had come to their answer. I believe our favorite captain was on my team, but he simply sat in a chair with his feet propped up on a desk reading the entire time. The instructor attempted to get him to participate several times with no success. I suspect the instructor thought he would make an example out of him, so he called him to the stage, handed him a piece of chalk, and told him to please outline the solution showing his computations. The blackboards were four feet tall and went across the entire stage. The captain immediately went to the blackboard and started writing gibberish. This blank, blank, blank, added to that yert, yert + this lump, equals this yert and this times that equals this. He proceeded to fill the blackboard from top to bottom in about six columns. The instructor tried to stop him several times and he would just say, "Patience major, you were the one who thought it was a good idea to have me present our teams solution, and I am getting to it." After filling the board, he wrote in large letters, our team computed the Sealift Command would have to move XX,XXX,XXX tons. And he sat down. The instructor checked his solution sheet, and it was the exact answer. The major just shook his head as he erased the board and called on the next team.

A few weeks after my brother Robert arrived in late August, I began drawing up plans for a two-story playhouse in my back yard. I worked on the plans for a few weeks and made up a list of materials. There was a lumber company that had gone out of business in Lawton, and the owner was a member of my Kiwanis Club. He told me there was a considerable amount of used lumber left there. He said it was full of nails, but I was welcome to get as much of it as I wished. My neighbor let me borrow his pickup truck for a couple of hours on three Sunday afternoons. Robert and I were able to get enough 4" x 4" and 2" x 4" lumber for the framing. We spent Sunday mornings pulling nails, while Kathy took the children to mass.

In early October I purchased a second car as we were really struggling to get things done with only one. Getting Robert to and from school presented the biggest challenge. I purchased a 1958 Buick Super with only 50,000 miles on it from an elderly lady who was the original owner. The inside was in great condition and it drove just fine, but the paint was blistered, and it did not look very good. The best part was that I only paid $200.00, according to my December letter to my parents. By the time Christmas vacation rolled around, I was ready to start building the children a playhouse.

We thought that word might spread to instructors that would discourage them from calling on our book reading classmate, but no such luck. Some weeks later we had another four-hour class. This requirement was to organize artillery support for combat for an operation where we were given 20 artillery battalions that had to be allocated between three attacking divisions based on a set of criteria handed out at the start of the class. Once again, we were being taught by a senior major that we had never seen before.

We were divided into four teams and given three hours to develop a solution and one hour to present and justify our solutions. Once again, our favorite captain joined his assigned team, sat down, put his feet up on a desk, and proceeded to read his book. The instructor tried to get the team to force him to participate with no success. When it came time to present the first solution, the major called on our favorite captain and said he wanted him to present the team's solution. The captain jumped to his feet, strode to the front of the room, and told everyone to move all the tables and chairs to the two sides of the room. The major asked what he was doing, and he told him this is a practical exercise, so we are going to do this visually. He next asked every student to get out a blank piece of paper and have a felt tip pen handy. He first had three students write the name of the three divisions and positioned them at the front of the room. He next started having students write down artillery battalion designations, and in some cases, he split battalions into two or three parts. He positioned each student in rows behind the three divisions based on which battalion was assigned to support which division. He asked each student to hold his sign up so the instructor could check the solution as he called it out. It took him about ten minutes to assign every unit to a student and he then proceeded to explain to the instructor what criteria had been used to assign each unit. The instructor was checking his printed solution and was simply speechless. Once again, it was exactly the school solution, and the captain went back to reading his book. To the best of my recollection, he was never called on again.

Sometime in August of 1968, we saw an advertisement in the local newspaper for a home for sale in Duncan, OK, a town just two miles north of Lawton. The ad described a large 4,000 sq ft, 5 bedroom, 3 bath home being sold with the furniture included. It had a separate guest house, a four-car garage, an office building, and a horseback riding arena with barn. It said it covered three city blocks on nearly four acres, and it had a polished marble driveway. It had no price but simply said, "MAKE AN OFFER." We called the listing real estate agent, and the next Sunday afternoon we went to look at the property. OMG, what a place! The house was two stories filled with antiques and it and the outbuildings sat on two city blocks that had been combined and the riding arena with stables was on a third block in a very run down area in Duncan. We were told that the owners were the three children of the founders of the Pace Oil Corporation, George L Sr and Mildred Pace. We were also told that the home was part of the father's estate and Mrs. Pace still lived in the home. We were told that George Sr. had made a will stating that the estate could not be settled until his youngest child was 21 years old. Unfortunately, just a short time later he died in an oil field accident the day after Christmas in 1950 when his youngest daughter, Nancy, had just turned seven. We were told the property had been for sale for a few years, but the three children could not agree on a price.

They had apparently had several offers, but none were accepted by all three children. I was able to determine that the last offer they had was for $50,000. We thanked the agent, said we would think about it, and we went home. I called my father and described the property and asked if he would co-sign a loan with my bank. He said he did not think buying such a property would be a good idea, as the upkeep would be expensive and I would likely struggle to afford it. We went back a few weeks later as no offers had been made, and we met Mrs. Mildred Pace. She was a very nice lady and we enjoyed meeting her. I took lots of pictures and sent them off to my father. He said he would consider coming down to look as it appeared to be a very impressive place. We had a number of family discussions about making an offer, but we put off making one.

On October 9, 1968, I wrote to my father stating, "I do not think I have ever felt worse about a deal. I found out from the real estate agent that they received an offer for $55,000 today. I expect they will accept it." I wrote off the property as being too expensive and believing it would likely be sold to someone else.

Back in our daily classes, our favorite captain got himself in trouble just a few weeks before we were to graduate. We had a Signal Corp 2nd Lieutenant for an instructor attempting to teach us about artillery communications. I have to say I thought it was one of the poorest classes we ever had, but it was only 50 minutes long. About 40 minutes into the class, our favorite captain laid down his book, reached into his pockets and pulled out two handkerchiefs (to this day I have no idea why he would have had two handkerchiefs). He proceeded to climb up on the table in front of him and started moving his arms to positions so that it quickly became apparent that his arms were in semaphore positions spelling out something. The instructor stopped teaching and asked the captain exactly what he was doing. The captain said, "Lieutenant, did they not teach you how to communicate with semaphores in your Signal Basic Officers Course?" The lieutenant allowed as yes, they did, but he was not sure he remembered enough to recognize what was being spelled out. The captain said, "So let me help you out" and proceeded to call out each letter as he signed it. "Y O U A R E T H E P O O R E S T I N S T R U C T O R W E H A V E E V E R H A D" and he sat back down. The lieutenant stood there for a minute, started crying, and left the classroom. We did not see the captain for a few days and when he returned, he no longer had a book, but continued to sit in the back of the classroom and pretend like he was listening. He graduated, and as if poetic justice was heaped upon him, he was assigned to the Artillery School as an instructor. I often wondered if he ever taught a class.

In April I sent a long letter to my parents. I wrote,

April 12, 1969

"As you can see from the pictures my playhouse project has turned into a small building withz three rooms. I have it all up and covered but I still have to put in the windows and have an electrician install lights and outlets. I bought a new car last week. I traded in my 1964 Oldsmobile Tornado on a new 1969 Mercury Monterey Custom station wagon. I paid only $2,000 difference and the sticker price was $5,150. You may remember meeting my fellow Kiwanian, Clinton Herring. He purchased

the local Lincoln/Mercury dealership, and he sold me the car at his cost as this was his first sale at his grand reopening. Having found out Kathy is pregnant again a few weeks ago, she really needs a bigger car to carry a bassinet and the three children. We are all well. Robert gets his grades next week and we hope he is doing better. He has six weeks left to the end of the school year and I hope when you folks come to pick him up, he will have successfully passed his junior year. My opinion is that he needs to join the army and grow up. He has not yet learned the meaning of the word responsibility. I know he cannot join until after his birthday, but if he falls in with the wrong crowd this summer he will never join. We had a call last evening from Mrs. Pace in Duncan. They still have not yet sold the place. She said she would very much like our family to have their home. She told me that it was her son, George Jr, a lawyer in Oklahoma City, who was the one that wanted more money, and she said I should call him and see what he thought would be a fair price. I called soon after hanging up and had a pleasant conversation with him. He said tax circumstances had changed for the family and they would now be willing to take less. He said if I could make a substantial down payment, in the $20,000 range, they would be interested in financing the remainder. I told him I did not have $20,000 but we would discuss what might be done. He would not tell me what price he would accept. This morning I called the real estate agent and asked her to draft an offer for $50,000 with a $10,000 down payment, the amount you agreed to lend me if we could make a deal. Will call you with what I find out."

The real estate agent drafted the offer. I had her include that the sellers would finance the $40,000 balance at 7.00% interest for five years (about half a percent under the going rate) and included that the offer was only good for 21 days. I sent a draft copy to my father, and Kathy and I signed it. Our agent took it to the family and within a few days the two daughters had accepted the offer, but apparently George, Jr, wanted either a higher sales price or more money for a down payment. The 21 days expired without a signature from George, Jr. and once again I wrote the sale off. The agent called me in the third week of May and stated she had all the signatures. I told her I was very sorry, but I had to leave for Vietnam in two weeks and there was no way I could make a deal and move my family in that time. I see on the internet, as of this writing, that the home is now listed as the "Pace Mansion."

We graduated from the Advanced Course on April 26, and we were placed on administrative duty pending our departure from Fort Sill. This meant that I had about six weeks to do whatever I wished. In preparation for another unaccompanied tour, my wife and I purchased two better quality tape recorders and a case of 50 three-inch tapes so we could send them back and forth without having to record over a previous tape. I still enjoy listening to these tapes as I have had them all digitized.

I did all I could do on the playhouse, but I never put the planned slide on the roof. By the time I left, however, it was very usable. The original drawings and photos are in the picture section of this chapter. I departed Lawton and reported to Travis Air Force Base in California on June 2, 1969, for my Military Airlift Command Flight to Vietnam

Two story playhouse I built in my back yard out of scrap lumber, hand drew plans for the building.

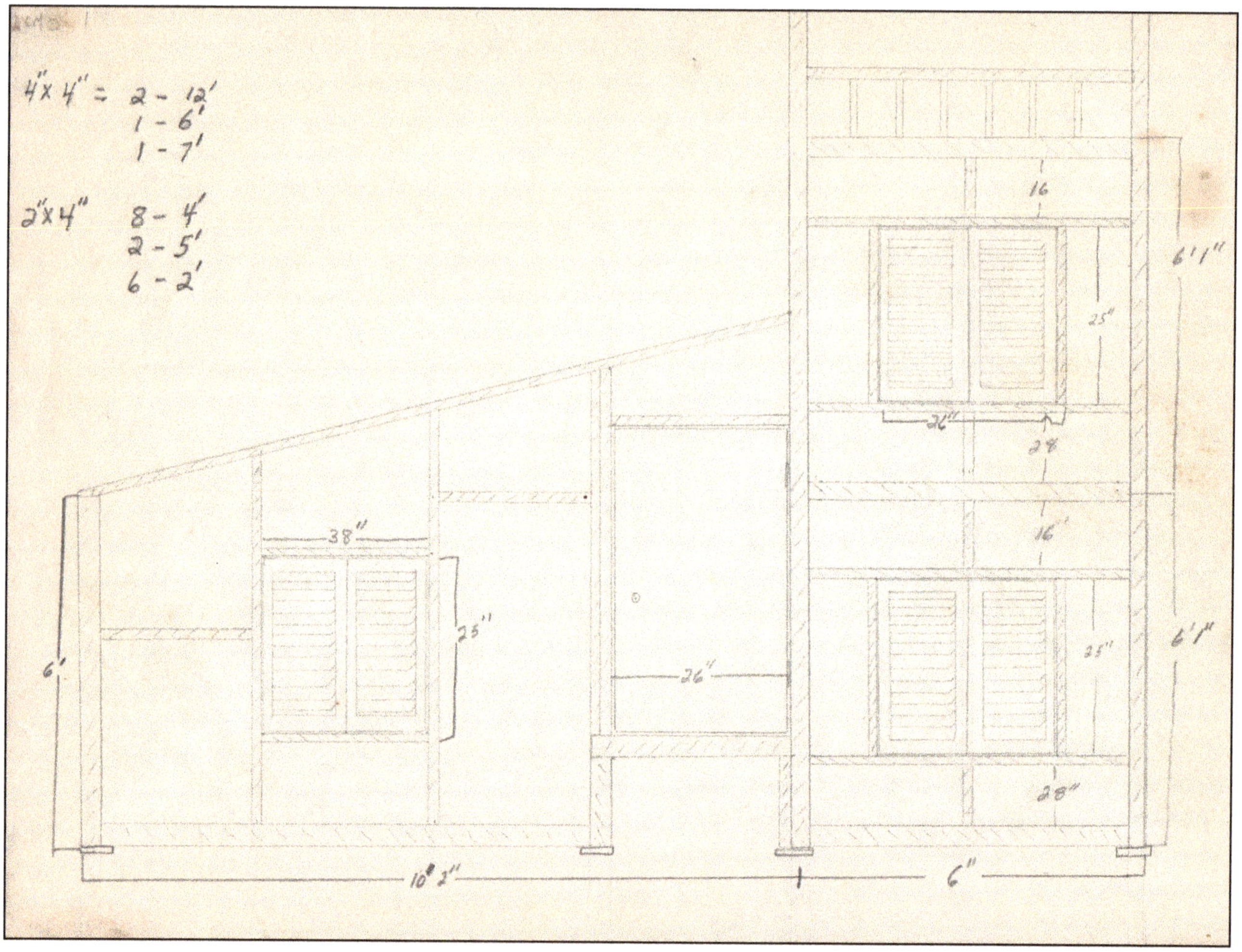

$4'' \times 4'' = 2 - 12'$
$1 - 6'$
$1 - 7'$

$2'' \times 4'' \quad 8 - 4'$
$2 - 5'$
$6 - 2'$

16

6'1"

25"

36"

24"

16'

6'1"

38"

25"

26"

25"

28"

6'

10' 2"

6"

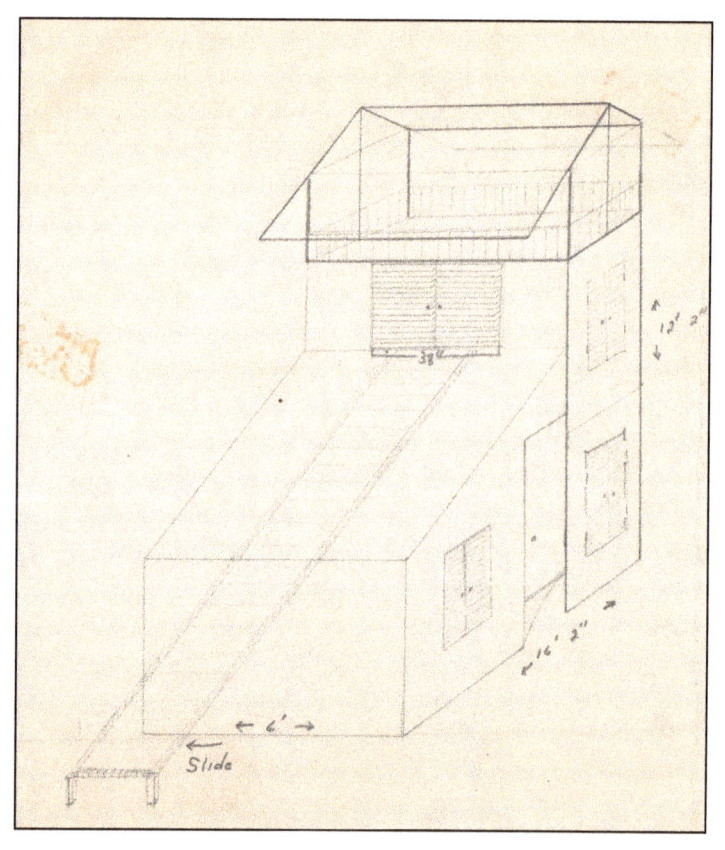

13' 2"

38"

16' 2"

6'

Slide

1964 Oldsmobile Tornado, Summer 1968, visiting Aunt Eileen in New Jersey.≠

1958 Buick Super purchased as 2nd car.

New 1969 Mercury Monterey with the new dealership owner, Clinton Herring, a fellow Kiwanian. Purchased as the first auto he sold just prior to my departing for Vietnam in May of 1969. I traded in my 1964 Oldsmobile Tornado.

GRADUATION CEREMONY
SNOW HALL AUDITORIUM
25 Apr 1969

OPENING SELECTION	77th Army Band
INVOCATION (Audience Stand)	USAFAS Bde Chaplain
INTRODUCTION	MG Charles P. Brown
ADDRESS	LTG John J. Davis
PRESENTATION OF DIPLOMAS	LTG John J. Davis
NATIONAL ANTHEM (Audience Stand)	77th Army Band
CLOSING SELECTION	77th Army Band

—7—

Graduation Booklet for FA Officer Advanced Course 1–69; July 1, 1968 to Apr 25, 1969

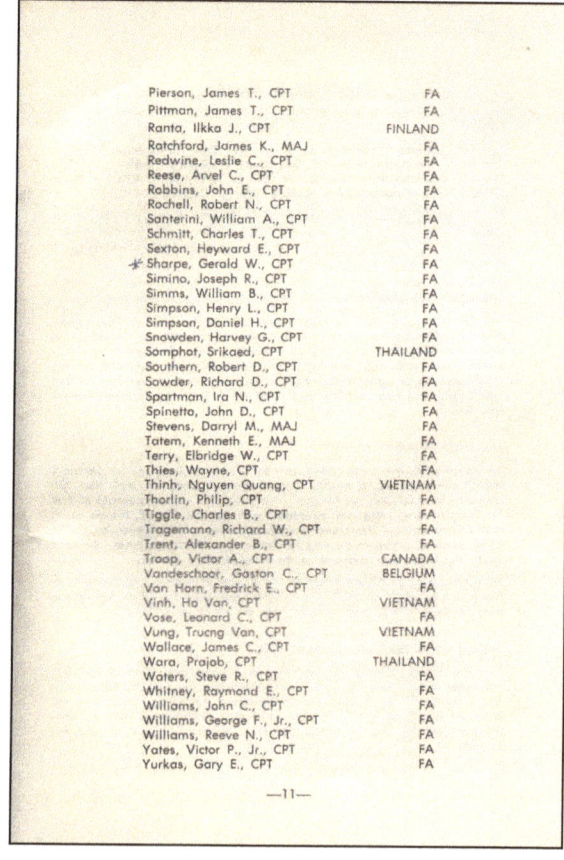

Pierson, James T., CPT	FA
Pittman, James T., CPT	FA
Ranta, Ilkka J., CPT	FINLAND
Ratchford, James K., MAJ	FA
Redwine, Leslie C., CPT	FA
Reese, Arvel C., CPT	FA
Robbins, John E., CPT	FA
Rochell, Robert N., CPT	FA
Santerini, William A., CPT	FA
Schmitt, Charles T., CPT	FA
Sexton, Heyward E., CPT	FA
Sharpe, Gerald W., CPT	FA
Simino, Joseph R., CPT	FA
Simms, William B., CPT	FA
Simpson, Henry L., CPT	FA
Simpson, Daniel H., CPT	FA
Snowden, Harvey G., CPT	FA
Somphot, Srikaed, CPT	THAILAND
Southern, Robert D., CPT	FA
Sowder, Richard D., CPT	FA
Spartman, Ira N., CPT	FA
Spinetto, John D., CPT	FA
Stevens, Darryl M., MAJ	FA
Tatem, Kenneth E., MAJ	FA
Terry, Elbridge W., CPT	FA
Thies, Wayne, CPT	FA
Thinh, Nguyen Quang, CPT	VIETNAM
Thorlin, Philip, CPT	FA
Tiggle, Charles B., CPT	FA
Tragemann, Richard W., CPT	FA
Trent, Alexander B., CPT	FA
Troop, Victor A., CPT	CANADA
Vandeschoor, Gaston C., CPT	BELGIUM
Van Horn, Fredrick E., CPT	FA
Vinh, Ho Van, CPT	VIETNAM
Vose, Leonard C., CPT	FA
Vung, Truong Van, CPT	VIETNAM
Wallace, James C., CPT	FA
Wara, Prajob, CPT	THAILAND
Waters, Steve R., CPT	FA
Whitney, Raymond E., CPT	FA
Williams, John C., CPT	FA
Williams, George F., Jr., CPT	FA
Williams, Reeve N., CPT	FA
Yates, Victor P., Jr., CPT	FA
Yurkas, Gary E., CPT	FA

—11—

Photo outside our garage, December 1968. Partial list of graduates, my name marked

Sherrin and Shelley at a picnic in a park at Fort Sill, OK, July 1968.

Sherrin and Shelley at the picnic and Sherrin with Aunt Eileen.

Sherrin and Shelley in our front yard and on our porch, August 1968.

Sherrin and Shelley carving pumpkins with me at Halloween 1968.

Sherrin's 3rd birthday party, October 1968.

Shelley in July 1968 and in February 1969.

Sherrin and Shelley with dresses made by their mother.

Sherrin and Shelley following their bath and in their Halloween costumes.

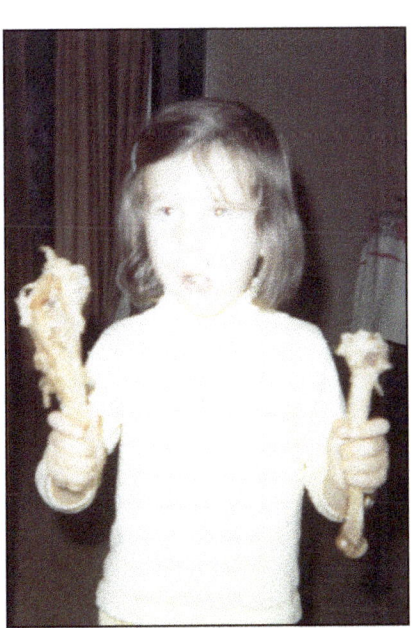

Sherrin with turkey legs and Sherrin and Shelley with Easter baskets, 1969

9. VIETNAM

CAMP GORVAD, PHUOC VINH
JUNE 3, 1969 TO MAY 25, 1970

TARGET ACQUISITION PLATOON LEADER
BASE DEFENSE OFFICER
(JUNE 3, 1969 TO AUGUST 17, 1969)

I ARRIVED AT THE REPLACEMENT CENTER IN SAIGON, VIETNAM ON JUNE 3, 1969, AND DEPARTED for Phuoc Vinh and the 1st Cavalry Division Artillery headquarters on June 5. On the morning I arrived from the replacement center, I was told that the commander had not decided where I would be assigned, so I should go to the dining facility and wait until the commander returned later in the afternoon. As a captain, I decided I would look around the tactical operations center, an underground bunker covered with logs and sandbags with the perimeter piled high with additional sandbags.

I quickly discovered that folks were moving furniture around and I asked what they were doing. I was told the DivArty had been tasked the previous evening to take over managing base defense from the commander of the infantry battalion assigned to provide protection for the sprawling base and its 650 helicopters. I met the captain who oversaw the soldiers on the perimeter, and I spent the morning riding in the back of his jeep as we visited each position dug in on the perimeter. In the afternoon, I helped set up the operations center including maps, radios, and status charts. When the commander, COL James A. Munson, returned on the afternoon of June 5, 1968, he called me into his office and said something like, "Sharpe, I am told you have spent the day reviewing the base defense and helping to set up the ops center. Since it now appears you know more than anyone else in my headquarters, I am going to appoint you as the Base Defense Officer in charge of 'AO Chief' (the name given to the operational area around the division headquarter base camp.) You will be filling the position of Target Acquisition Platoon leader responsible for the radars and crater analysis of the division artillery headquarters. You will report directly to the DivArty S2, MAJ Charles Snyder."

I met the other officers in the S2 that included CPT Jack Garven, whom I replaced. He moved up to assistant S2, and about a month later he replaced MAJ Snider as the S2. CPT Richard "Dick" Murphy was the radar officer. He had taken over the job about a month before I arrived, and he told me recently that I was one of his instructors at OCS as he was a graduate of OCS Class 16-67. He turned out to be so good at managing the radars for the DivArty that he stayed in the position for much of his tour. He was a great help to me later in the tour. There were two other captains in the S2 section, but I do not remember their names.

The Division Base had recently been renamed Camp Gorvad after a Lieutenant Colonel Peter Gorvad. He was the Commander of the 2nd Battalion, 12th Cavalry, and he was killed on March 8, 1969, while directing the defense of Landing Zone Grant against an attack at 3:30 am by 1,000 North Vietnamese soldiers. The battalion tactical operations center (TOC) had been dug into the ground and covered with logs and sandbags just like ours, but it took a direct hit from a 120mm rocket with a delayed fuse that penetrated into the TOC before exploding. Everyone inside the bunker died, but the defenses held, and 157 enemy bodies were counted the next morning.

Having served in a Target Acquisition Battalion in my first assignment after graduating from OCS, I was very comfortable overseeing the target acquisition radars assigned to the division artillery. We had two AN/MPQ-4A counter battery/counter mortar radars used in the Camp Gorvad base camp, each with a crew of six and a large number of small AN/PPS 5 ground surveillance radars attached to every fire base and some deployed on the perimeter of our base camp.

Each radar had an assigned crew of two personnel. This radar was a lightweight, man-portable radar used to detect and locate moving personnel and vehicles day or night under virtually all-weather conditions. The radar was rugged enough to withstand rough field handling, but repair was done only at a designated maintenance facilities, so as soon as one became nonoperational, it had to be swapped out. This practice would later cause me much grief.

Being in charge of a whole infantry battalion commanded by a lieutenant colonel was new to me. It took a few days for me to sort out what I could and could not do, and when to fall back on my boss, Major Snider, who was the intelligence officer, and/or when to raise issues up the chain of command to COL Munson, the DivArty Commander. We settled into a comfortable routine of me going out to inspect the perimeter defense during both day and night with the company commander that was on the week-long rotation of manning the perimeter. We depended on his soldiers to keep us safe. Fortunately, the battalion had several experienced officers and NCOs who had actually fought the North Vietnamese out on firebases. That had built bunkers that would withstand overhead fire from rockets and mortars. They also installed chain link fencing in front of their bunkers to explode rocket grenades before they reached the soldiers inside. In a letter home on June 18,1969, I described the base defense as follows,

In order to defend our camp we have 45 bunkers and 15 towers around the four-mile perimeter. Each perimeter bunker is manned by three men. We have barbed wire four feet high and six feet

REMINISCENCES OF A LIFE WELL SPENT

wide. The rolls are stretched out to form the barrier and they completely surround the camp. We also have mines, grenades, grenade launchers, and machine guns available in and around every bunker and tower.

Crater analysis was also new to me. The division base camp was attacked by rockets or mortars two or three times a week. As soon as the first shell hit, several things would happen at once. Two attack helicopters were either in the air or sitting on a pad with their engines running and they would immediately get into the air and/or begin searching for the distinctive flashes of light that were created by the firing. If they could spot flashes, they would attack the area around the flashes hoping to catch enemy soldiers leaving the launch site and a command of "check fire" would be sent to the artillery. If they saw no flashes after a minute or two, the attack helicopters would race away from the search area after having been told "check fire," and the artillery would start firing at previously identified firing locations.

At the same time my job was to grab my helmet, flak jacket, and compass and head for the door of the tactical operations center. If the shells we hitting close, I would simply run toward where they were landing trying to locate a crater, the hole left in the dirt where the rocket or mortar landed. If it was not close, I would jump in the jeep that was kept outside the door of the TOC. I would try to get to where the shells were reported as landing as quickly as possible.

Once I found a crater, I would identify it as a "mortar" or "rocket" as they came from much different distances away from the base camp and made very different holes in the ground. If it was a rocket, I would look for a piece of the rocket to determine if it was either a 122mm or 140mm, as they also came from different ranges and looked different. It was then my job to go to where the rocket exploded and sight back on the furrow in the ground the rocket made before it exploded and give the back azimuth or direction from which the rocket came.

As soon as that information was known, the attack helicopters were called away if they had not already done so and artillery would start firing all around the possible launch sites. Mortars were harder to determine a back azimuth as they sometimes fell almost straight down and exploded. With a little experience examining craters, however, I learned to figure out the back azimuth, as it always had a little bit of an angle. The same firing sequence was used with mortars. Artillery would start firing shortly after I sent a back azimuth which gave the attack helicopters time to get out of the line of fire. This was a very dangerous job as sometimes the enemy firing would go on for a couple of minutes. There was a distinctive sound of an incoming rocket or mortar which gave you time to dive into the same crater you were evaluating and wait until the shell exploded, then jump back up and get back to work. Fortunately, mortars and rockets almost never landed in the same place twice and I never got a scratch.

Near the end of my second month in Vietnam, the big news in the world was that the United States was going to try to land men on the moon. We were fortunate that the Armed Forces Network broadcast television and radio to bases around the world. We had a small television in our tactical operations center, and on July 16 we were able to watch the liftoff of the Apollo 11 moon mission

COL(RET) GERALD W. "JERRY" SHARPE

aboard a Saturn V rocket. Aboard were the mission commander, Neil Armstrong, and crewmen Buzz Aldrin and Michael Collins. We saw frequent clips on television and heard radio transmissions from the crew of what was called the command module over the next four days as they headed for orbit around the moon. On July 20, the crew reported they were preparing to board the lunar lander which they had named "Eagle."

Neil Armstrong and Buzz Aldrin transferred to Eagle, and they separated from the command module and headed for the surface of the moon. About an hour later, streaming on live television, they reported "Eagle has landed." Neil Armstrong was the first to leave the lunar lander in his space suit and as soon as he planted his feet on the moon's surface, he uttered his now famous words, "One small step for man, one giant leap for mankind." A short time later Buzz Aldrin joined him on the surface with cameras recording everything happening. They named their landing spot "Tranquility Base." Michael Collins continued to orbit the moon in the command module. They ended up spending 21 hours and 36 minutes on the moon. They gathered samples of rocks and soil and loaded it onto Eagle.

In perhaps the most watched portion of the trip and the most worrisome, Eagle lifted off the moon's surface and rendezvoused with the command module. Neil Armstrong and Buzz Aldrin transferred back to the command module and Eagle was jettisoned. (There are some conspiracy reports that say that "Eagle" may still be orbiting the moon.) After 30 orbits the command module fired its rockets to leave moon orbit and headed for home. As they approached earth, weather reports indicated there was a storm brewing in their designated landing spot. Fortunately, they had an alternate site selected and the US aircraft carrier Hornet was on site for recovery operations. At first light, on the morning of July 24, under the watchful eye of four large Sea King helicopters just 13 miles from the Hornet, the module splashed down in the Pacific Ocean. There was immediate concern because the module landed upside down. Fortunately, this had been considered and the crew activated flotation devices on one side of the module that flipped it right side up. There was great concern about bringing unknown contaminants home from the moon, so the crew had to go through days of decontamination and examinations before they could appear in public and be reunited with their families. It was a great time to be an American as the moon landing and the return of the astronauts safely to earth was viewed as winning the space race with Russia.

A number of things were happening at home during my first six months in the country. On August 12th I received a letter from Dad telling me about his brother Woodrow who lived in North Carolina. On Saturday, August 9, "Woody" and his wife Fern drove up to Norfolk, VA to help their daughter Diane and her husband Mike move into a trailer, as Mike was stationed in the Navy there. They returned home Sunday afternoon. Dad wrote, "When nearing home on Sunday, Woodrow complained of chest pains which he thought were due to something he had eaten or indigestion. These pains grew worse as they arrived home, and he asked Fern to take him to the hospital. Woodrow lay down on the back seat of the car and was dead by the time they reached the hospital." I later found out that all of the family assembled in North Carolina for the funeral on Tuesday, August 12, 1969.

In a letter from my mother, I found out that my brother Stanley had married his first wife, Melody Key Abbot, on Saturday, September 6, 1969, at 1:00 pm in a very nice outdoor ceremony at the Pillar of Fire cabin in Eldorado Canyon. Mother wrote,

The rush of the stream acted as music to accompany the I Do's. After the ceremony, Melody's folks went home to finish preparing for a reception scheduled for 4:00 pm. We had a picnic lunch after the ceremony for the Dallenbachs, Truitts, Stanley, Melody, and some of their friends. The reception was very nice, and Stanley and Melody received a number of very nice gifts. They have rented a small apartment in Denver on Logan Street but hope to find a home in the country to rent.

On October 8, our son Sean Raymond was born in Lawton, OK. This was the only time in my career that I had to go on two unaccompanied tours in three years and I would only serve one more such "short" tour before my retirement. On November 22, 1969, my brother Charlie married Sandra "Sandy" DeJohn at our home in Westminster. They had a small reception following the ceremony. Needless to say, I was unable to attend.

ASSISTANT S2, 1ST CAV DIVARTY
(AUGUST 18,1969 TO OCTOBER 23, 1969)

A few days after the astronauts returned to Earth, I was appointed to take the place of the DivArty Assistant S2 who was rotating back to the states. I had spent nearly three months of day and night work on base defense, and I participated in both the morning and evening briefings seven days a week to talk about our base defense status. The assistant S2 served as the senior night duty officer in the tactical operations center, so I turned over both of my jobs to new officers. Base defense was taken over by a new captain, and my job as target acquisition platoon leader was turned over to an incoming captain, David "Dave" Maule.

On August 25, 1969, I began working the 7:00 pm to 7:00 am shift. Working a full night shift meant rearranging my meal schedule. While the dining facility served a midnight meal, I often found myself too busy to leave the tactical operation center. Fortunately, we kept a good supply of cases of c-rations. Each case had 12 meals in cans and packages in small boxes (see a photo of a case of 12 boxes in the picture section). There were 12 meals to a case with such appetizing titles as Chopped Ham & Eggs, Pork & Rice, or Ham & Lima Beans—none of which I would eat. Every box had some type of fruit or desert, and a small folding can opener called a P38 (See pictures). I would open a new case and find the Pork & Beans meal that had sliced peaches, pound case, blackberry jam & crackers, chewing gum, and crackers. I would sometimes open the box of Pork Slices and take out the canned apricots or the Chopped Ham and Eggs and remove the fruit cocktail and fruit cake. I kept a supply of these items in my desk and would snack on them through the night.

As the senior night duty officer in the tactical operations center, I coordinated several activities that required either planning or execution while the senior staff was off duty. This included coordinating resupply missions to fire bases that would take place early in the morning, approving requests for artillery fires from division artillery cannons, usually for what was called H & I fires (harassing & interdiction). These fire missions would be on areas the enemy might use to move at night or places where the enemy might congregate to prepare for an attack on a fire base.

I was also responsible for replacing ground surveillance radars on a fire base if one became non-operational during the night. This was a challenging task, as the helicopter that delivered the radar was very vulnerable to enemy fire while landing, unloading, loading the broken radar, and then taking off and returning to Camp Gorvad. We would send a pair of attack helicopters with each mission to attack anyone who thought it was a good idea to fire at the transport helicopter or fire onto the fire base while the radar was being delivered and the broken one retrieved. Fortunately, we never lost a helicopter in these night operations while I was in the DivArty.

My major responsibility was coordinating artillery fires and directing attack helicopters during a rocket or mortar attack on Camp Gorvad. We had a well-defined set of policies and procedures for who would do the attacking and who would not. In mid-October 1969, we had a rocket attack just before first light. The two attack helicopters were on the ground, and they immediately took off. I told the artillery to check fire until the attack helicopters could scan the area. Within a minute or two of the attack, the division artillery commander, COL Munson, came into the TOC and sat down in front of the large map we used to plot all activities. A few moments later, one of the attack helicopters reported seeing flashes and sent the coordinates in. I passed the coordinates to the fire direction center with instruction to work up fire missions around those coordinates but to remain check fired. Just as I told the attack helicopters to "Roll hot" which meant attack the area where flashes had been seen, the DivArty Commander jumped up and said, "No, check fire the helicopters and let the artillery fire." I am not sure where I got the nerve to do so, but I turned to the commander and said, "Sir, there can only be one person coordinating this attack." I took off my head set and held it out to him and said, "Sir, if you want to take over, please do so, but, if not, please sit down and let me do my job." He sat down. As soon as the helicopters made one strafing run with machine guns blazing on the suspected firing location, I told them to move away from the area and I instructed the artillery to commence firing, which they did for the next 15 or 20 minutes. By the time the artillery finished, I turned to find the commander and other senior staff officers had apparently all gone to breakfast.

Much to my surprise, a lieutenant colonel I had never seen before came up to me and introduced himself as LTC Morris Brady. He said something like, "Young man, you seem to be the only one in this tactical operations center that knows his ass from a hole in the ground. How about you explain to me what just happened."

I proceeded to walk him through our policies and procedures on who had priority on attacking a suspected target and the criteria I used to decide whether the attack helicopters or the field artillery would have priority in firing. He listened intently, asked a few questions, and then left the TOC. I found out when I came to work the next evening that LTC(P) Morris J. Brady was to become the new DivArty Commander the next day. I also found out that he had been the battalion commander of the attack helicopter battalion in the division just two years earlier. LTC Brady had little patience with briefers that were not sure about what they were talking about. He would chew their ass and tell them that they had better get their sh-t together or they would be looking for a new job.

I never had that problem, but often argued with him as to whether I had made the right decision over something that had happened during the night. I soon found out that much of the arguing was over things that he felt I was just too junior to make such a decision, and I should wake up either the operations officer or himself before I made such a decision. I always used the same argument. I would say to him, "Sir, if I had awakened you, would you have made a different decision than I did?" He would say, "No, but that is not your decision to make." I would then say, "Well sir, if and when I make a decision you disagree with, you let me know and then I will be happy to start waking you up. But if I am making the same decision that you would make then I am simply NOT going to wake you up, Sir." Sometimes that sufficed and other times he would come back with some other reason, and we would argue some more. After one long argument, the DivArty XO, LTC Howard Guffy, called me into his office and told me, "Captain, you cannot talk to a full colonel the way you do." COL Brady was promoted to full colonel shortly after his assumption of command. I said to him, "Sir, you can talk to COL Brady any way you want, but I have no intention of changing the way I talk to him. If he does not like the way I talk to him, I am sure he will let me know."

After having served nearly five months in the DivArty, at the morning briefing of October 24, I briefed Col Brady that I had once again sent out a radar without waking him. He simply exploded and told me to stop the briefing and get my ass off the stage as he had talked about this for the last time. He said something like, "Go to bed, I will deal with this later." I went and had breakfast and went to bed. I had a hard time getting to sleep because COL Brady had ended a few officer's careers, and I suspected mine was headed in the same direction.

I finally got to sleep, but someone woke me up at about 4:00 pm and said the commander wanted to see me. I quickly got dressed and went to his office. His clerk went into the commander's office and then came out and told me to take a seat as the commander was busy. I had sat outside his office for about two and a half hours when he came out of his office, and without so much as looking at me, told his clerk he was going to the 7:00 pm briefing. My brain was coming up with all the bad things that could happen to me, to include being sent home and kicked out of the army for what could be called disobeying a direct order. About 8:30 pm the colonel returned and once again went into his office without looking at me or speaking to me. He left his door open and a couple of minutes later he shouted, "Sharpe, get your ass in here."

I immediately went in directly in front of his desk, came to attention, and saluted saying, "Sir, Captain Sharpe reporting as ordered." He said, "Stand at ease." I did and he just looked at me for a minute or two without saying a word. I was getting more worried as every second passed. Finally, he said, "Sharpe, do you think you could be the division artillery intelligence officer, S2? I just fired MAJ McCrary." I stared at him for a minute or so and then just exploded!

I said something like, "G_d Damn, Son of a Bi_ch - colonel. Did you just let me sit outside your office for five hours believing my career might be over, and now you are asking if I think I can do Major McCrary's job?" He smiled and said, "I just wanted to make sure the next time I tell you to do something you will do it." I then said something to the effect of, "Well colonel, I have news for you, if you think I am going to change the way I do my job because you have tried to scare me for five hours, I am not. But yes, as long as you understand that, I am more than ready to be your S2." He said, "OK, get your ass out of here and get to work."

S2, 1ST CAV DIVARTY
(OCTOBER 24, 1969 TO MAY 25, 1970)

So, on October 24, 1969, I took over as the S2. MAJ Thomas "Dan" McCrary, who I replaced, was on the lieutenant colonels list for promotion. He was a West Point graduate, class of 1956, and I found out recently from COL (Ret) "Dick" Murphy, my target acquisition officer, that Colonel Brady was apparently kind to him on his efficiency report as he went on to command a battalion at Fort Bragg and also retired as a full colonel. I went back to working 18+ hour days. I was now responsible for a section of three captains that included "Dick" Murphy, "Dave" Maule, and Arthur "Butch" Williams, as well as four sergeants and several enlisted personnel. One of the enlisted personnel on the night shift with me before I became S2 was Sgt Gary Enzfelder. He had been a Private First Class and did such a good job that we got him promoted to SP4 and then Sergeant in just eight months. We have stayed in touch over the years and he reminded me in a recent letter that one of his memories include Col Brady getting very mad that a trail being used by the enemy that had been reported the previous two morning was left out of the briefing. He noted that he had felt bad about leaving it out of the notes he prepared for me, but that I did not seem to let Col Brady's tirade bother me. I once again had the additional responsibility of overseeing base defense as well as producing intelligence our assigned infantry battalion could use to try to find the enemy soldiers before they could attack us.

A few weeks after I assumed my new role, the commanding general of the division was wounded in a mortar attack at 3:00 one morning. A piece of shrapnel nicked the side of his head just as he was sitting up in bed in his trailer. At the morning briefing it was reported that the CG was irate and told COL Brady this has to stop. He said something to the effect that he had given us an entire infantry battalion to clear AO Chief of enemy personnel and yet they continued to attack us. He also stated that if we could not solve this problem, he was going to either get a new DivArty commander or have some other organization take over base defense.

When COL Brady returned, he told everyone at the morning briefing what the CG had said and then turned to me and said, "Sharpe, figure out how to put a stop to this. Tell me what support you need, and I will get it for you." All I could say was, "Yes, Sir." Immediately after the briefing I got everyone in my section together and told them what the CG had said. I told them we were going to go back and search our records for the past 12 months and make a list of the type of weapon, date, time of day,

day of the week, weather, and the phase of the moon for every single attack. We were going to put that on a spread sheet, separate the data into three sets by type weapon, and then plot the location of every known and suspected firing location. Finally, we were going to look for patterns by day of the week, time of the night, phase of the moon, and/or weather at the time of each attack. I said I wanted this done by the following morning's briefing, if not sooner. I also told them to simply forget sleeping until this is complete.

We finished gathering the data and plotting the known or suspected firing location of each attack and had the data color coded by the three type weapons used to attack us shortly after the evening briefing concluded. A little after 9:00 pm, I sent everyone to bed except my assistant S2. He and I spent the next six hours poring over the data looking for patterns. By 3:00 am we had concluded that there were distinct patterns we had perhaps not been aware of before. Mortar attacks almost always occurred between 2:00 am and 4:00 am. Virtually all of them occurred in the four or five days a month when there was the least moonlight, and interestingly enough, they almost never attacked during rainstorms. By far the most interesting result was that there were only four locations where we suspected virtually all the attacks originated.

Without waking up COL Brady or our operations officer I had the LTC infantry battalion commander and his operations officer awakened, and I asked them to come to our TOC as soon as possible. They arrived about 4:00 am and were very unhappy to find that a captain had gotten them out of bed and no one senior to me was present from the DivArty staff. I reminded them I was the staff officer in charge of base defense, and I asked them to please take a seat. I also told them that if they were still unhappy after I finished the briefing, I would awaken either the commander or our operations officer or both and have them come to the TOC.

It took me just 15 minutes to brief them on what we had found. I suggested they take this information back to their headquarters and create a plan to establish ambush positions around these four sites for the next three or four nights, that happened to qualify as low light nights, to see if they could capture the mortar being used and to come back and brief the CG at our 7:00 am morning briefing on their plan. They said they could do that, and this looked very, very promising. I then asked if they wanted me to wake up the CO or the operations officer and they said they did not.

At the 7:00 am briefing I told COL Brady I had a special briefing on mortar attacks and wanted to know if he wanted me to go first or after everyone else briefed. He said first, do it now. I condensed the briefing to about ten minutes and then introduced the infantry battalion commander who outlined an operational plan to put out 12 ambush parties each afternoon for the next four days, have them dig in and hide, and then try to detect of any movement in the vicinity of the four possible firing locations between 2:00 am and 4:00 am. COL Brady approved the plan and told everyone that he did not want one word of this plan briefed at division. To make a long story short, on the second night, the enemy mortar crew was ambushed, and all five enemy soldiers were killed. A helicopter was sent to pick up the mortar and COL Brady had it presented to the CG at the morning briefing with a big red bow attached. As the story goes, second or third hand at best, this did not go anything like COL Brady thought it would.

He assumed the CG would be pleased at the quick solution to the problem. Instead, the CG was irate that this plan hadn't been briefed to the division. He supposedly warned that if anyone ever again took it upon themselves to do things in his area of operations without his knowledge, there was going to be hell to pay. COL Brady was told to have someone remove the damn mortar from the stage and he would talk to him in private when the briefing was over. When COL Brady returned, he did not say one word about any unhappiness at division, instead he asked that everyone in the S2 section and all the staff on duty assemble in the TOC. After we gathered everyone, COL Brady told us that this was perhaps the best intelligence staff work he had ever seen in his career. He congratulated the staff on a job well done and then directed that he wanted a similar analysis done on the rocket firings. I told him he would be pleased to know that it was already completed, and while we had planned to brief it at the morning briefing, we could do it at the evening briefing if he wished. We presented the second analysis that evening, but we were not able to come up with specific locations to target as the rockets were almost always fired from new locations. I am pleased to report that we did not have another mortar attack while I was in Vietnam and only had two more rocket attacks.

About a month after capturing the mortar, an attack helicopter reported killing a North Vietnamese soldier hiding in a large open field. The helicopter crew landed and retrieved a large pouch the soldier had been seen carrying. Apparently, he was a courier as the pouch was filled with reports to higher headquarters and letters from the soldiers operating in AO Chief. I kept translated copies of a few of the letters but cannot find them. In essence they reported that in the previous month something had changed in AO Chief because they could hardly move without being bombed or strafed. They reported soldiers being burned alive by 55-gallon drums of napalm being dropped by helicopters. They reported on losing their only mortar and the entire crew. They reported that every time they went out to fire rockets it seemed enemy soldiers were waiting for them. In one letter home a soldier reported he was starving because he and a number of his fellow soldiers were afraid to leave their underground bunker stating they had sent people out to get food and they never came back. These letters and reports were sent to division and were briefed to the CG. He was reported to be very pleased with them. I considered them as validation of the great work we were doing in managing the base defense of Camp Gorvad.

In December 1969 we had a visit by a team from the Department of the Army, Military Personnel Center. There were two officers from the Artillery Branch who interviewed the 20 plus field artillery officers assigned to the DivArty. They had reviewed our files before arriving and then divided us up into two groups to discuss what our next assignment might be. The major I met with indicated I had been doing well taking college courses part-time, but that to have a successful career I needed to get my bachelor's degree. He said he thought it would be a good idea for me to apply for civil schooling. He suggested that when I filled out my "Officers Assignment Preference Statement" that he handed me, that I should indicate my number one choice of assignments following Vietnam should be to attend civil schooling. I filled out the form and prepared a two-page request for civil schooling, and on December 31, 1969, I mailed the request to the Artillery Branch.

LTC Guffy, the DivArty XO, and my rating officer was scheduled to return to the USA the beginning of the 3rd week of January 1970. This required him to write efficiency reports on the six individuals he rated. The actual transfer date was January 14 when he turned over his duties to his replacement LTC Jack G. Callaway. I had some concerns about what LTC Guffy would write as he did not always appreciate my candor. He wrote a very nice paragraph and rated me top in all the blocks except one. He gave me a 98 out of 100 in his comparative ranking of all officers he knows. I was pleased to learn however that COL Brady had rated me 100 out of 100 and also gave me the best possible ratings in all categories. COL Brady wrote the following glowing report.

Captain Sharpe has performed his duties as the Division Artillery Intelligence Officer in a truly outstanding manner. Although his position calls for a major, Captain Sharpe was selected because of his maturity, extremely professional abilities, and a rare gift of being able to get the job done regardless of the degree of difficulty. He has repeatedly and consistently demonstrated a capacity for achievement that equals or exceeds any officer I have ever known. Through his drive and unusually high standards, he has organized his staff section into what surely must be the most effective organization of its size and type in the army today. Captain Sharpe's duties far exceed that of the normal Division Artillery S-2 in that, in its present situation, the Division Artillery's responsibilities include infantry ground combat operations within its own area of responsibility. This has caused Captain Sharpe to expand his efforts into far more detailed and tactically oriented areas of infantry combat operations. He has met this requirement by developing facilities for extensive Prisoner of War interrogations and intelligence assimilation. He has established firm contact with all adjacent and superior headquarters seeking a more complete exchange of information. His effectiveness has been such that he is recognized as an authority of note within his area of interest. Captain Sharpe has done this while continuing to perform in a peerless manner the normal function of a Division Artillery S-2. His target lists are always detailed, current, and accurate and are available on short notice. His survey teams and survey information center are the most active and efficient in Vietnam. The spirit, courage, and dependability of his crater analysis teams is widely known. In short, this is the most effective captain I know in the army today. His maturity, professionalism, and highly developed military talents are such as to set him apart from his fellow officers. I strongly recommend that Captain Sharpe's abilities and potential be recognized by promotion ahead of his contemporaries and attendance at the Command and General Staff College at the earliest possible date."

On January 27, 1970, I received a letter stating that the Artillery Branch had received my request, they had reviewed it, and had determined that I qualified for consideration. They said I would receive an invitation letter soon with instructions on how to proceed. Sure enough, on February 26 I received a letter outlining what paperwork and transcripts were required. I was also instructed to submit an updated "Officers Assignment Preference Statement." I had already requested transcripts, so I was able to almost immediately send in everything they asked for. Two weeks later I was notified I should apply for admission to my first choice, the University of Oklahoma, at Norman. I applied and I received an acceptance letter from the University on April 15. Finally on May 4, I received an instruction letter stating I would be sent to the University of Oklahoma for what was called the "Bootstrap Program" for the period June 5, 1970, until December 29, 1971.

REST AND RECREATION (R&R), HAWAII, (FEBRUARY 18, 1970 TO FEBRUARY 24, 1970)

The one bright spot in our year in Vietnam was that we received a week of R&R. The slang term for the week was I&I or Intercourse and Intoxication. For married soldiers we were allowed to go anywhere we wished that was serviced by the Military Airlift Command as we were flown free to and from the location aboard military aircraft. Day one and seven were travel days with five nights of actual R&R.

I chose Hawaii, by far the most popular place, but some chose to go to Thailand or Hong Kong. There were two in-country R&R spots for single soldiers, one at Vung Tau beach, about 60 miles southeast of Saigon, and China Beach, about eight miles south of Da Nang. Single soldiers could also take five days in Bangkok or at least five other overseas locations, including Manila in the Philippines.

I had heard about an all-inclusive resort where you paid one price for everything. It was on the Big Island and was called the Kona Village Resort. Kathy and I arrived and departed Honolulu just hours apart the third week of February. Guests all arrived at the resort on Sunday afternoon by charter bus from Kona or the Kona Airport and we all went home on Friday. Upon arrival, the entire staff of the resort was lined up to meet us and one by one we met each person who told us their name, asked ours, and then welcomed us to the resort. From that moment until we left, we were called by our first names by everyone. It was amazing as an hour later one of the beach boys saw us coming down to the beach they would say, "What kind of beach chair would you like, Jerry and Kathy?" or a bartender would say, "Kathy, can I get you a drink?"

There were three different kinds of thatched roof cottages: ones that were out in the lava field, those that faced the beach, and ours that were built up into the palm trees so no one could see into your cottage, and it was very private. Each was luxurious with a king size bed, great bath, and comfortable sitting room. There were common areas in the resort with games and entertainment, a community store, a great dining hall that served food day and night, and if you wanted to go on a picnic, they would pack a basket with a selection of items you could pick. The standard rate per night for a couple was $75.00 but they offered service members on R & R a rate of $42.25. It was a wonderful week.

Back in Vietnam we had a wonderful homemade shower enclosure. It had a cement floor, four-foot-high plywood walls, and a three-foot screen above the plywood to keep the mosquitos and flies out. It had two shower heads and outside was an air force wing tank, mounted on high stilts, that was divided in half top and bottom. (Unfortunately the pictures at the end of the chapter were taken just before our wing tank was installed.) Water was stored in one 55-gallon drum and fuel oil in another. There were two water lines from the water drum, one to the top half of the wing tank and the second to the shower to provide cold water. You had to climb up a ladder and turn on a valve to let water go to the top half of the wing tank and a second valve to let fuel oil go to the bottom of the wing tank. We would light the fuel oil and it would heat the water as it flowed across the steel plate and out the other end to the hot water pipe. It was primitive but ingenious and allowed us to take a two or three-minute hot shower. When finished we had to climb back up the ladder and turn off the two valves.

One of the main roads on Camp Gorvad was perhaps five feet higher than the shower, with a bank sloping down to the shower. We would watch trucks pass and pray the wind was blowing away from the shower or clouds of dust would blow down and fill the shower enclosure.

One afternoon, at about 5:00 pm, I was taking a shower by myself and watched a long line of trucks pass by on the road above me. Fortunately, the wind was blowing away from the shower, carrying the dust to the far side of the road. I was facing the road and observed the next to last truck stop right above the shower. A soldier jumped out of the truck and rolled what was clearly a hand grenade down the bank and it hit the back wall of the shower. I immediately dived flat onto the concrete away from the back wall, expecting to hear an explosion at any moment. When nothing happened after about 15 seconds, I jumped back up and saw the truck start to move away to catch up with the others. I grabbed my towel, wrapped it around me, and ran to the tactical operations center next to the shower where the colonel's driver was sitting in his jeep. On my way I passed an officer and yelled at him to turn off the shower and keep people away until I returned as there was a live grenade behind the shower. I jumped in the jeep and told the driver to follow that long line of trucks that was making a huge dust cloud as it moved. We caught up with the dust and saw that the trucks had turned into the engineer battalion compound.

As we pulled in, the trucks were parking all over the lot. The drivers and assistant drivers were going into the dining facility for supper. I saw a lieutenant and briefly explained what had happened and asked him to get me the engineer company commander. Just a few minutes later an engineer captain showed up and again I explained what had happened without saying I knew it was the next to last truck. I asked him to get all the drivers and assistant drivers out of the dining facility and have them pair off and line up in the order of the trucks they were driving. With the help of his first sergeant and a couple of other officers they finally got everyone lined up in order from the first truck to the last. I asked the company commander to accompany me as we walked down the row stopping at the next to last pair of soldiers. I asked them which soldier was the assistant driver. After one said he was, I asked him to step out of the rank. He was a private first class, and I said, "Soldier, explain to me exactly why you thought it was a good idea to get out of your truck, throw a grenade down an embankment, and then drive away?"

He made a feeble attempt to deny it, but I told him I had seen him from five feet away. He admitted he had done it and explained to us that they had been on the grenade range all afternoon and had one grenade left over. Rather than turning it over to the officer in charge, he and the driver decided to just bring it back to camp and turn it in there. As they were driving back, he said they concluded that if they showed up with an extra grenade, they were going to be in big trouble, so they decided to just "Get rid of it." I turned to the company commander and said this is now your problem. Please have your first sergeant follow me back to the division artillery compound and retrieve the grenade. I never did find out what they did to the soldiers, their NCO, and the lieutenant in charge, but spending an hour in the sun in only a towel and my shower flip flops was no fun.

There was a French 75mm cannon next to the flagpole that was fired with blank ammunition every morning and evening with the raising and lowering of the flag, much as I did as a second lieutenant at Fort Sill. Unfortunately, one morning, the shell blew up inside the weapon, killing the two soldiers that were doing the firing. I was appointed to do an investigation of the incident. After taking statements from everyone involved, having the explosive ordnance detachment examine the weapon and what was left of the shell, and having them examine the remaining stock of blank shells, I could not reach a conclusion as to why it blew up. I had possession of the remainder of the shell and over the years have used it as an ash tray on my desk. See the photo of it at the end of the chapter. It now resides in my display cabinet of Army memorabilia.

In February of 1970, COL Brady asked me to prepare an article that would describe "Nighthawk," my idea to combine a Huey transport helicopter and a Cobra attack helicopter to operate at night in support of infantry squads and platoons conducting night operations. The Huey was armed with two 7.62-mm machine guns and a 40-mm grenade launcher. It also had a powerful infrared light and scope that turned night into day for the crew. Additionally, the Huey mounted a 50,000-watt xenon searchlight to help locate enemy positions. The Cobra was armed with 76 2.75-inch rockets with a variety of warheads and fuses. The article was first published in the weekly *First Team* newspaper and then was published in the March 1970 edition of *Army Digest* under my byline. A copy of the *Army* magazine article is in the photo section.

On April 26, 1970, MG Elvy B. Roberts, CG of the 1st Cavalry Division received orders from LTG Michael S. Davidson, CG of II Field Force to prepare a plan for a coordinated attack within the next four days to go across the Cambodian border to kill and capture North Vietnamese, who had used the area as a sanctuary for years. We had bemoaned the fact that for years North Vietnamese soldiers could flee across the border and then give our attack helicopter pilots the finger knowing we could not fire at them. We later found out that President Nixon had approved planning on April 22, but he directed that the only people in Vietnam to be told were commanding generals until they were notified planning could start. MG Roberts assembled a top-secret planning staff of 14 individuals from the division, six from the DivArty including myself and our S3 operations officer, MAJ Tony Pokorny. We were joined by six individuals from II Field Force and two individuals from the ARVN Third Airborne Brigade, with three airborne battalions. We worked virtually without sleep until the morning of April 28.

MAJ Pokorny and I were responsible for determining what artillery would be needed, where it should locate, and how much ammunition would be needed. We gave this information to the logistics folks so they could figure out how to get it transported to where it would be needed. We also prepared targeting plans based on the numerous aerial photographs that we were given of the area.

Early on the morning of April 28, the division was further directed to be prepared to commence operations within 48 hours. A combined task force would make the initial assaults into Cambodia. Command and control of this operation was to be given to the ADC for Maneuver of the 1st Cavalry Division, BG Robert M. Shoemaker. Additional U.S. and Army of the Republic of Vietnam staff joined our small planning staff at this time, and we prepared the final plans for the operation. In preparation for the assault, the planning group was moved from Camp Gorvad to the 1st Cav's 3rd Brigade headquarters in Quan Loi, just 40 miles from the Cambodian border and about the same distance NW of Camp Gorvad.

The concept of the operation was that Task Force Shoemaker, consisting of the 3d Brigade with one mechanized infantry battalion and one tank battalion, under operational control of the ARVN 3d Army, with an ARVN Airborne Brigade, and the 11th Armored Cavalry Regiment, would conduct air assaults and ground attacks into the "fishhook" of Cambodia. Following an intensive preparation phase of B-52 strikes, tactical air strikes, and then artillery bombardment, the ARVN Airborne Brigade would air assault into the area north of the objective to seal off escape routes and begin operations moving to the south. Simultaneously, a task force would attack north across the Cambodian Border with the 3d Brigade on the west and the 11th Armored Cavalry Regiment on the south and east. All elements would then conduct search and interdiction operations to locate and exploit enemy lines of communication and capture cache sites of weapons, ammunition, and supplies in the objective area as well as capture or kill North Vietnamese soldiers putting up resistance.

In the early hours of May 1, six flights, each of 20 or more B-52 bombers, dropped their heavy ordnance on hard targets within the primary objective area. The last bomb went off at 0545. Fifteen minutes later an intense artillery preparation began with the priority of fires directed at the proposed landing zones of the ARVN Airborne Brigade's objective area in Cambodia. D-day had arrived.

At 6:30 am the 11th Armored Cavalry Regiment began its movement from the northwest of An Loc toward the border. At the same time, a 15,000-pound bomb with an extended fuse designed to detonate about seven feet above the ground was dropped to clear the jungle at landing zone EAST. This was followed fifteen minutes later by a similar drop at landing zone CENTER. Shortly after first light, the Forward Air Controllers began directing tactical air strikes on pre-planned targets, shifting to the ARVN Airborne Brigade's objective area during the period from 7:00 am to 8:00 am.

The 1st Squadron, 9th Air Cavalry, began aerial reconnaissance operations early on May 1st and by 7:40 am had established contact. Five North Vietnam Army soldiers and their 2½-ton truck became the first recorded casualties of the operation. At 8:00 am the 1st Squadron, 9th Cavalry conducted a landing zone reconnaissance which was followed ten minutes later by the combat assault of an ARVN airborne battalion into landing zone EAST. The landing zone was secured and became a fire support base when six 105-mm howitzers and three 155-mm howitzers from the 1st Cav were inserted shortly thereafter. During this air assault, the 11th Armored Cavalry Regiment had moved out of their staging area and crossed the border, moving north. In the 3d Brigade area, C Company, 2d Battalion, 47th Infantry (Mechanized) crossed the Cambodian border at 0945, followed by elements of the 11th Armored Cavalry Regiment to the east, which crossed approximately fifteen minutes later.

An ARVN battalion of the airborne brigade began its combat assault into Objective "B" at 9:46 am in a 42-ship helicopter lift supported by 22 Cobra gunships and reached landing zone CENTER by 10:05 am. Another ARVN Airborne Battalion completed its combat assault into Objective "A" on the west.

Contact was immediately established with a panicked North Vietnam Army force of approximately 200 men. The Cobras supporting the contact expended most of their rockets and machine gun ammunition on groups of 10 to 30 North Vietnam Army men fleeing the area in a dozen directions. It was apparent that tactical surprise had been achieved during the combat assaults as there were no reported instances of .51-caliber ground-to-air firing.

During the afternoon of D-day, two companies of the 2d Battalion, 7th Cavalry, made a combat assault into Objective X-RAY in the northern portion of the 3d Brigade area of operation. This movement had been tentatively planned by General Shoemaker, and due to the relatively light resistance throughout the area, he ordered its execution as the final combat assault of D-day.

The 1/9 Cav had a field day catching small groups of NVA trying to evade, resulting in a record total of 157 NVA killed by helicopter. TAC air also had a record setting day by putting a total of 185 sorties on hard targets which resulted in an estimated 100+ NVA KBA in the ARVN Airborne AO alone. Among the ARVN Airborne forces, the 5th Battalion performed well with 27 NVA killed, and 8 prisoners taken during the day. The prisoners were later identified as members of the 250th Convalescence Battalion; the 50th Rear Service Group; and the 1st Battalion, 165th Regiment, 7th NVA Division. The 3d Company, 3d ARVN Airborne, made the first significant cache discovery of the invasion at 5:20 pm when they found a large medical cache of up to 6,000 pounds. The cache included the finest in modern surgical equipment and had been imported from Western Europe via Air France, possibly through Phnom Penh. The ground contact of Company H of the 2/11 ACR was the highlight of the 11th ACR operations during the day. After passing through a regimental-size base camp, a large enemy force was encountered in trenches to the north. The ensuing battle left 50 enemy dead versus 2 U.S. KIA, the only U.S. combat fatalities of D-day.

As friendly forces moved across the border in both airborne and mechanized assaults, the North Vietnamese quickly found themselves surrounded in their bunkers by ARVN forces flown in by helicopter on the north and by US forces on the south. They reported in writing, after the war ended, that surrounded, they waited until nightfall and then with security provided by the 7th Division, broke out of the encirclement and fled north to unite with the COSVN in Kratie Province in what would come to be known as the, "Escape of the Provisional Revolutionary Government (PRG)." Trương Như Tảng, then Minister of Justice in the PRG, recounted that the march to the northern bases was a succession of forced marches, broken up by B-52 bombing raids. Years later Trương would recall just how, "Close [South Vietnamese] were to annihilating or capturing the core of the Southern resistance—elite units of our frontline fighters, along with the civilians and much of the military leadership." After many days of hard marches, the PRG reached their northern bases. The column needed many days to recover and Trương himself would require many weeks to recover from the long march.

The next few days of operations were characterized by a continuation of maneuvers begun on D-day. The enemy made strenuous efforts to avoid contact and to determine the extent and placement of the Allied forces. His command and-control apparatus was completely disrupted and he was caught off guard and ill prepared. The North Vietnamese High Command scattered in twos and threes, and a large exodus of trucks going in all directions was noted by helicopter pilots from the 1st Squadron, 9th Cavalry.

US and ARVN forces found North Vietnamese supply bases and caches stretched along the entire length of the Ho Chi Minh Trail. One located by U.S. soldiers sprawled for well over a mile and consisted of at least 18 structures, including a mess hall, barracks, training facility, and hospital. In the final accounting after the incursion, MACV reported that the amount of supplies captured or destroyed in Cambodia was 10-times greater than the amount captured during all of the previous year in South Vietnam. The list included, among other things, 18 million rounds of ammunition, tens of thousands of weapons of every type, 435 vehicles, and 700 tons of rice. The allies also claimed that 11,350 Communist troops had been killed and 2,300 had been captured or had surrendered.

One unusual job I had as the DivArty S2 was to personally approve the taking home of war souvenirs. In the months before May 1970, I signed perhaps six to ten requests a week to take home a North Vietnamese rifle, pistol, bayonet, or some other gear captured on the battlefield. A soldier was allowed to take home only one weapon. Within a couple of days of the Cambodian invasion I started getting 20 to 30 requests a day. I quickly learned that virtually every helicopter coming back to our base camp from Cambodia had many more than one weapon per person aboard. The crews would bring many, and in addition to bringing weapons for everyone in their unit, a black market quickly developed. The policy allowed soldiers to take a weapon home, so no one thought it was unusual, but I ended up having to set aside an hour in the morning, one in the afternoon, and one in the evening with the understanding if you were standing in line at the end of the hour, you had to return another time. I do not know how many I signed, but I do know that I still kick myself for not accepting one of the many offers to bring me a North Vietnamese SKS rifle.

As I approached the last week of my tour in Vietnam, I began asking COL Brady at the morning briefings when he was going to name my replacement and allow me to go to Bien Hoa, the large base next to Tan Son Nhut Air Force Base, where tradition was that you went and got four days to out process, do some shopping, and relax for your last few days in country. His answer was always the same, "Just keep doing your job, Sharpe. You will be on your plane as scheduled." This went on through day five, four, three, and finally day two, the morning before my Pan American flight was to depart at 10:30 am the next morning. I asked COL Brady if he did not think it would be a good idea if I spent the last day briefing my replacement. He gave me the same answer but said, "Come to my office after the briefing." When I reported in, COL Brady handed me my efficiency report that had perfect numerical scores and a written paragraph describing how I performed my duties. He said, "Sit down and read this and see if you have any questions."

When I reported in, COL Brady handed me my efficiency report. I noted that both LTC Callaway and COL Brady had rated me 100 out of 100. All other blocks were the best they could be. LTC Callaway had written an entire extra page with simply glowing comments. Some of what he wrote follows:

Captain Sharpe's position is unique with respect to his multiple responsibilities and greatly exceeds that of any remotely similar circumstance with which I am familiar. Considering his age, rank, experience, and level of professional education, I judge that his talents and capabilities are unique and of the most outstanding nature. He has an extremely quick, perceptive, imaginative, and analytical mind which enables him to execute all tasks quickly and thoroughly. His recommended solutions are always practical and workable. He is an outstanding briefer and his ability in this role easily exceeds that of any other captain, and most of the majors that I have observed in recent years. The enthusiasm and dedication with which he approaches each task is impressive and productive. He is a dynamo of energy and appears to be almost tireless. Regardless of the type or variety of pressures to which he is subjected in a crisis, when we all depend on his performance for success, and time is critically important, he has repeatedly demonstrated he has the ability to produce that which is so direly needed."

COL Brady wrote in his section:

Captain Sharpe is one of the most efficient and effective officers of his grade that I have ever known. He has performed the duties of the 1st Cavalry Division Artillery S2 in a truly outstanding fashion; this despite the fact that his duties have been drastically modified to include the intelligence efforts to support tactical maneuver units and the complete responsibility for managing some 31 surveillance radars assigned or attached to the Division. In accomplishing these diverse tasks, he has demonstrated a maturity and judgement far beyond his years. He is capable of performing to the most exacting standards without supervision. Captain Sharpe has the rare talent of grasping a concept and producing a full-blown plan which he supervises through complete execution. Captain Sharpe has accomplished all of his tasks in an outstanding manner while continuing to produce unequaled results as an artillery intelligence officer, His attitude, dedication, and courage are exemplary. He is an articulate speaker with a commanding presence. This is an unusually quick officer of high intelligence and unlimited potential. I recommend that Captain Sharpe be promoted ahead of his contemporaries and attend Command and General Staff College at the earliest possible date."

After reading the full report, all I could say was, "Thank you sir."

At the evening briefing on May 24, I again asked what the plan was for my departure, and COL Brady assured me that after I gave the morning briefing, I should have my bags packed and I would get on my flight as scheduled. To be honest, I was doubtful that there was some magic that could be worked that would see me taking the hour's drive south to Bien Hoa, out process, and get on the plane as scheduled. On the morning of May 25, I was the first briefer and when I finished, COL Brady came up on the platform and told everyone that, "The DivArty is losing the best staff officer I have ever met." He said that a brand-new major would be arriving later in the day to be the next S2, and before everyone should come up and shake my hand in farewell, he had one more task to perform. Orders were read for a Bronze Star Medal, an Army Commendation Medal, and an Air Medal. After pinning on the medals and shaking hands with everyone, COL Brady told me, "Sharpe, the DivArty SGM has completed all of your out processing, and he will accompany you on my helicopter to catch your Pan Am flight."

We loaded my duffle bag and one small carry-on bag and took off in COL Brady's helicopter for the 25-minute flight to Tan Son Nhut Airbase. Much to my amazement, we landed on the tarmac just a short distance away from the large Pan Am airplane that appeared to already be loaded as there were no soldiers going up the stairs into the airplane. The SGM and I got off the helicopter that carried a one-star red plate on both sides as the position of DivArty Commander was authorized to be a brigadier general. I asked the SGM where I had to go to out-process, and he said you simply take your bags and go get on the airplane. I shook his hand, grabbed my bags, and started walking toward the stairs of the airplane.

About halfway to the stairs, I was intercepted by a major who said, "Where the hell do you think you are going, Captain?" I said, "I am going to get on that airplane and if you have any questions, you can ask the brigadier general who owns that helicopter." About that time the sergeant major arrived and showed the major some piece of paper as I walked on toward the plane and then up the stairs into the airplane. I worried as I stowed my bags and found a seat that some military policemen were going to come aboard and escort me off, but apparently whatever paperwork the sergeant major had satisfied the major that I was authorized to board the airplane.

Thus ended my year-long stay in Vietnam. My plane landed about an hour late on the 26th at Travis Air Force Base in California. By the time I cleared customs and out processed I had less than four hours to catch my airplane from San Francisco to Oklahoma City. I had to search for a taxi willing to take me the 65 miles south to the San Francisco Airport. I found one and told him I would pay him double if he could get me there with two hours to spare. He made it and I was the last one to board my flight back to civilization and to my follow-on tour at the University of Oklahoma.

The Phuoc Vinh District Headquarters building. By 1995 it was converted to an elementary school.

1969

1970

25 years later, 1995

2023

I am standing in front of my sleeping quarters that had ammunition boxes filled with sand outside.
Bottom two photos show what happens when mortar shells explode nearby protecting us inside.

The early version of our shower, before the elevated wing tank. Where a grenade hit the back side.

The entrance gate to Camp Govad above and I am standing by an AH-1 Cobra attack helicopter.

OH-6A helicopter known as a "Loach" in which I spent a few hundred hours. The post chapel.

The 1st Cavalry Division base camp called Camp Gorvad. I was in charge of base defense.

View of the runway and of the district headquarters, taken from the door of a helicopter.

Left, the fortified post theater and exchange, and right, the service club.

Photos of the battery of 175mm guns at Camp Gorvad. Could fire 30+ KM or 17 miles.

Left is the battery of 8" howitzers. Most accurate artillery weapon. Range 16.8 km or 10.5 miles.

Right is a battery of 155mm howitzers. Range of 14 km or nearly 9 miles.

Left a battery of 105mm Howitzers at Camp Gorvad. Used on every firebase. Range 7.1 miles.

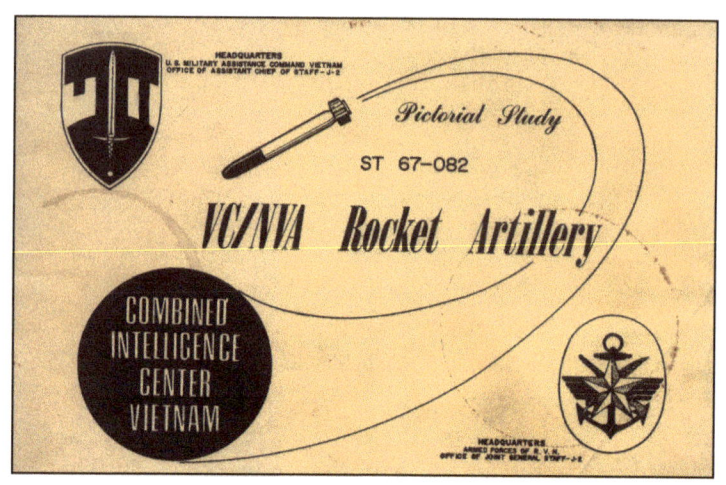

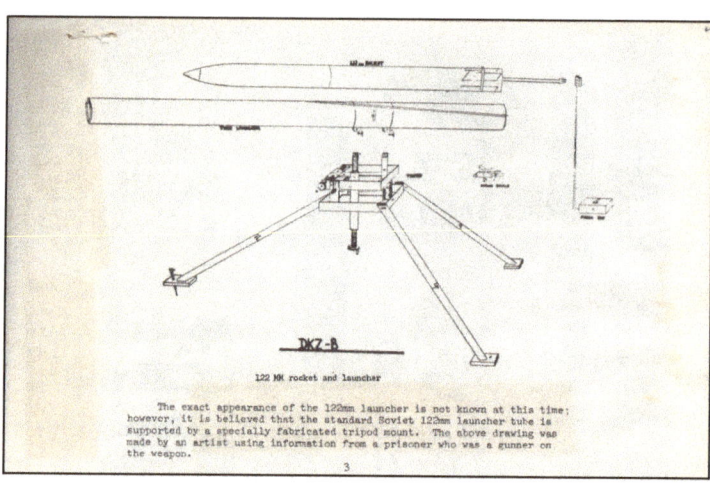

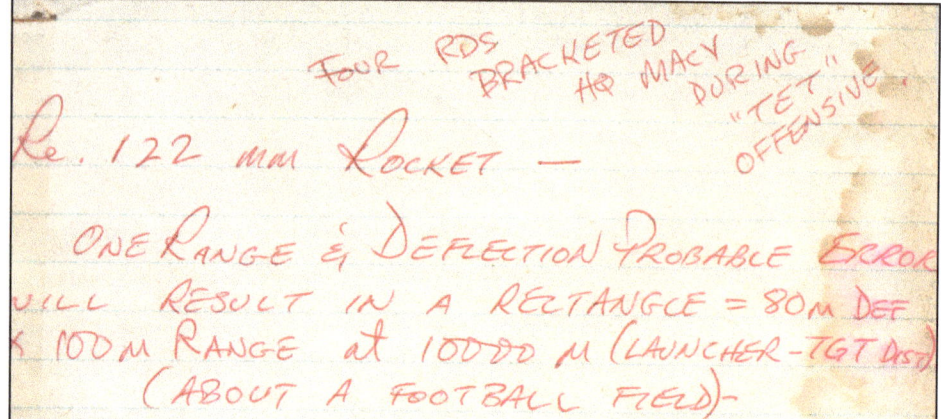

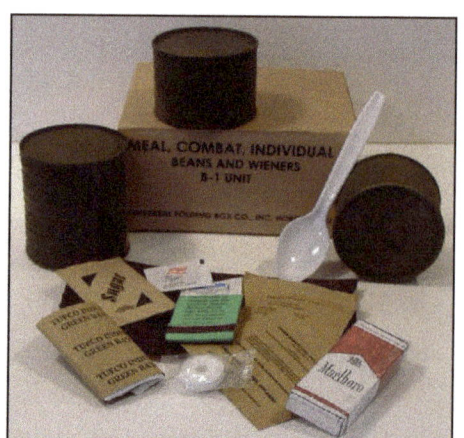

My reference book on North Vietnamese artillery. I filled it with addon notes. C-ration meal box.

A case of C rations with 12 meals. Some were much better than others. I gathered fruit cans.

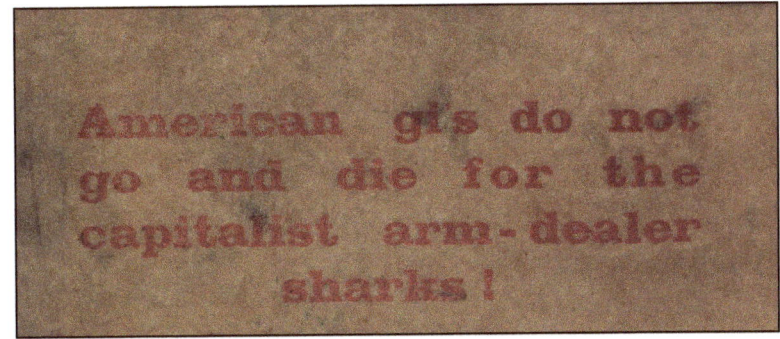

The P-38 can opener. There was one in every box of C rations, but we wore one on our dog tag chain. The North Vietnamese littered our area with propaganda leaflets. I had many, but only found one.

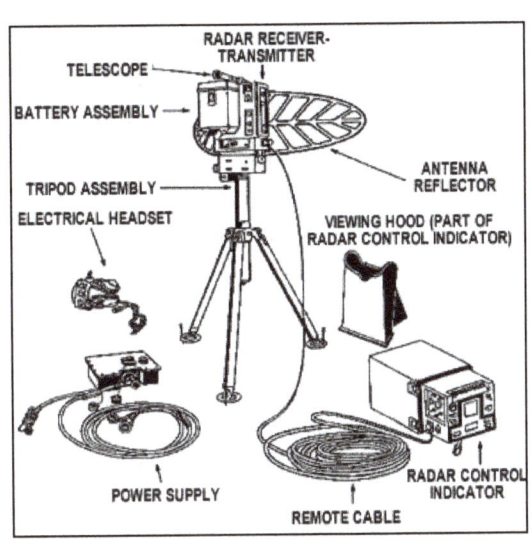

Left is the Q4 counter battery, counter mortar radar. It could detect and compute the point of origin of shells but was little help with rockets. The right two are the AN/PPS-5 Ground Surveillance Radar that caused me no end of grief in replacing them on firebases at night.

My ticket to fly aboard a military aircraft from Saigon to Honolulu, Hawaii for R & R.

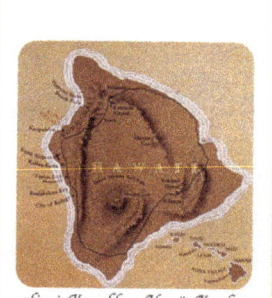

at historic Ka'upulehu on Hawaii's Kona Coast

KONA VILLAGE
P. O. Box 366, Kailua-Kona, Hawaii
Phone: Hawaii 324-0155 After May 1st, Call 325-5555
Cable Address: COCONUT

OWNED AND OPERATED BY
ISLAND COPRA & TRADING COMPANY, LTD.
A SUBSIDIARY OF SIGNAL PROPERTIES

Hawaii's Favorite Hideaway

KONA VILLAGE RESORT

a long-ago South Seas village,
reborn on Hawaii's golden Kona Coast

I awoke quite early, made coffee in my
Maori cottage and took it out on the lanai.
The dark was a smashing blue-black and
the air a delicious, clean cool...

Mornings insinuate in gold-shafts down
slopes of the guardian-volcano, Hualalai.
Days are palette-smears of color—a billion
sun-sparkles on cobalt sea; brilliant splashes
of hibiscus and bougainvillea; the copper and
green tones of cocopalms and shadelacings
through feathered algarobs.

A whipped-cream surf froths an uncrowded
beach and tumbles into tide pools. Plovers
and stilts strut on the sand. Finches flit and
mynahs gossip along coral pathways. Mallards cluck commands over a glassy lagoon
and small fish ruffle its surface at feeding
time.

There is a Hawaii-feeling here you can rub
between your fingers. It is a very real spirit
—a kind born only in solitude, rooted in
benevolent beauty. For Kona Village is more
a *re-creation* than a fine resort. It is this that
sets it apart.

KONA VILLAGE RESORT
Hawaii's hideaway resort

a long-ago South Seas village, reborn on Hawaii's golden Kona Coast

INFORMATION & RATES — 1970-71

Accommodations
Seventy thatched-roof hales—spaced for privacy—stand on stilts beside ocean, beach, lagoon and gardens. All are designed in spacious hale styles of the South Pacific outside but interiors have every-modern, wall-to-wall luxury of finest resort hotels. Sandy paths lead from all cottages to the crescent-shaped swimming beach and airy New Hebrides Long House. Here, meals are served and guests gather for informal, Hawaiian-style evening entertainment.

Our Setting and How to Find Us . . .
Kona Village has some 12,000 acres separating it from the world. But the small community itself clusters around a central sandy beach 15 miles from Kailua-Kona on the Big Island of Hawaii. It's close to Honolulu (150 air miles), yet seems centuries away in its remote setting at Kaupulehu, once a thriving Hawaiian metropolis.
You can take a Royal Hawaiian Airlines flight directly to Kona Village and land on our private air-strip. Or, take Aloha or Hawaiian Airlines to Kailua-Kona. Then travel by limousine, or rent-a-car, to our village by the sea. Drive 17 miles from the new Keahole Airport to our gatehouse on Highway #19. Turn left, and 7 miles of private road leads right into Kona Village.

What to do besides snooze in the sun . . .
Sail, or outrigger. Swim in the ocean or waterfall-fed fresh water pool. Join guided hikes to Hawaiian petroglyphs and ancient Hawaiian worshipping and shelter sites; use our snorkels and fins to see underwater coral reefs; enjoy nightly entertainment . . . all if you wish. There's no charge for recreation facilities* and no "appointed time" for you to partake.

Our beachcomber hales have every comfort . . .
Our smallest hales are 460 square feet big, with either king-size bed or extra long twins. Carpeted; private dressing room and bath; sliding glass walls opening onto pleasant lanais with ocean, lagoon or garden views and settings; electricity, of course, and coffee-makings, but no TV. Telephones are at the office if you really must contact the outside world!

What to expect in shopping services . . .
Like the entire village, our Island Copra & Trading Co. general store is thatched-roofed and airy. Its selections include film, sundries, swimwear and Hawaiian wear. It even rents baby sitters!

What to bring besides a swim suit . . .
Come casual and pack the same. Hikes call for sturdy, comfortable shoes and some evenings are cool, so bring a sweater or light jacket. There are no dress-up occasions and aloha shirts or muumuus are always appropriate. You don't need a raincoat — we have only 6" of rain per year at Kaupulehu.

* Except for charter fishing boats and hunting expeditions.

Dining and Cocktailing . . .
Our oceanside New Hebrides Long House provides three deluxe meals each day (American, European or Hawaiian cuisine). Now and then special meal events, such as a real Hawaiian luau, are held. Cocktails are available at the poolside Ship Bar or in the spacious Samoan Long House.

Rates are Hawaiian-style — very gentle . . .
Charges for two include your private hale, three sumptuous meals daily and use of all recreation facilities. Sorry, cocktails aren't included. But we promise — you won't mind a bit.
Daily rate for hale at Full American Plan (all meals included):

Single Occupancy	$40	$50	$60
Double Occupancy	$55	$65	$75	$85
Third person in room		$20 per day extra	
Each child under 10 years		$10 per day extra	

Rates are subject to 4% state tax and change without notice.

A Honeymoon Special! Of course . . .
Only $199 for two includes 3 nights and 4 days in a deluxe private hale overlooking the ocean; all meals; champagne and Mai Tai on arrival; airport transportation both ways from Kailua-Kona; sales taxes and tips extra. Additional nights $65.
The Honeymoon Special is only available when reserved and prepaid in advance. Sorry, no refunds on unused portions . . .

Reservations? They're instant — EASY!
To confirm reservations, you or your travel agent call — toll free in the USA — the phone number 800-AE 8-5000 — seven days a week and ask for reservations at Kona Village, Hawaii. It's the American Express Space Bank Reservation Service. Kona Village is a member for your convenience.

Or from this area:

	Call:
CANADA	
Montreal	514-866-7696
Toronto	416-486-4850
EUROPE	
Dublin	770-735
London (Telex: 23683)	01-493-4020
FAR EAST	
Bangkok, Thailand	865-144
Hong Kong	H-249-141
Kuala Lumpur	28578
Manila, Philippines	874-941
Singapore	94066
Tokyo, Japan	214-4721
Auckland, New Zealand	32009
Osaka, Japan	Call local telephone information
Sydney, Australia	277-852
Taipei, Taiwan	571-261
Kona Village, Honolulu office	923-4924

KONA VILLAGE RESORT
at historic Ka'upulehu on Hawaii's Kona Coast
Owned and operated by Island Copra and Trading Co. Ltd.
Write P. O. Box 366, Kailua-Kona for our free brochure.

ALOHA R&R HAWAII

[signatures]

PRESIDENT

HAWAII VISITORS BUREAU

We chose an "All Inclusive" resort to spend our R&R in Hawaii. For one price they provided everything other than alcoholic drinks. They even prepared picnic baskets if you chose to go of exploring. It was a beautiful resort and the staff treated R&R couples with a special hospitality.

Postcards and flyers from the Kona Village Resort.

Two photos we took or had taken during our stay.

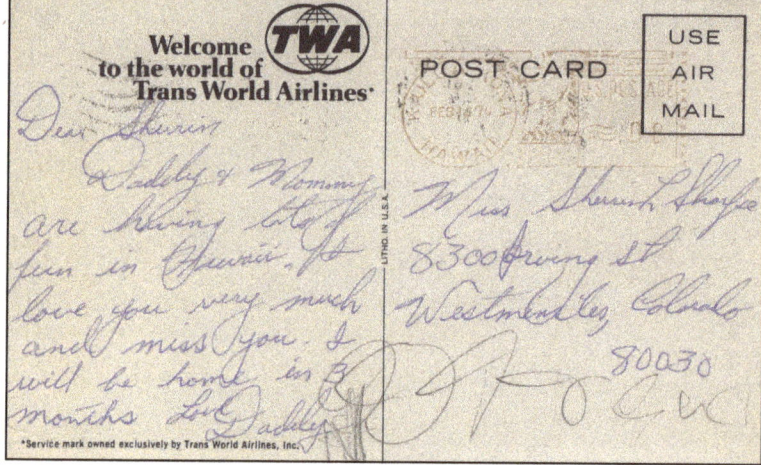

Two more souvenirs from our stay and the postcard we sent Sherrin from Honolulu.

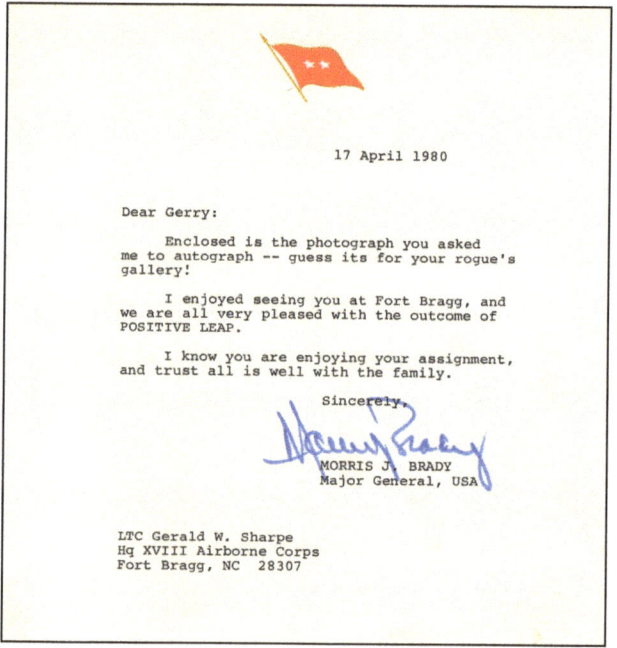

17 April 1980

Dear Gerry:

Enclosed is the photograph you asked me to autograph -- guess its for your rogue's gallery!

I enjoyed seeing you at Fort Bragg, and we are all very pleased with the outcome of POSITIVE LEAP.

I know you are enjoying your assignment, and trust all is well with the family.

Sincerely,

MORRIS J. BRADY
Major General, USA

LTC Gerald W. Sharpe
Hq XVIII Airborne Corps
Fort Bragg, NC 28307

Left: The photo I received from MG Brady. He was a tough boss, but I learned a great deal from him. Inscription says: "To the finest intelligence officer in the Army. The real boss of AO Chief"

Nighthawk

CPT Gerald Sharpe

Cobra gunship joins Nighthawk search ship to support 1st Air Cavalry Division ground units.

A hail of bullets and rockets streaking down a beam of light is part of the 1st Air Cavalry's surprise for Charlie operating during darkness.

It is a DivArty team known as Nighthawk, composed of a Huey helicopter from Battery E, 82d Artillery, and a Cobra gunship from the 2d Battalion, 20th Aerial Rocket Artillery's Battery B, located at Camp Gorvad.

This team mounts two 7.62-mm mini-guns, each capable of firing 4,000 rounds a minute; a 40-mm grenade launcher, delivering 150 grenades in 30 seconds; and 76 2.75-inch rockets in a variety of weights, warheads and fuzes.

A powerful infra-red light and scope turn night into day for a keen-eyed target detection crewman, who also has at his disposal a night observation device, which needs only starlight to operate effectively.

Besides these two modern electronic detection systems, the crew uses the 50,000-watt xenon searchlight to locate and mark enemy positions.

The moment an infantry squad spots movement, word is flashed to the Nighthawk. In a matter of seconds, the unit briefs the pilots and marks the positions with a light that can only be seen from the sky. Quickly, the Nighthawk begins searching the area until the target detector picks up movement on his scope. The pilot of the Huey banks to the left while the co-pilot informs the Cobra pilots circling above.

CAPTAIN GERALD SHARPE is assigned to Head-quarters, 1st Cavalry Division (Airmobile).

As the searchlight beam hits the ground, the Skytroopers below relay their positions in reference to the light before the Huey's radio crackles with—"Nighthawk Two, this is Dodge Six, you are clear to engage."

The mini-gun pours a constant stream of fire and lead down into the small beam of light. Moments later, the Blue Max overhead releases a burst of mini-gun fire and a salvo of rockets from pods tucked under its sides. The light goes dark, but infra-red scopes continue to probe the target area for an assessment of damage.

The mission is completed. For Nighthawk, it means that the enemy has been caught and stopped in the darkness before he can strike at positions on the ground.

The article I wrote that was published in March 1970 edition of Army Magazine

Just like in Turkey, we exchanged tapes every week or so. It was more convenient than letter writing.

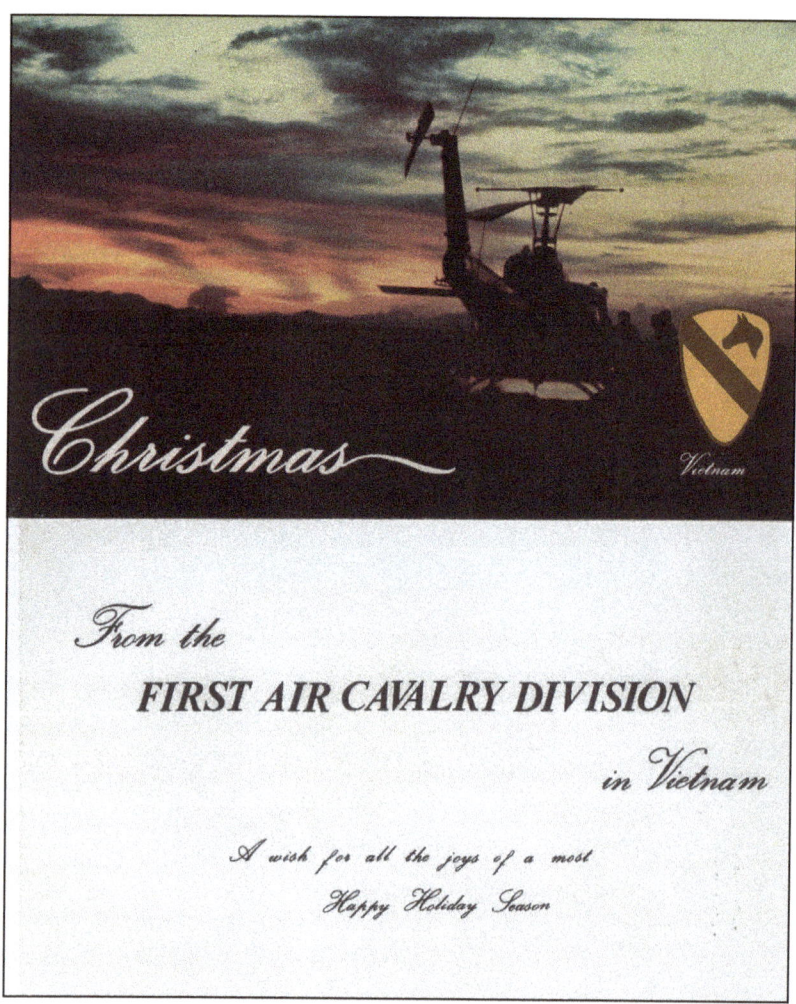

We were offered the opportunity to purchase boxes of Christmas cards. I sent out many. Above, right is the remains of the artillery shell that exploded in the breach killing two soldiers. We had little use for Vietnamize money but I brought a few bills home as souvenirs.

The Kennedy grandparents came to visit and Sherrin celebrated her 4th birthday.

Sean was born October 14, 1969 and Kathy took photos on the first day home.

Photos taken with two different Santas when Sean was six weeks old.

Picture taken at Sean's baptism. Photo at my parents' home before Kathy departed for R&R

Shelley and Sherrin playing in a local park.

Children visiting Mexico while on a trip to El Paso.

10. UNIVERSITY OF OKLAHOMA

NORMAN, OKLAHOMA
MAY 26, 1970 TO JANUARY 3, 1971

I FLEW DIRECTLY TO OKLAHOMA CITY AND RENTED A CAR FOR ONE DAY TO DRIVE TO NORMAN to register for classes. I was directed to check in with the full colonel who was the professor of military science in charge of the ROTC program. He told me there would be ten active-duty army officers attending and gave me instructions on how to register. My wife had rented a partially furnished home in Norman and had left a key under a flowerpot, as she would not arrive until the next day. She had stayed in our home in Lawton, near Fort Sill while I was in Vietnam. I registered for classes and drove to the house. It was within walking distance to the university, so the next morning I simply walked to my classes.

My first class of the day was in a 200-person classroom. I arrived early and there were one or two army officers outside — I could tell they were army officers as they were the only ones with virtually shaved heads like mine. I introduced myself to them and about that time an extremely good-looking young lady wearing no bra came up and started talking to me. She looked like a *Playboy* bunny. As we walked into class, she followed me and sat down next to me. She flirted a little, and after class as we walked out, she told me she was looking for someone to live with rent free and in return she would sleep with me. I told her as attractive as that sounded, I doubted that my wife and three children would appreciate another roommate. She smiled, said no problem, and went and introduced herself to another army captain. They lived together for as long as I was there.

Kathy and the three children arrived the second morning. I saw my son Sean for the first time. Both Sherrin, now almost five, and Shelley, almost four, had grown like weeds and of course I had not seen Kathy since R&R in Hawaii in February. Over the next week we arranged for a moving company to bring our furniture and belongings from Lawton to Norman. We decided not to rent out our home right away, so we hired a neighbor boy to mow our grass and water the lawns. We visited Lawton about once a month during our time in Norman.

1970 was the height of the anti-war movement across the USA, and demonstrations and marches were being held on the OU campus once or twice a week. I made a personal decision that because the Army was paying for my education, I would wear my uniform every Friday. At first, I would hear comments like "baby killer" or "war monger" as I walked to class. After a few weeks, however, individuals in my classes would talk to me and tell me that I did not act like a baby killer or fit any of the other terms military personnel were called on campus. I took the opportunity in every class to tell my classmates that I had just spent a year in Vietnam and the way the news media often described soldiers was simply not true from my personal observation. I explained that we were simply doing what the government asked us to do, what they trained and paid us to do, and we all did it as fairly and ethically as we could. Even a few of my professors who had been very vocal about being anti-war at the start of the semester told me that maybe they had a wrong opinion of army officers and soldiers if most of them were like me.

After I had been in school for about two months, I got a message from the same full colonel, head of the ROTC that had explained how to in-process, that he wanted to see me. I went to Col Ladd's office that day after class. He told me he wanted me to stop wearing my uniform on campus, as he was having enough trouble with anti-war students disrupting his training. I asked him if he was giving me a direct order to stop. He said no he was not, but he would appreciate my cooperation. I explained to him that I thought I was doing more good by wearing my uniform than I was causing trouble, as I felt I was getting the opportunity to change students' minds about Army officers. I said I planned to continue my Friday patriotic demonstrations but if I perceived it was inflaming anti-war students I would consider stopping. I never stopped.

After having been in Norman a few months, my wife Kathy met a lady in a women's social sorority called Beta Sigma Phi, Delta XI Chapter (It is now known as the XI Beta Rho Chapter). She met some of the members and they explained that Beta Sigma Phi International (ΒΣΦ) was a non-academic sorority with more than 100,000 members in chapters around the world. It was the largest organization of its kind in the world. Founded in Abilene, KS, in 1931 by Walter W. Ross "For the social, cultural, and civic enrichment of its members." The organization now has about 200,000 members and has chapters throughout the United States and Canada, and in 30 other countries that provide social and cultural opportunities for women of all ages, usually beginning in their twenties or thirties, allowing them to stay active and build friendships that last a lifetime. It is primarily a social organization, but some chapters do community service work. Kathy agreed to join and over the next year we attended many social functions and held at least two at our home. One project I agreed to carry out was the making of a child's folding playhouse out of 5/8" plywood. It was dubbed "The Gingerbread Doll House" and I have a copy of the flyer in the pictures following this chapter. I worked on it evenings for perhaps a month and then it was displayed at one of the meetings. Members were given two-dollar raffle tickets and I forget exactly how much was raised, but I believe it was over $500.

Between the summer and fall sessions, I decided to let my beard grow out, which I had never done before (see pictures). Kathy was not thrilled with my decision and almost daily lobbied for me to shave it off. On August 11, 1970, we made a trip to Lawton to check on our home on 50th Avenue. The temperature was hovering around 100 degrees and we got in and out of the air-conditioned car a few times on the trip down, while there, and on the trip back to Norman. We arrived back at our rental home a little before supper time and I laid down on the couch and took a short nap.

About 5:00 pm I awoke to find the right side of my face paralyzed. I could barely talk as my lips did not seem to work correctly and my facial muscles clearly were not working. Kathy got a neighbor's daughter to come babysit for us and she drove me to the Air Force hospital at Tinker Air Force Base, about a 20-minute drive from Norman. After checking in I was taken to an examination room, and for the next 20 to 30 minutes there was a parade of doctors coming in asking me to try to do this or that, but none seemed to be able to tell me why my face was paralyzed. Finally, the full colonel hospital commander and his deputy, a lieutenant colonel came in and started through the same routine – can you blink your right eye – no! Can you smile – no! after a few more of the same questions the door flew open and a major said in a rather loud voice, "Get your butts out of here, now!" Much to my shock and amazement the two colonels left without so much as a word.

The major introduced himself; I regret I cannot remember his name. My first words were, "Damn major, I've never heard anyone talk to two colonels like that and then have them scurry away without so much as a word." He simply said something to the effect, "Young man, they have no clue as to what is wrong with you, but I believe you have what is called Bell's Palsy." He went on to add something like, "And this is your lucky day because I am a reserve officer on my two weeks of active-duty training here, and my specialty at the Mayo Clinic is Bell's Palsy."

He proceeded to tell me that no one knows exactly what causes the disease to attack a person, but quite often it is associated with getting in and out of air conditioning. I confirmed that was exactly what I had done all day long. He went on to explain that the condition attacks the facial nerve and causes a sudden weakness in the muscles on one side of the face. He said that in more than 90% of the cases the weakness is temporary and significantly improves over some number of weeks. He further explained that as I could tell the weakness makes half the face appear to droop. Smiles are one-sided and the affected eye resists closing.

Now he said, "That is the good news, but the bad news is that in about ten percent of the cases the facial nerve that passes through the skull behind the ears swells up to the point that without intervention the nerve dies, and you are paralyzed for the rest of your life." He added, "Unlike that parade of doctors that were in here who had no clue what was wrong with you, we will ensure you are *not* going to be paralyzed for the rest of your life."

He went on to explain that every hour for the next 24 hours he would perform connectivity tests by placing a light electric current behind my ear on the facial nerve and see if he received the pulse in my face. If he received a pulse all was well. So, we began hourly tests. 6:00 pm, 7, 8, 9, 10 11, 12, and 1:00 am.

At 2:00 am the major said, "Sorry, but it appears you are part of the 10%, as I am not getting any connectivity. I have an operating room reserved and we are going there now." He went on to explain, "What I will do is cut behind your ear, separate your ear at the point where the facial nerve comes through the skull, and I will then carefully enlarge the hole with a drill allowing the facial nerve to expand or swell up without it being pinched by the bone. I will then sew your ear back in place and we will go back to conducting connectivity tests until late afternoon." I woke up about 4:00 am as the major was conducting his second test. He told me all was well and if nothing changed by afternoon, I could go home the following day and in six to eight weeks the paralysis should disappear completely. That is exactly what happened. Except for a scar behind my right ear, I have no evidence I ever had Bell's Palsy. Kathy, for years, would ask if I wanted to "Get Bell's Palsy again" if I went a day without shaving. In her mind it was not the getting in and out of air conditioning that caused my condition, but rather payback for having grown a beard. Not because of that, but I have never grown one again.

I truly enjoyed my classes at OU. The business management classes were my favorite, as I was learning something new about how to be a better manager in virtually every class. One instructor required us to read an article written by a very well-known time management specialist, William Oncken, Jr. entitled "The Theory of Monkey Management." A few years later, in 1974, Oncken summarized his article in the Harvard Business Review, and it has become part of management theory taught in business schools ever since.

As an army officer I found the theory relevant and eye opening. I tried to practice it throughout the remainder of my working life. Oncken concludes that managers should view every problem as a monkey on the shoulder of the person who has the problem. He states that most managers have no concept of what the subordinate is doing when he or she comes into their office. The subordinate will begin to explain the problem he or she has and wants some advice or suggestions as to how to solve the problem. As the discussion progresses, the monkey starts to climb off the shoulder of the subordinate and get onto your shoulder if comments are made like, "Let me think about a solution," or "Put that paper in my inbox and I will get back to you with some ideas."

The manager's goal, according to Oncken's "Monkey Theory," is to minimize or eliminate subordinate-imposed work and develop initiative in subordinates. He wrote that managers should learn how to get their subordinates to look after their own monkeys. He suggests statements like, "Please describe the problem as best you can." The manager should listen to see if the statement of the problem is clear. Then say something like, "What do you recommend be done to solve the problem?" Listen carefully, perhaps offer a suggestion or two, then say, "That sounds great. Please let me know what progress you make on solving the problem." The idea is to clearly decide on whether or not this is a problem that can be solved by the subordinate and to then issue guidance and ensure the "monkey" is firmly astride the subordinate when he or she leaves your office. Oncken used the analogy of the monkey on the shoulder to illustrate how managers spend much more time solving problems for their subordinates than they even realize. Unfortunately, many workers try to pass their problems over to other people, particularly their supervisors, to deal with, for various reasons, including but not limited to, fear of failure or the inability to take responsibility.

I was very fortunate to have a retired US Army full colonel as another of my management instructors. I rwegret I cannot remember his name. In one of his classes, he told us about the management style he used as a commander. He told us he liked to call it the "Whirling Dervish" style. He said he had adopted the name from a Turkish religious group that whirls around in a constant frenzy of action. He said that during duty hours he was constantly on the move visiting soldiers, NCOs, and officers in their workplaces. He would spend only a few minutes asking a soldier or NCO a couple of questions. If he was satisfied with the answers, he would compliment the person answering, and if he was not satisfied, he would tell the individual he expected a better explanation the next time he stopped by.

He said this method allowed him to evaluate every part of his command daily, identify high performers, and write down in his notebook those who needed additional training or counseling. He carried a notebook and let the person see him making notes which allowed him to review his activities at the end of the workday. He would return to his office at supper time, review his notes, and make a list of those he needed to counsel and compliment. Usually that was the NCIC or the officer in charge of the soldiers, NCOs, or junior officer who did both poorly and did well. He told us that one of the things he found lacking as a junior officer was any positive feedback. He said his commanders were quick to "chew his ass" but seldom said a positive word.

He explained to us that subordinates need positive feedback just as much as they need criticism. This brought to mind my experiences with 1LT Hogan in Germany and Cpt Floyd Whithead in OCS. I do not believe either ever said a positive or encouraging word. Later in my Army career I read a number of articles about a management style called "Managing by Walking Around" or "MBWA" written by the famous management author Tom Peters, author of *In Search of Excellence* and *A Passion for Excellence* plus 15 other books. I quickly recalled the professor who taught me the lessons I used throughout my army career and in my rise through the corporate world after retirement. Peters in short wrote, "MBWA is about listening, facilitating, teaching, reinforcing values, and giving feedback. These things can only be accomplished by means that are visible and tangible. A senior manager's ten-minute visit with a neophyte salesman — in the field, in his territory, far from the headquarters — is relayed within minutes around the entire organization." Peters suggests that managers should publicize the fact that they plan to spend more than 50% of their time out with employees, customers, and clients. I highly recommend anyone wanting to improve their management effectiveness to read his article, particularly the last page where he gives eight suggestions on how to immediately become more effective as a manager.

I had grown up reading series books like *The Hardy Boys*, *Nancy Drew*, *The Bobbsey Twins*, *Tom Swift*, and many more. I have continued my love of reading books throughout my life. After I became an officer, I developed a habit of visiting Goodwill and Salvation Army Thrift stores whenever I had the opportunity. They sold hard cover books for ten cents each, and I started buying children's books. By the time we moved to Norman I had perhaps 60 or 70 children's series books and numerous children's books we had purchased for Sherrin and then Shelley.

We visited an antique store in Lawton before I went to Vietnam, where I found and purchased my first four-section Globe Wernicke glass fronted, stacking oak bookcase for $75.00. Otto Wernicke first began building quality furniture in 1899 and went on to become the seller of the bookcases of choice for laws firms and private homes all over America. They came in solid oak and mahogany veneer. Not long after I arrived at OU, I saw an advertisement in a local paper that a law firm in Oklahoma City was selling an entire floor's worth of mahogany Globe Wernicke bookcases with seven sections and a top and bottom for just $50 a set. I had seen an oak one with just four sections for $250 so I knew this was a real bargain.

My wife and I knew we did not have money to spend on bookcases, nevertheless, the next day we went to look. The address given was a high-rise office building. We parked and went in to see the receptionist. She told us the law library was on the fifth floor and it was locked, but she gave us a key. She told us to let her know if we wanted to buy some and she would have someone put our names on the stack or stacks and we could come back to get them.

There must have been a hundred or more stacks of bookcases in the room. We spent the next hour and a half unstacking and restacking bookcases to get five near perfect sets. We paid the lady using some of our rent money and we hired a man to get them moved home for us. It took us a few years to fill them, but, because of this purchase, I started seriously collecting children's series books.

I found out from a magazine article that virtually all the books I had been reading were written by one man, Mr. Edward Stratemeyer. He started writing books in 1899 and quickly discovered there was a huge demand for children's books. He formed a syndicate, hired writers to prepare manuscripts based on plot outlines he gave them, and he would then review and modify them, and finally published more than 100 series under more than 50 pen names during his lifetime. We were fortunate that throughout my career the Army moved the bookcases and hundreds of books numerous times for us as part of our military housing benefit.

Oklahoma football was a big deal on campus. Every student received a season ticket and attendance was more than 60,000 for most home games. In 1970, 1971, and 1972, the head coach was Chuck Fairbanks, and his offensive coordinator was Barry Switzer, who would become head coach in 1973. My first season as a student saw the team post a 7-4-1 record losing to the number two and three teams in America – Texas and Nebraska as well as Oregon State and KS State. We ended the season ranked 20th in the nation and tied Alabama 24 – 24 in the Astro-Bluebonnet Bowl in the Houston Astrodome.

Of note during the 1970 season was Barry Switzer's introduction of the "wishbone" offense in the fourth game of the season. The 1971 season saw Oklahoma perfect the wishbone offense and win 11 games, losing only to the number one team in the nation, Nebraska, in a hard-fought battle 35 – 31.

Nebraska went on to become the national champion with Oklahoma being number two. Of note was the team's offense. They broke nine national offensive records including highest average points scored – 45 per game and most average rushing yards ever achieved at 472.4 per game. Seven of those records stand today as team records. One game in particular was quite controversial. The team had beaten KS State 32 years in a row from 1937 to 1967 and then lost the next two years. In 1968 the team lost in Manhattan, 21 to 59, and then in my first year at OU in 1969 we lost 14 – 19 in our stadium. Before the game in 1971, there was a lot of trash talk about them kicking our butt for the third year in a row as we returned to Manhattan. Unfortunately for them, the final score was 75 to 28 in our favor, and instead of putting the second and third strings in in the fourth quarter, the first team stayed in and scored three more touchdowns. There was a lot of complaining about "running up the score," but the only thing the team had in mind was revenge for the two previous seasons. The team played on New Year's Day in the Sugar Bowl in New Orleans and beat Auburn 40 – 22.

As I previously mentioned, I joined my first Kiwanis Club in Lawton. It was a very large club with more than 125 members. I saw several articles in the local paper about "The Big Red Kiwanis Club" of Norman shortly after I enrolled in school, so I contacted one of the members, Robert Flagler, who invited me to a meeting. Robert was in the OU Law School Class of 1970, became the Kiwanis Club President the next year, and later he became my divorce lawyer. I was invited to join and found that, unlike Lawton, it was a very small club with only about 25 members. We met weekly and carried out a few service projects. I was almost immediately asked to become Chairman of the Youth Services Committee that looked for ways to help underprivileged children.

Within a few months of joining, I was elected to the board of directors and later became vice president of the club. In August of 1971, I was asked to help found a Key Club in Norman High School, which is like a junior Kiwanis club. We needed to sign up 20 high school students to get the club chartered. On 22 September I drafted a letter to Kiwanis International requesting literature and informing them who my replacement would be as the Key Club Activation Chairman. I have since learned that the Big Red is now known simply as the Kiwanis Club of Norman and the Key Club in the high school is still active to this day. One of my last activities before I turned over the job of organizing a Key Club to another member was to speak to the entire student body of Norman High School. The following is the text of the speech I gave on November 3, 1971.

A CHALLENGE TO THE YOUNG MEN OF NORMAN, OK

If I asked you to name a period in history when young people were reveling; when the "Jesus Movement" was growing; when baseball and football stars caught the headlines; when talk of revolution ran through the land; when attitudes towards sex and marriage were liberalizing; and when discrimination, race, and crime were on everyone's mind, what would you say! Today? Yesterday? The 1960s? All these answers could be correct. But I have in mind the 1920s.

Babe Ruth, Red Grange, and Jack Dempsey reign as sports heroes, evangelism is at new heights, sex and petting are the newfound subjects of discussion — much to the dismay of parents and the delight of students; young people are caught up in prohibition, shorter skirts, new dances, and even women's liberation.

Why do I compare the 1920's with today? Because in Sacramento, CA, in May of 1924, an organizational meeting was held that has not been forgotten to this day. On that date a new club was organized in the Sacramento High School. The club came from a simple idea, an idea that could have been tried years earlier, and an idea that was to be the steppingstone for thousands of other clubs to follow.

What kind of club was it that would multiply from 11 members to 100,000 members throughout the United States and Canada and would still be with us 47 years later? The club was, and is, a junior service club. It was, and is, patterned after a club much like some of your fathers, uncles, brothers, or perhaps neighbors belong to today. These clubs have many names, and you may know some or all. Lions, Rotary, Elks, Jaycees, and in this case, the Kiwanis Club.

The club was there, and is here, to be unlike any other one in this high school. This uniqueness stems from three things, its sponsorship, its aims, and its programs. First the club is unique because it will be sponsored by your local Kiwanis Club composed of professional and businessmen of this community. Through this sponsorship and the associations developed, you will learn more about your city and how it functions. And you will have the opportunity to gain knowledge of various business and professions, one of which you may, because of this association, choose as your way of life.

Secondly, the club will be unique in its aims of developing initiative, leadership ability, and good citizenship practices through work in and for the community. These qualities are encouraged of each member accepting responsibility for the club's administration and activities. And third, since the club will be independent of the sponsoring Kiwanis Club, you will direct and control its programs and activities.

The organization of a new club faces the same challenges today that were faced in 1924. These challenges are apathy, resistance, competition, even jealousy, but perhaps most of all, a lack of understanding as to what the organizational priorities behind the club are to be.

It was for this reason that I was asked to speak today. I have been associated with Kiwanis since 1966 and with Kiwanis Key Clubs for about the same amount of time. In my profession as an Army officer, I have had the opportunity to see many young people coming directly into the service from high school and one can readily spot those soldiers that have had some opportunity to develop their leadership ability through this type of organization. This opportunity is available to each of the young men here today. The challenge is here, the doors to opportunity are open, and each of you could do something to make this world a better place in which to live.

Each of you knows the watchwords of today are pollution, poverty, drugs, and the environment. All these things, and more, are subjects that are challenged by an organization such as the

Kiwanis Key Club. The key to success, and that is why it is called a Key Club, is that we believe the true key to success for young people is the early development of leadership.

For those of you who desire to know more about this organization; for those who are willing to accept a challenge; are willing to work hard; are ready to take just a little bit of your time away from your car, sports, or friends; and just contribute a little bit of your time toward improving Norman, these are the individuals we invite to join the new Key Club of Norman High School.

You can see some of the young men sitting here behind me that have already accepted the challenge. The application to formally charter the club will be sent out in the near future and will contain the names of the charter members of this organization. Just as in 1924 there was an opportunity to become a charter member of the Sacramento High School Key Club, so to today you have an opportunity to become one of the first Key Club members in Norman High.

From this beginning the limits of your achievements and those of students to follow are bounded only by the efforts you are willing to expend. So, today I challenge you to take up this leadership, to assume this responsibility, to join the Norman High School Key Club, and to work to improve your school, your community, and your nation. Thank you.

After receiving orders in October to go to Fort Benning, GA, and being assigned quarters on post, we began discussing what to do with our home in Lawton. We decided to rent it out. We went to visit John Jones and Lon Parks, the owners of Parks Jones Realty in Lawton, They were the ones who had sold us our home. After meeting with them we decided to let them manage our property. They indicated that they could find military renters rather quickly as there was a shortage of rental properties in the area. I should point out that more than 50 years later, John Jones is still managing the company.

Shortly before I graduated, I was selected by the members of the Big Red Kiwanis Club to be their Kiwanian of the Year for 1971. I was presented with a very nice plaque. Two weeks after the speech I gave at Norman High School, our fourth child, Terrence William Sharpe, was born on November 15, 1971. A week later, our second daughter, Shelley, came down with chicken pox but fortunately none of the other children contracted it. I graduated with my bachelor's degree in business management on December 27, 1971, and a week later we packed up our household for a door-to-door move to our assigned quarters on Fort Benning, GA.

Beta Sigma Phi Sorority Party at our home.

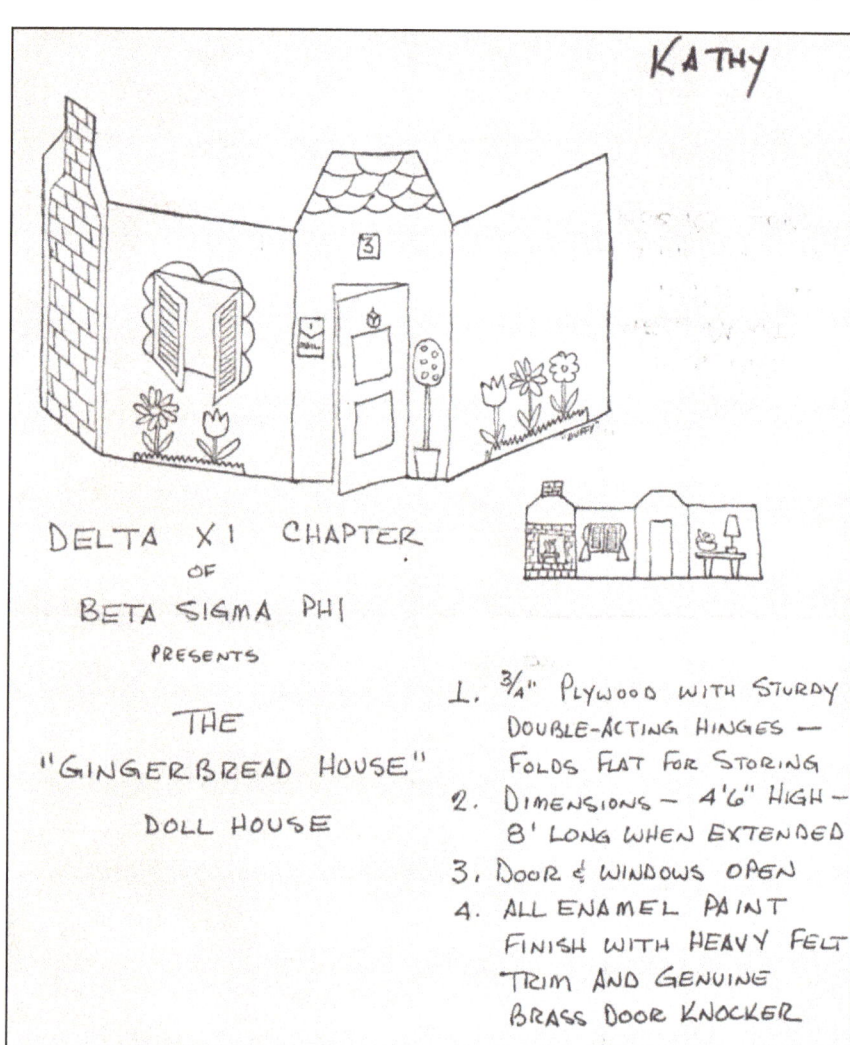

KATHY

DELTA XI CHAPTER
OF
BETA SIGMA PHI
PRESENTS

THE
"GINGERBREAD HOUSE"
DOLL HOUSE

1. ¾" PLYWOOD WITH STURDY DOUBLE-ACTING HINGES — FOLDS FLAT FOR STORING
2. DIMENSIONS — 4'6" HIGH — 8' LONG WHEN EXTENDED
3. DOOR & WINDOWS OPEN
4. ALL ENAMEL PAINT FINISH WITH HEAVY FELT TRIM AND GENUINE BRASS DOOR KNOCKER

Building the Gingerbread House.

Beta Sigma Phi Sorority fund raising project. I built the doll house and Kathy did the painting.

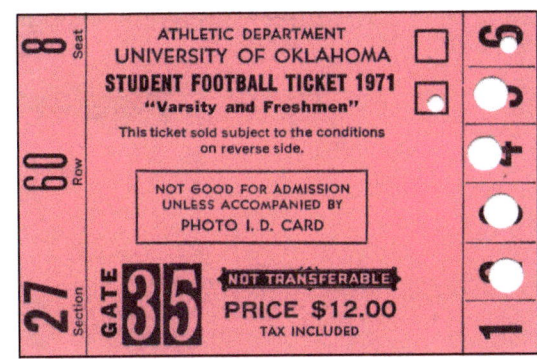

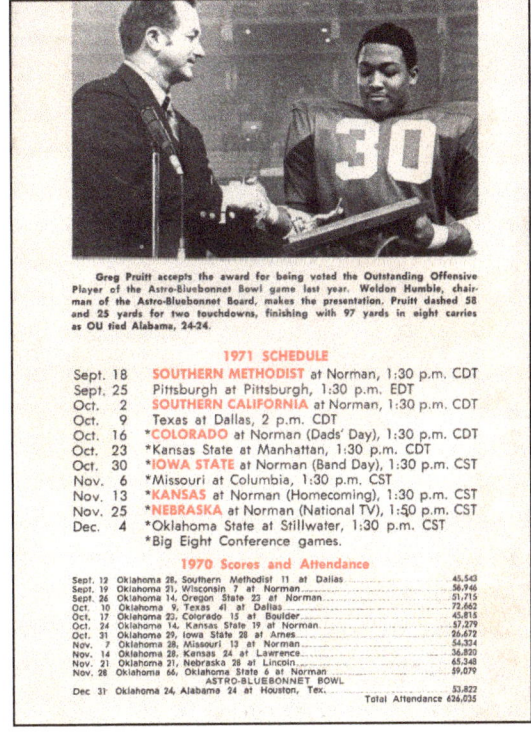

OU Football Media Guides for 1970 and 1971. Football season tickets.

Oklahoma University Football Stadium following a 1970 game, Taken from a news helicopter

Walter Conkite, September 1937, calling OU football games. Me, bearded, before Bell's Palsy

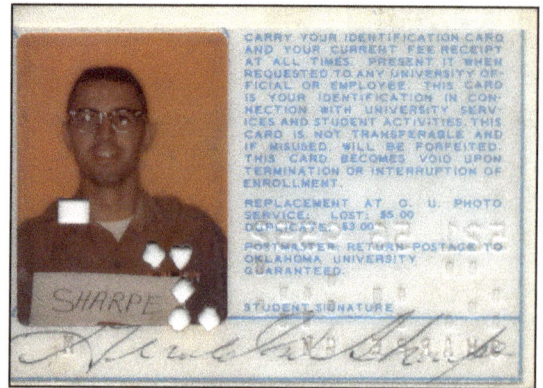
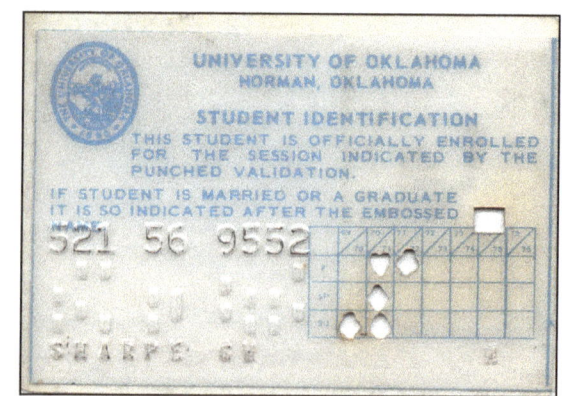

My Oklahoma University Identification card

TOM STEED
4TH DISTRICT, OKLAHOMA

Congress of the United States
House of Representatives
Washington, D.C. 20515

November 19, 1971

Capt. and Mrs. Gerald W. Sharpe
1230 Leslie Lane
Norman, Oklahoma 73069

Dear Friends:

Since a new baby has recently arrived at
your home, I believe the enclosed government publications, "Infant Care," and
"Your Child From 1 to 6," may be helpful
to you.

Let me take this opportunity to express
my best wishes and my hope that the baby
will have a happy and useful life.

Sincerely yours,

Tom Steed

TOM STEED, M. C.

TS:ag
Enclosures

Note from our local congressman on the birth of Terry.

Captain and Mrs. Gerald W. Sharpe
proudly announce the birth of

TERRENCE WILLIAM

on

15 November 1971

7 lbs., 11 oz.

Birth Announcement of Terrence W. Sharpe.

University of Oklahoma Association

Makes Known To All Men By These Presents That
GERALD WILLIAM SHARPE

Has Completed The Requirements For The Degree Of
BACHELOR OF BUSINESS ADMINISTRATION 1972

At The University of Oklahoma
And Has Accordingly Been Awarded The Degree
With All The Honors, Rights and Privileges
Pertaining Thereto.

The back of my OU Association card.

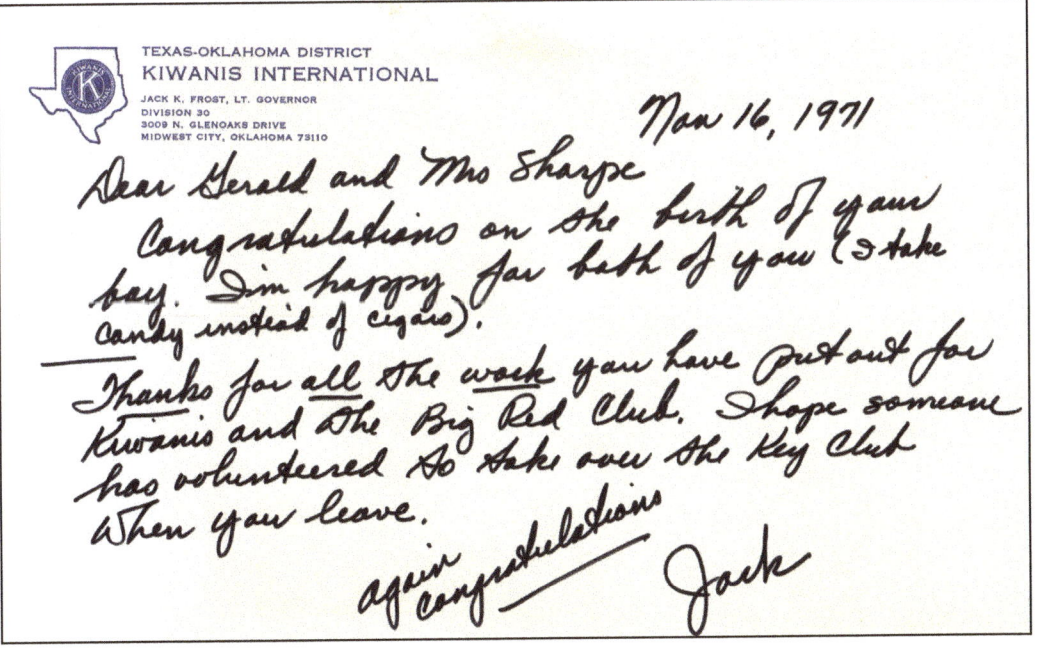

Note from the Kiwanis District Governor on the birth of Terry.

TEXAS-OKLAHOMA DISTRICT
KIWANIS INTERNATIONAL

JACK K. FROST, LT. GOVERNOR
DIVISION 30
3009 N. GLENOAKS DRIVE
MIDWEST CITY, OKLAHOMA 73110

Nov 16, 1971

Dear Gerald and Mrs Sharpe
Congratulations on the birth of your
boy. I'm happy for both of you (I take
candy instead of cigars).
Thanks for all the work you have put out for
Kiwanis and the Big Red Club. I hope someone
has volunteered to take over the Key Club
when you leave.

again congratulations
Jack

ALUMNI ASSOCIATION

This certifies that

GERALD WILLIAM SHARPE
Is a Life member of the

**University
of Oklahoma
Association**

Jack J. Ledbetter

Executive Director

My OU Association card.

  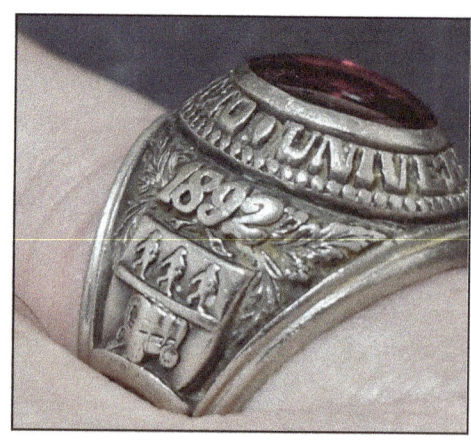

Class ring purchased to commemorate my graduation.

Student protesters were out on the campus almost daily. Colonel Ladd, (left) the ROTC Commander, asked me to stop wearing my uniform to classes, but I declined, believing it was a positive thing. There were anti-war demonstrations on campus often, but I was never caught up in one.

Sanders, Samuel, A&S, '71, Waldo, Ark.
Saunders, Becky, Kappa Delta, Pharm, '71, Oklahoma City
Schiff, Shirley, Ed, '71, Duncan
Schopflin, William, Sigma Chi, Engr, '71, Kansas City, Mo.
Schuepback, Charlotte Braly, Ed, '71, Killeen, Tex.
Schug, John, Delta Tau Delta, Bus, '71, Oklahoma City

Schultz, Larry, Engr, '71, Great Bend, Kan.
Scism, Stanley, A&S, '71, Tecumseh
Scoggins, Jeff, Sigma Alpha Epsilon, A&S, '71, Edmond
Scott, Warren L., Bus, '71, Oklahoma City
Sebert, Ginger, A&S, '71, Tulsa
Sharpe, Gerald, Bus, '71, Lawton

Shelden, Michael Von, Bus, '71, Oklahoma City
Sheppard, Michael, A&S, '71, Midwest City
Shipley, Darlene Baldwin, Ed, '71, Wister
Shipely, Frank, Engr, '71, Poteau
Smuller, Margaret, Ed, '71, McAlester
Shumard, Doug, Bus, '71, Jackson, Miss.

Student photos from the annual yearbook.

KIWANIS INTERNATIONAL

KIWANIAN OF THE YEAR AWARD

PRESENTED TO
CAPTAIN GERALD W. SHARPE

BY THE
KIWANIS CLUB OF
THE BIG RED, NORMAN, TEX-OKLA DIST.

1971

Award for establishing a "Kiwanis Key Club" in the Norman High School. It is still an active Key Club.

Oklahoma University annual yearbook.

My first Globe Wernke bookcase.

The Oklahoma State Regents for Higher Education
acting through the

University of Oklahoma

have admitted

Gerald William Sharpe

to the degree of

Bachelor of Business Administration

and all the honors, privileges and obligations belonging thereto,
and in witness thereof have authorized the issuance of
this Diploma duly signed and sealed.

Issued at the University of Oklahoma at Norman, Oklahoma on the
twenty-third day of December, A.D. nineteen hundred and seventy-one.

For the State Regents:

Chairman

Secretary

Chancellor

For the University:

President Board of Regents

President of The University

Academic Dean

University of Oklahoma, Bachelor of Business Administration diploma.

Sean playing in our moving boxes in our Norman, Oklahoma rental house, June 1970.

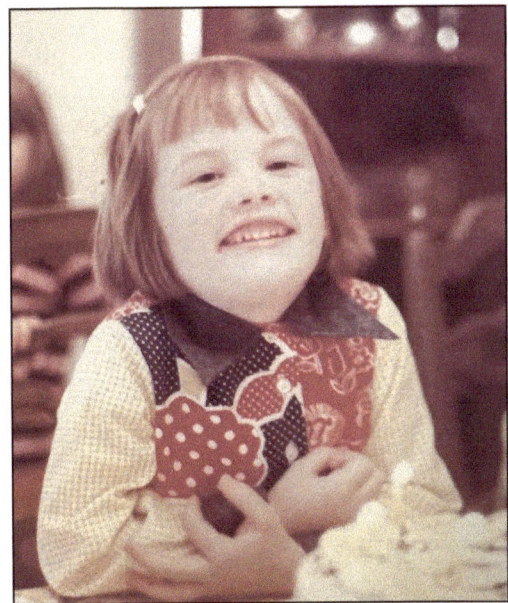

Shelley's fourth birthday and Shelley at home.

Sean playing in in front of our fake fireplace and asleep in his highchair.

Sean trying out a piano. Children sitting on front steps of our Norman home.

Sean's second birthday and standing by his toy camel from Turkey

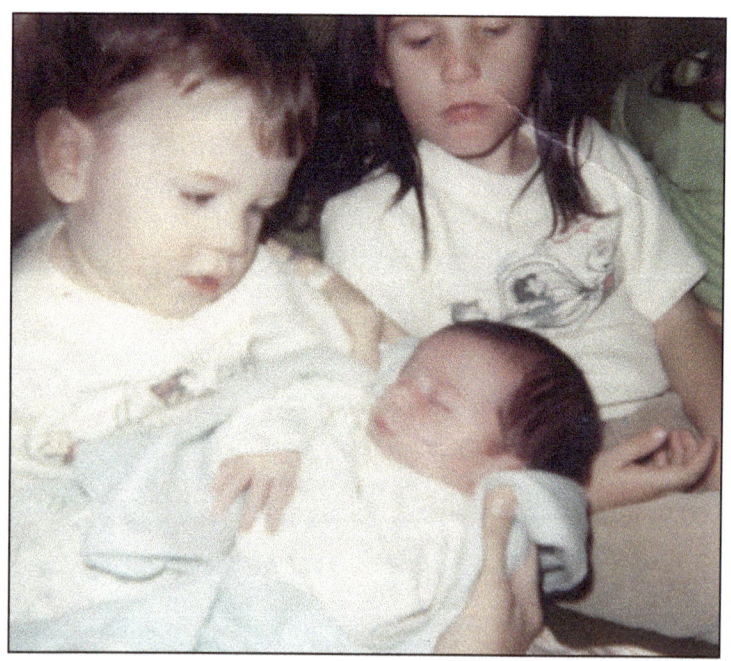

Sean and Shelley holding their brother Terry for the first time

Kathy with all four children together at home for the first time.

ACKNOWLEDGMENTS

I COULD NOT HAVE WRITTEN THIS AUTOBIOGRAPHY WITHOUT THE LOVING HELP OF MY SISTER
Eileen and her husband Ron Fournier who have contributed countless hours to editing, finding family photos, and offering suggestion on how to improve my story telling. One of Eileen's greatest contributions was researching and writing the three family appendices included at the end of the book. Likewise, my uncle Ray Ritchie, even though he is 95, reviewed and edited early drafts of this first book.

My four children not only reviewed what I wrote and offered suggestions, but recalled events in their early lives that I had completely forgotten. They combed through photo albums and provided many of the pictures from their childhood.

Arthur "Art" Jacobson, my high school classmate, has contributed many of the photos in the high school chapter of this book. Art, Willard Staats, and I met at Willard's house two years ago and shared many stories. Willard, who recently lost his wife, and Art remain among my best friends.

As I mentioned in my dedication, I would like to acknowledge the contributions of the following Pillar of Fire ministers to my life and early education:

To Reverend Albert Wolf and his wife Thelma, who became like second parents to me. I spent more time at their house than my own. He ran the farm and had infinite patience in teaching me how to do things. He taught me to drive at an early age and trusted me to drive trucks and tractors about the farm. I loved the man, respected him, and enjoyed my time working for him.

To Reverend Ray Sharpe who taught me how to do maintenance on plumbing and electrical switches, how to patch holes, to trim trees, to use ropes and pulleys to lift heavy objects, how to do a basic survey, and in general, how to do a multitude of other things.

To Reverend Reuben Truitt who taught me how to swim

To Reverend Robert Dallenbach who showed me how to take photos, critiqued what I did, and taught me how to be a radio announcer for classical music programs on the church radio station, KPOF

To Reverend William Staats who ran the church garage and taught me about basic maintenance on vehicles,

To Reverend Overk, whose first name I never knew, who taught me how to lay brick and concrete block while he was helping dad build the extension on our house.

Bob and Mark Babcock, and Matt King of Deeds Publishing, have shown patience and understanding in shepherding me, a first-time author, through the publishing process. They have done a marvelous job of producing this book.

Last but not least, my partner, Myla "Mae" Figueroa has read and edited every word of this book multiple times. She has a wonderful understanding of the English language and has offered suggestions for improvement in every part of the book   I could not have done this without her loving help and assistance.

ABOUT THE AUTHOR

GERALD W. "JERRY" SHARPE IS AN 83-YEAR-OLD RETIRED US ARMY COLONEL. BORN TO OLD-TIME Methodist ministers, raised in an Amish like environment, and educated in church schools. He enlisted following two years of college at Western State College in Gunnison, CO. Jerry became an outstanding enlisted soldier and was offered the opportunity to attend Artillery OCS where he was a distinguished graduate. He subsequently served tours at or in Fort Sill; Turkey, Vietnam; Fort Benning, Fort Bragg; Germany and Korea. He was married twice for 22 years each and raised four children. He has 12 grandchildren and eight great grandchildren. He became active in civic affair and received numerous local, state, and national awards. His career was marked by excellent performance at every rank up to full colonel. He received his bachelor's degree at Oklahoma University and his master's degree from Shippensburg University while attending the US Army War College. He was poised to be a general officer, but a new Chief of Staff of the Army wanted younger generals, effectively ending his and hundreds of colonel's careers. He retired in 1989, worked for several large corporations, and is living in paradise with his partner, Mae, in the Philippines.

APPENDIX 1

L to R: Sunday, Ed, Mary Jane, Ivy, Woodrow, and Bessie

SHARPE FAMILY HISTORY IN THE PILLAR OF FIRE 1926–1940

THE PILLAR OF FIRE IS A METHODIST-BASED CHURCH organization founded in 1902 by Alma White, the first woman to become a bishop in the United States. By the late 1920's the church owned schools, colleges, radio stations and missions throughout the U.S and in London, England. Bessie Sharpe and five of her six children arrived at the church headquarters in Zarephath, NJ on Oct 8, 1926. A picture possibly taken the same day

Bound Brook school, auditorium and housing.

Having suddenly and traumatically lost his father at the tender age of five in the "Spanish flu epidemic" of 1918-1920, Sunday had to grow up quickly and assume many responsibilities along with his older siblings on the family tobacco farm. After they left the farm and were separated, he ever after missed the lost family closeness and was often drawn back to his early North Carolina home. He was very close to his nurturing, encouraging mother, even though they were separated for long periods of time. As the youngest boy in the family, he looked up to his older brothers and yearned for approval. He became very serious-minded and emotionally reserved. As a father of five boys, he probably unconsciously tried to recreate the family farm environment of everyone working hard together. He instilled in his children a strong work ethic and did not spoil them, even though he left much of the discipline to Thelma. He was highly competitive, especially with his sons, and yet he was intensely proud of them.

Mary Jane and Sunday, probably in 1927

Zarephath campus: Liberty Hall on right, Main Bldg. in front

Above: Sunday and Bessie (early 1930's).
Below: Clockwise from left: Hazel (1930), Elton & Nellie Gray (1930); Ivy (in middle) in LA with two co-workers (early 1933 before the earthquake); Sunday and Mary Jane in Cincinnati (1933)

shows the youngest child, Mary Jane, with a big smile, perhaps because it was her eighth birthday. Sunday was just twelve and looks very unhappy and about to cry, probably wishing he were back home in NC.

The family was immediately separated. The two older children, Ivy (17) and Ed (almost 16), lived in the school dormitories and attended high school on the main campus. Bessie, Mary Jane, Sunday, and Woodrow (almost 14) were assigned to live in a three-story building in Bound Brook, about three miles from the campus.

At the August 1929 Camp Meeting, Bessie was reassigned to Wilkes-Barre, PA, while Ivy was "consecrated" as a deaconess and spent the next year at Za, probably taking college courses. She and Mary Jane (age 12) appear in the Apr. 1930 census living in the "Academy Lodge" (probably what later became known as the Main Building). In the same census Edward (19) and Woodrow (17) are shown living in Liberty Hall, the boys' dormitory. They are both listed as "ministerial" students.

Around this time, Sunday's unhappiness must have begun affecting his behavior. Any student who got in the slightest trouble, (even for so much as being caught smiling at the opposite sex) would often

be "sent off the place", and Sunday at some point was sent to live with his mother. The April, 1930 census shows Bessie and Sunday living in Wilkes-Barre, PA. Bessie is classified as a "roomer" with the occupation of "missionary". Sunday's age is listed as 13 although he was actually 14. Often part of the punishment for misbehaving was a denial of schooling, so he may have missed a whole year of school. We don't know where he finally began high school in 1931, but his problems must have continued, because in 1932 he relocated to North Carolina to start the 10th grade, living with his oldest sister Hazel, who had two small children of her own—Elton (6) and Nellie Gray (3).

The first family member to leave the church was Ed in February of 1931. According to Burton, his departure was caused by a disagreement with the church over his interest in Gertrude Speer who became his wife two years later. Before leaving the Pillar, Ed completed two years of college and Gertrude finished four years of high school (according to the 1940 census).

In June of 1930 Ivy was sent to Detroit for the summer and then on to Cincinnati for the school year. She served in Huntington, WV during the summer of 1931. When Ivy left Cincinnati, Bessie was assigned to take her place for five months. Then Bessie returned to Wilkes-Barre where she remained for several years except for visits to NJ and NC. On Aug. 31, 1931 Ivy departed for Los Angeles where she spent the next 2½ years.

Mary Jane was sent to the Pillar school in Cincinnati in Sep. of '32 at the age of 14, while Sunday was attending school in NC. After one semester, Sunday relocated to Cincinnati to finish the 10th grade. Both Sunday and Mary Jane returned to Za in the summer of 1933. Sunday completed his senior year of high school at Za and graduated in June of '34. Mary Jane, however, did not make it through the year. She probably got in trouble over a boy and was sent to Wilkes-Barre on Feb. 23, 1934. She returned to Za for her brother's commencement, no doubt along with Bessie.

Ivy returned to Za from LA in April, 1934. The whole family was together for the first time in years when Sunday graduated from high school. However, Ivy was under a severe mental strain, possibly from a broken relationship or maybe

Hazel, Gertrude & Ed outside Hazel's house in NC, May, 1934

Gertrude, Bessie, and Mary Jane holding Burton on Ed's front steps in NJ, 1936

Mary Jane, Ed, Nellie Gray, Ivy and Hazel in NC, 1937

Roy Brice

Ray B. White

Bishop Alma White

from a trauma suffered during the Mar. 10, 1933 disastrous earthquake in Los Angeles. She had a breakdown of some kind and was committed to Greystone Park Mental Institution on Aug. 22, 1934 (Sunday's 20th birthday). She would remain there for one year. This was a very difficult time for the whole family, especially since Alma White told the family members that they needed to get "prayed up" and become "fully consecrated" to the work before Ivy would get better. Both Woodrow and Sunday "prayed through to victory" while Ivy was in the hospital and Sunday felt a call to be a preacher.

In Sep. of 1934 Sunday and Mary Jane returned to Cincinnati as workers. Mary Jane at age 17 had not yet graduated from high school. At the same time Woodrow, who had graduated from high school, was sent to the Colorado headquarters as a painter (according to the 1940 census). On Jan. 1, 1935 Sunday began his first diary. According to his entries, Mary Jane left Cincinnati for Za on Aug. 21, 1935, the day before Sunday's 21st birthday. A week later he received word that his mother and Mary Jane were leaving the Pillar for North Carolina, taking Ivy out of the hospital to go with them. Ivy and Mary Jane never returned to the church.

Mary Jane returned to New Jersey within a few months and finished high school while living with Ed and Gertrude. She then got a job at Bell Telephone as an operator. By April, 1940 she was sharing an apartment on the main street of Summit with a pharmaceutical bookkeeper. One month later on May 30, she married Don Cruver, and the couple moved into an apartment in Plainfield. Ivy probably stayed in North Carolina for several years. Around 1939 or 1940, she returned to New Jersey and lived with Ed and Gertrude in their new house in Gillette. She got to know Theede (Ted) Rystedt, who she married in 1946, while living with Ed. She would pick up the milk and eggs they purchased from Ted's farm a couple of blocks from Ed's home. When the war started in 1942 she moved into her own apartment in Summit after she got a job at Ciba Pharmaceutical Company which was making medical supplies for the military. We don't have a record of when Bessie returned to the Pillar, but it was probably within a year or so. In 1940 she was assigned to a church in downtown Denver as caretaker. The census for that year shows she had completed just one year of high school.

After high school graduation in 1934, Sunday was sent to Providence, RI for the summer to learn how to do missionary work from a Pillar missionary, Roy Brice. Missionary work consisted of selling subscriptions to Pillar magazines, talking to people about the work of the church, and soliciting funds. The Pillar operated as a "volunteer" organization in that no one received a salary. A portion of all funds taken in were to be used for the upkeep of the local "home/church" and expenses for the missionaries. The balance of the money (along with a report of the entire sum) was to be sent weekly to headquarters. From Providence, Sunday moved to the Cincinnati school campus in Sep., 1934, where he served as a "jack of all trades" in addition to being a missionary in training.

Sometime toward the end of 1934, Sunday was saved under the preaching of Ray B. White, Alma's son. Afterward in a letter to Ray Sunday wrote: "You have had a greater influence over my life than any other person I know." Ray wrote to Sunday "It does me good to know that you have had a touch on your soul, and that you feel the call to preach...Keep at it boy; there is a future for you." In a Dec., 1934 letter, Ray smoothed over a denial of Sunday's request to come see his sister Ivy with: "We are all much encouraged about your spiritual activities and development...Sunday, you are on the highway to sanctification. I have great faith for your future." Around the same time Bishop Alma White also wrote to Sunday saying that Ray had told her a year earlier that he knew God had called Sunday to preach. Without giving any details, she expressed appreciation for how Sunday had "helped to straighten up things concerning Samuel Ingler" (the son of a missionary couple). She continued, "Yes, Sunday, I feel sure the Lord has called you to preach and will do what I can to help you in any way to make good." These letters are the earliest in Sunday's collection of over 400 letters from his early days in the Pillar. He obviously treasured the letters, looked up to the writers, and would hold Bishop White to her promise to help him on numerous future occasions.

The Pillar of Fire grew out of the "holiness movement" that arose in the 19th century among Protestant churches in the U.S. This movement was characterized by a doctrine of sanctification as a second distinct "work of grace" after salvation, as well as by a strict separation from worldly culture and values. They also believed one could lose salvation by backsliding. Most other Protestants believe that the believer is "secure" because all his sins have been covered by the sacrifice of Christ and that sanctification is a life-long process of spiritual refinement as the in-dwelling Holy Spirit directs more and more of a person's attitudes and life. Unfortunately, the Pillar doctrine led to people never being sure about what state of grace they were in, whether saved, backslidden or sanctified. Thus some spent years praying over and over to obtain a gift from God that would take away their sinful thoughts and tendencies. This made for very insecure Christians with an inward focus on their own minds and hearts, instead of an outward focus on sharing the Gospel of faith in Jesus with a sinful, "worldly", and often downtrodden world.

Alma White was a charismatic leader and influenced many to follow Jesus. But her ministry (as we read in Sunday's diary) probably over-emphasized negative "battles" against "worldliness", Catholic Romanism, and modernistic ministers, denominations and schools. She was self absorbed and needed to control the church and everyone in it. She as much as preached that the Pillar was the only organization promoting "true religion". This self-absorption and need to dominate and be in charge was passed down to her son, Arthur, who led the church after Alma and Ray died in 1946. One of his primary objectives was preserving "his Mother's work" and so the organization became somewhat "backward looking" instead of moving strongly forward. It has been gradually shrinking ever since.

The church always emphasized "missionary work" but the financial needs of the growing organization led Alma to promote heavily the necessity of raising funds for her many expansions and building projects. The course of Sunday's life work was pretty much set on June 16, 1935 when she had a private meeting with the "older workers" in Cincinnati stressing how the temporal (financial) and spiritual go together. Sunday ominously noted: "She as much as said Norman & I had to go out in the missionary field." Norman Sillett would become Sunday's close companion and best friend on missionary trips for the next thirty years or more. Norman was easy-going with a wry sense of humor. Sunday was a take-charge person who had no difficulty asking for free hotel rooms or anything else. They were equally successful at soliciting even though they both tended to be reserved.

Up to that point Sunday had been working with four or five different older ministers on one or two day missionary trips. But over the following three months Norman and Sunday launched out on several multi-week trips totaling nearly 3,000 miles. They stayed in different hotels almost every night which Sunday usually got donated. They sold hundreds of Pillar magazines and sent hundreds in donations back to Zarephath. There were occasional opportunities to talk to someone about spiritual matters, but the primary focus was distributing magazines and collecting donations. Sunday proved very adept at asking for donations of materials, services and money. He seems to have enjoyed the travel and the opportunities to talk to people, but he also felt the time spent on the road took away from his calling to be a preacher.

In the midst of these trips came the departure of Bessie, Mary Jane and Ivy for North Carolina. Sunday was in turmoil about what to do. He was torn between his loyalty to his biological family and his family-like loyalty to the church and his co-workers. He felt he didn't have any choice but to leave the church. He tried to find a job somewhere in Cincinnati and when that didn't work out, he left Cincinnati for a month and travelled to NC to be with his family and look for a job. He had just turned 21 and was, no doubt, ready to find a mate and start his own family. He had gotten in trouble over several girls in past years and there was someone in NC he had been writing to. Part of the reason he went to NC was to see if M.W. (her full name doesn't appear) was seriously interested in him. Perhaps fortunately (for us children) his reception from M. W. was "rather cold", probably hastening his return to Cincinnati on Nov. 15th.

Sunday was very guarded about revealing too much in his diary. Evidently his emotional repression and introspective nature only increased as he matured. After all the trauma and change in his life in 1935, he doesn't mention any of it in his summary of the year. Instead he lists nine news items from the year and only one personal memory of note: "I have done quite a lot of traveling and have seen a number of sights." If Sunday kept a diary in 1936, it does not survive. He worked in Cincinnati until September 1937 when he returned to Za to begin college. He completed five semesters before his studies were interrupted by the blowup over his relationship with Thelma Ritchie.

APPENDIX 2

RITCHIE FAMILY HISTORY

Abovel left: Hugh Hunter Ritchie, age 18, in Mobile, AL, 1913. Above right: Marion Jane Lewis, about age 12, in Mobile AL with a neighbor child (about 1911).

THELMA EDITH RITCHIE WAS BORN ON SEP-tember 29, 1919, in Mobile, Alabama. Her father, Hugh Hunter Ritchie, was born in 1895 near Mobile. Hugh's father was Mortimer Ritchie (1851-1932), a farmer and sawmill worker. The Ritchie family had been in Alabama for at least two generations. Mortimer and his wife, Evie, had ten sons in twenty years. Most of them helped with the family farm as youths, worked in sawmills in their twenties, and later worked in the shipyards. Hugh was the seventh child.

Hugh's grandfather, James Uriah Ritchie (1813-1874) was born in So. Carolina but was living in Alabama at least by 1826 when his father (also named James Uriah) died. James married Theresa Emma Sobieski Traywick in 1835 at age 21. They had twenty children in thirty years, 17 boys and 3 girls. The oldest boy died in the battle of Gettysburg. The ninth child died in the Battle of Spanish Fort, outside Mobile near the end of the Civil War. Mortimer (Hugh's father) was the 13th child. Theresa was six years younger than James Uriah. She bore twenty

Marion and Hugh Hunter Ritchie, taken in Monroe Park in Mobile, perhaps on their wedding day, Oct. 18, 1918, when Marion was 18 and Hugh was 22.

Above: Thelma, Aunt Mary and Hunter taken in Belle Isle, Detroit, MI in the summer of 1922. Below: Aunt Mary, Marion holding Irene and Hugh. Hunter and Thelma as young children – probably taken at Zarephath camp meeting in 1924.

1924: Marion, Irene, Hunter and Thelma at Zarephath.

children and yet outlived him by 18 years, living to age 87.

Thelma's mother, Marion Jane Lewis, was born July 23, 1899 in Chicago. Her father, Edward Wallace Lewis (1870-1960), was born in Alba, Pennsylvania. He was always called Wallace or Wally. He married Gertrude Kellow (born in Scranton, PA) in 1894. They moved to Chicago, IL very soon after, and their first three children were born there: Richard (1895), Leon (1898), and Marion (1899). In 1900 Wallace is listed as chief clerk of the RR. Around 1904 the family moved to Mobile, AL because they had relatives near there. In 1905 a son, Gayle, was born and in 1908 their son Leon died. Gertrude Kellow died in Aug. of 1917. In the 1920 census Wallace's occupation is listed as blacksmith.

Marion and Hugh met in Mobile, AL. They married on Oct. 18, 1918. Hugh was listed as a carpenter on their marriage license. Thelma was born almost a year later on Sep.29, 1919. In Jan. 1920, the census lists "Hugh Richey" as a battery helper in the shipyard. They moved to 418 Bishop St., Houston, TX sometime that year and Hugh Hunter, Jr. [always called Hunter] was born there on Dec. 13.

Marion's unmarried aunt, Mary Lewis (Wallace's sister), joined the Pillar of Fire in 1921 at the age of 34. In the summer of 1922, the family visited her in Detroit, MI. She must have been serving there as a missionary at the church and school. She was probably encouraging Marion and Hugh to join "the work". But in 1923 they were again living in Mobile when their third child, Irene, was born on Nov. 23.

Hugh and Marion probably joined the Pillar of Fire, in Aug., 1924, at camp meeting. A picture of the young family with Mary Lewis in uniform was probably taken at that time because Irene looks about nine months old. The Ritchies lived at Zarephath for at least a year. There is a picture of Marion and the three children taken during winter outside the Assembly Hall at Zarephath. It must have been taken in early winter of 1924 because Irene looks about a year old.

In Feb. of 1925, Marion had a fourth child, Calvin, who lived only 17 days. He is buried at the Zarephath cemetery. The Ritchies were probably at the first camp meeting Bessie Sharpe attended in Aug. of 1925. We don't know if Bessie brought her children along, but if so, it's possible Sunday and Thelma could have seen each other as young children. Camp meeting was always the time in the P of F when "appointments" were made to the many branches of the church.

The Ritchies were probably assigned to Denver, CO after the 1925 camp meeting. On May 5, 1926, a fifth baby, Esther Mae, was born and lived only four days. The death certificate shows the family living at 1845 Champa St., Denver (the P of F headquarters in Colorado). The family remained in Colorado for about six years. Some of that time they served at the P of F church in Colorado Springs (226 S. Weber St.). On April 15, 1929, a sixth child was born there. Ray is the only surviving Ritchie sibling. He is now 95 and lives in Hemet, CA.

The census taken on April 29, 1930, records Hugh, Marion and the four children living once again at Champa St. in Denver. Hugh was listed as a missionary. Three months later, on July 23 (her birthday) Marion had her last baby, who unfortunately was born dead. In 1930 and 1931 Hugh served as pastor of the Pueblo, CO church at 819 Court St. In early 1932, Hugh and Marion left the Pillar and returned to Alabama where they lived at 55 N. Holmes in Prichard, a suburb of Mobile.

Thelma began the second half of the fifth grade at the Prichard School in 1932. She completed sixth grade the following year and in 1934 entered Murphy High School —a brand new re-

Above left: Thelma probably at graduation from seventh grade in 1933 at age 13. Above right: Thelma in Feb. 1938 as a high school senior at Alma Preparatory School (age 18).

Ritchie Family at Zarephath: Hunter, Thelma, Ray, Hugh, Marion and Irene; probably summer of 1937.

Left: Thelma as a senior. Right: Thelma at high school graduation.

gional school in downtown Mobile. By 1937 the family had moved to 205 W. Main in Prichard and Hugh was listed in the city directory as a [house] painter.

For unknown reasons in the summer of 1937, the family returned to the Pillar, and Thelma entered her senior year at Alma Prep School at Zarephath, NJ. The three older children lived in the school dorms, and Ray, age 8, lived at the Bethany Boys Home run by the Weaver family. He was separated from his parents and siblings and describes the home as an orphanage. Hugh worked as a painter in various church locations in NJ and PA. Sep. 1937 is the same time that Sunday returned to Zarephath and entered Alma White College as a freshman.

Thelma graduated from high school in June of 1938. She entered Alma White College in the fall. For the next two years she took a full load of classes, went to summer school, had various duties, did missionary work, and sang in the girls' choir. Most stressfully, she taught young children every morning. She wasn't given any training for this or much support from the administration. She had a difficult time controlling the class. From her 1940 diary we know she had a terrible time in the classroom. When she sent a misbehaving student home, she got "jumped on" by Mrs. Gertrude Wolfram who told her she needed to change her methods and apparently recommended not punishing the children.

She tried this approach, and the results were even worse. She had a bit of a breakdown and fortunately received some encouragement from Sister Della and Miss Kubitz. She went back to the class with a new determination to make the children mind. "When they saw I really meant business they settled down some."

Alma White College Freshman Initiation Banquet, Oct. 20 1938. Above: Thelma is the second girl from the left. Below: Sunday is the fourth male head from the left (partially obscured by Carolyn Staats); Thelma is the second girl from the front with dark hair parted in the middle. This is the first and only picture with Sunday and Thelma in the same room before they were married.

The story of Thelma and Sunday's courtship and marriage will be the next installment of the story.

SOME EARLY HISTORY OF THE LEWIS FAMILY

Marion's father was Edward Wallace Lewis (called Wallace) and his father was Edward Livingston [E. L.] Lewis (born in Carbondale, PA in 1843). E.L.'s family moved to Troy, PA when he was a boy. Enlisting at age 18, E.L. served in the Union army as a drummer. Some letters survive that give details about his service. They will be uploaded. He married Minerva Crandall in 1877, lived in Troy, PA until around 1883, and had eight children in twenty years. Two died in childhood. The family moved to Chicago sometime before 1884, probably at the same time as Wallace. E. L. and his son Leon are listed as "stationary engineers". One of his sons died on his wedding day in Chicago in 1894.

The 1900 census shows the E. L. family lived at 6438 Normal Ave in Chicago. The Wallace Lewis family lived about 25 miles away at 9327 Victoria Ave, probably where Marion Jane was born. Soon after 1900, both families moved far south, a pivotal decision leading to the meeting of Marion and Hugh Ritchie. The 1910 census shows that E.L's family moved to Loxley, AL just east of Mobile where he became a farmer. Also in 1910, Wallace's family lived in a suburb of Mobile very near the Hugh Ritchie family farm, and Wallace worked as a blacksmith. This is probably the time and location when Hugh and Marion met. In 1917 Wallace's wife died and in 1918 Marion married Hugh Ritchie. In the 1920 census Wallace had moved to a Mobile house two blocks from the river and was working as a boiler maker in the ship yard.

By 1920 E. L. (age 76) was back in Troy, PA living with his unmarried children, Leon and Mary. Mary joined the Pillar of Fire in 1921, but she had evidently been supporting the work by 1918, when her name appears in the P of F magazine as having pledged $100 toward the mortgage on the Zarephath property. She faithfully served in Detroit and other Pillar missionary homes until May 25, 1935, when she was assigned to Colorado. She spent the rest of her life there, dying in May, 1974, at age 89. Her brother, Wallace joined the Pillar around 1935. In that year he was living and working as an office clerk at the Jacksonville, FL church and missionary home.

Left: Thelma at Zarephath. Right: Thelma at Zarephath with Jessie Cubbedge who later married Thelma's brother, Hunter.

APPENDIX 3

COURTSHIP AND MARRIAGE OF SUNDAY & THELMA

IN AUGUST, 1937, WHEN THELMA RITCHIE arrived at Zarephath, NJ for her senior year of high school, Sunday Sharpe moved from Cincinnati to ZA to begin his college years. They were both boarding students whose parents were far away. They were also both "southerners" and became attracted to each other. In those days it was forbidden for boys and girls to spend time together. Much of the courting went on clandestinely on the "back road" from the campus to Millwood along the river. Here, one June day in 1938, after graduating from high school, Thelma and Sunday had a memorable romantic moment.

Ten months later (probably near the same spot) on April 17, 1939, Sunday proposed and Thelma accepted. Thelma's parents were in Jacksonville at the time and did not arrive back at Zarephath until sometime in July. Sunday must have waited till they arrived and then asked Hugh Ritchie for his daughter's hand. Thelma was almost twenty and Sunday was almost twenty-five. Hugh thought Thelma was too young and needed

Above: Sunday at Zarephath. Below left: San Francisco Missionary home. Below right: Sunday inside the home in 1940.

to finish college. The church leaders found out and called Sunday into a meeting on Aug. 1, 1939, where eight people ganged up on him. A.L. Wolfram told Sunday he had "broken the rules" by getting engaged without telling the church. Sunday replied that he didn't know any other way to find out if the girl liked him. He thought he was doing the right thing. He also said others who had gotten married had broken the rules. He was told he could go to the missionary field, but he replied that that would be "absurd" unless he had some assurance that this would be worked out. Later he was told by Alma White that he could go to the Providence home and stay until after camp meeting. This he refused to do and so turned in his credentials. Before Sunday left, Thelma told him she had decided not marry him. After leaving, Sunday secretly sent a letter trying to get her to change her mind. Thelma refused to read the letter and sent a reply, berating him for leaving as soon as he could not get what he wanted. This must have changed his attitude because he shortly returned to the work and took an assignment in San Francisco. The meeting notes and Thelma's letter come from the Pillar archives (link).

After Sunday left for California, Thelma was so stunned and dazed she hardly knew what had happened. She threw herself into her studies and made A's and B's. But she couldn't get him out of her thoughts and longed for him so much she thought she would "go crazy".

She was under a great deal of stress from taking a full load of college classes in addition to teaching young children every morning, and struggling to keep them under control. On April 25 she noted that "the out-of-doors always has a wonderful effect

Sunday, likewise, threw himself into missionary work, preaching, reading, devotions, and music, in addition to the mundane chores of missionary home and car maintenance. He took a college correspondence course in Biography and started weekly piano lessons from a local teacher.

His primary work was collecting donations of money and materials. He traveled far and wide from his home base, often spending one or two nights out. One week he took in over $100, and in one day he took in $65 which "broke all records" for him. He often wrote letters to Alma White and sent her money, perhaps aiming to prove he was a valuable contributor to "the work". He also wrote articles for the Pillar of Fire magazines.

On March 30, 1940, Sunday noted "Here it is almost the end of March which makes seven months that I have been in California which in some ways seems more like seven yrs. Still, I feel that the time has been well spent, and I have learned a few things which I would have been ignorant of otherwise. What the outcome will be I do not know, but I am walking the chalk line."

Sunday had been kept informed about Thelma by Norman Sillett, his friend and missionary companion. Norman's sister, Margaret, was Thelma's best friend. He became convinced that Thelma still had strong feelings for him and so decided to broach the subject of marriage to Thelma in a letter to Alma White sent on May 20. We don't have his letter, but Alma's reply praised his "marvelous success" in California and told him to work it out with Thelma. Sunday immediately sent Thelma a letter, relaying the news and delicately asking if there was any chance of reconciliation.

ZAREPHATH, N.J.

May 22, 1940

Rev. Wm. Sunday Sharpe,
    San Francisco, Cal.

Dear Bro. Sharpe:

I received your letter. Now in regard to your future, I have no comment to make one way or the other. I say it is up to you and Thalma. I feel I should be left out of the whole matter, and out of your and Thelma's decision regarding it.

The Lord has wonderfully used you there in San Francisco and I could not go against light and say for you to go here or there, as long as He has given you the marvelous success that you have had there in behalf of the work.

Now settle it between yourselves and leave the church and me clear out of the picture. Now what else could you do if you were in my position, that would satisfy all concerned?

Yours sincerely,

Alma White

on me. I feel soothed and quieted, and very close to God when out in the woods with nature. People and the cares of life seem to fade away, leaving me peaceful & filled with wonder at the sight of God's majesty & glory. Somehow the pain of being parted from Sunday is eased, too, and I think only of our dreams fulfilled and believe that someday they will be! It has been almost eight months now since he left, and I still care for him as much as I ever did. In fact, I am more sure than ever of my love for him; still, I am praying only for God's will to be done."

Thelma's father Hugh had objected to the marriage at first partly because he thought Sunday might leave the church, taking Thelma with him. But as time went by, he became convinced that Thelma knew her own mind and heart. In mid-May Thelma confides to her diary that her parents were at last reconciled to the marriage.

On May 29, Thelma received Sunday's first letter in nine months: "I was so nervous & excited that I could hardly get it out of the envelope. Oh, I don't know what to say or how to express all that letter meant to me. I can only say that he still loves me and still wants me to marry him. The best part of it all is that the bishop said it was up to us & we were to settle it between ourselves."

Thelma immediately responded professing her enduring love. A flurry of letters ensued about the details of their union. The Bishop ruled out a New Jersey wedding, so Thelma ultimately packed her trunk and on June 29th boarded a train in New Brunswick for a three-day journey through Chicago and Denver to Oakland, California. (The following are quotes from Thelma's diary.)

Thelma and her grandfather, Edward Wallace Lewis, in Chicago.

Sunday and Thelma at their wedding. Miss Garretson (with the couple below) performed the ceremony.

June 30: Arrived in Chicago this am at 8:30. Miss Hollander met me at the train. Spent day at missionary home. Bishop White, Dr. A.K. & wife were there, having left ZA yesterday also. Saw Grandpa [Lewis] after about sixteen years [since she left Alabama]. Spent a wonderful day with him. He certainly has a heart of gold and I thought the world of him from the moment I saw him. Left Chicago at 5:30 on Zephyr train for Denver. Was a wonderful train & I enjoyed myself immensely.

July 1: At 9 a.m. I got to Denver. Spent a memorable day with Mrs. Sharpe [Bessie, Sunday's mother], Aunt Mary [Wallace's sister], and Helen Frey visiting many of the places I knew as a child. Worn out by evening. Took bath & went to bed early.

July 2: Had a splendid rest last night-- slept about ten or eleven hours without waking once. Left Denver this a.m. at 9 on the D&RG (Denver and Rio Grande) Railroad. Wonderful time viewing the gorgeous sights from the train. Very comfortable. Saw Pikes Peak, Moffat Tunnel, Colorado River, Gore Canyon and many other beautiful sights.

July 3: Arriving at the dock, Sunday met me and once again after almost a year, I felt the comfort of his arms. Am supremely happy although very tired.

July 9 (Tues): This is the great day. Sunday & I were married at 1 o'clock in the parlor by Miss Garretson. Had cake & ice cream after ceremony. About 11 or 12 people present. After eating Mrs. [Emma] Wolf & Sunday & I went to Park where we took some pictures. Came back, then started on our honeymoon. Spent the night in a nice hotel in

Friday, July 12: Today was a most interesting one for Thelma and I. We left the hotel about 9:00 a.m. and went to a restaurant for a bite to eat. Then we drove northwest into the country, came to a farm house near which we parked the car and from there took a long walk into the dense forest and along a rippling stream lined with round pebbles. We watched the fish and threw rocks into the water. At noon we went to the farm house and the old lady was very kind and gave us a quart of milk as a part of our lunch. In the p.m. we drove to another place and read some from "The Cabin" by White, went for another walk then drove back to Eureka.

Saturday, July 13: We left the Inn rather early and drove northwest into Oregon to the Caves, arriving there about 3:00 p.m. The caves were declared a national monument in 1912 by President Taft and a highway has been constructed through a Mt. pass for 20 mi., leading to the entrance. Of course, we paid a 55¢ fee each and went through. We traveling for a mile and a quarter through narrow passageways, up ladders and into large rooms, we saw some beautiful sights of foundations of stalagmites and stalactites. Millions Chamber was the most beautiful with its Tulip-shaped formations in a cone-shaped room that rose up sheer many feet to a point at the top. Many- colored lights played on their formations.

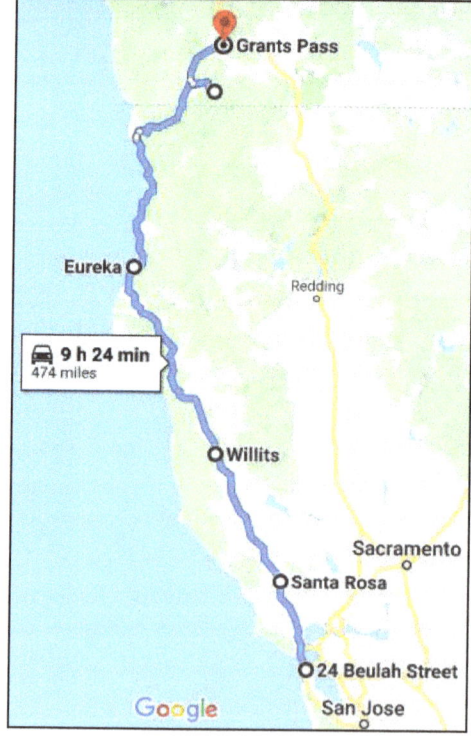

Above: Entries taken from Sunday's diary.

Left: Map of the honeymoon trip taken by Sunday and Thelma in July, 1940.

Santa Rosa. Both tired but very, very happy. It all seems like a dream. Yet it is so natural to be with Sunday. Never dreamed it would be that way.

July 10 (Wed.): Continued north on our trip today. It was grand driving along early in the morning when the air was so fresh and the sun so bright. In afternoon we stopped under a shady tree that was some distance away from the highway & rested. Read some in our diaries and talked a long time. Got to Willits around five p.m. where we spent the night. Got a room in the Van Hotel.

July 11 (Thurs.): Left Willits early this a.m. Ate a bite in the car just outside of town. Day was beautiful & we saw the wonderful redwoods. I had never seen them before & I was fascinated beyond words. I looked & looked until it seemed I couldn't get enough of their beauty. Once we stopped and rested for a while under their great towering shelter. Spending night in Eureka.

July 12 (Fri.): A wonderful hiking in the woods outside of Eureka. Never spent a lovelier day in my life. The woods in some spots were a literal paradise. In afternoon we drove to a place where we rested for awhile then read some & finally took another walk, this time up a steep hill. Got stung by a wasp while trying to get some fragrant flowers. Spending tonight also in Eureka.

July 13 (Sat.): Traveled on to Oregon this morning. Arrived at Caves around 3 pm. Spent an enjoyable afternoon exploring the passages & chambers that were filled with beautiful formation—stalactites & stalagmites. Left around 5 or 5:30. Got to Grants Pass an hour or two later. Got a cabin at the Riverside Motel Court. Bought some groceries for supper & for tomorrow.

July 14 (Sun.): Stayed around the cabin most of the day. Was a very hot day & we almost melted. Rested in afternoon. Had a malted milk down town this p.m. then took a little drive out to a Memorial Park.

July 15 (Mon): Left Grants Pass early this a.m. to drive the 450 miles home. Day was hot & we were tired when we arrived home in afternoon, but thankful we had made it safe & glad to be back home. Enjoyed reading letters & cards & opening several gifts that were waiting for us. Went to bed early.

July 16 (Tues.): Washed clothes this a.m. while Sunday cleaned the car. This afternoon, rested some. Spend a full day getting settled. Like our room very much. [They lived in one room on the first floor of the missionary home at 24 Beulah St, San Francisco.]

APPENDIX 4

U. S. ARMY

Recruiting & Career Counseling JOURNAL

MARCH 1966

ON THE WAY!

OFFICER CANDIDATE SCHOOL
FORT SILL, OKLA.

BECOMING AN ARTILLERY OFFICER

By 2ND Lt. Joseph A. Rollo
Information Office
U. S. Army Artillery and Missile Center
Fort Sill, Okla.

> First in a series of articles reporting on highlights of a typical artillery OCS class from the first week of training to completion of course and graduation as commissioned officer.

Candidate Robert L. Stein, a middleclass guide, gives helpful suggestions to members of Class 8-66 prior to processing for entry into OCS the next day.

Candidates are issued "basic load" of literature. They will rely heavily on the text books as they progress in the 23-week Officer Candidate course.

One hundred and thirteen soldiers, ranging in rank from private to warrant officer and representing 34 states, reported into Officer Candidate School, Fort Sill, Okla., and attained the rank of "candidate" with one goal in mind: completion of the course and a commission in the U. S. Army as a second lieutenant.

The group was assigned to Delta Battery, commanded by Capt. Floyd D. Whitehead, a 1960 graduate of West Point. Five tactical officers are also assigned to Delta Battery to help in the supervision of the candidates.

First Lieutenant Francis E. Israel Jr., executive officer, has a Master of Science Degree from the University of Knoxville.

The four platoon leaders are 1st Lt. Gerald W. Sharpe and 2nd Lts. John R. Treanor, John C. Gambaccini and Pat J. Razook. Lieutenant Sharpe is a graduate of OCS and attended Western State College in Colorado.

Lieutenant Treanor graduated from St. Mary's University, San Antonio. Lieutenant Gambaccini graduated from Kent State University in Ohio. Lieutenant Razook holds a Master of Science Degree in Student Personnel and Guidance from Oklahoma State University.

FIRST WEEK

The new candidate class, Number 8-66, began its processing Sunday, Oct. 31 for the 23 week long class.

Members of Class 8 spent most of their first hectic week being welcomed to OCS and receiving the tools and equipment which they will use during the next six months. They received their texts and instruments on Tuesday and were amazed at the voluminous manuals and gunnery equipment they will have to master.

In a ceremony Thursday afternoon Col. Charles E. Howard, the OCS commandant, presented the class guidon to Officer Candidate Battery Commander Marvin C. Williams, Danbury, Tex.

The class guidon bearer was Candidate Daniel Russel III, Glenarden, Md. The guidon will be presented to the honor graduate at the completion of the course.

On Friday morning, Col. Robert C. Williams, deputy assistant commandant of Sill's Artillery and Missile School, welcomed the class to the school. He told them that units at Sill fought to get graduates of OCS because they were the "best-trained artillerymen available."

After the welcoming address, the candidates attended their first classes. The instruction was given by the Artillery Transport Department and covered the general policies of the Army's maintenance program. Duties of the driver were stressed and the responsibility each individual has for the equipment assigned to them.

Candidates learn the Army from the ground up so that they will become familiar with the jobs of the lowest grade enlisted personnel before taking command as an officer.

As lowerclassmen during the first seven weeks of the course they learn the duties of the "worker." Later, during middle and upper class, they will learn the duties of the non-commissioned officer and junior officer, respectively.

The week was completed with an inspection in ranks by the TAC staff on Saturday morning. If the candidates had not learned what OCS was all about during the inspection they had a good indication of the standards expected of them.

With a long first week behind them, the candidates can look forward to many classes and a rigid schedule of activities to keep them always on the alert so they will learn how to function under pressure.

Two inspections and one written exam highlighted Class 8's first academic week of OCS training. After a week of processing, the 113 candidates began a rigid schedule of classroom instruction.

Although subject to inspection 24 hours a day, formal inspection of personal appearance and individual weapons was conducted on Saturday, Nov. 6 by the battery commander, Capt. Floyd D. Whitehead, and his platoon leaders.

SECOND WEEK

Class 8 received its first written exam from the Artillery Transport Department of the Artillery and Missile School on Tuesday. The exam covered the instruction the candidates had received on the Army's maintenance program.

On Wednesday, the candidates were finally given a chance to conduct their own inspections of various Army vehicles.

A junior officer is often assigned the task of motor officer and is responsible for the condition of the equipment assigned to his battery. The instruction on Wednesday gave the class practice in knowing how to determine the readiness of a vehicle.

An integral part of OCS is the close relationship each candidate has with his platoon leader. On Wednesday afternoon, each candidate was counseled by the officer responsible for his training. The candidates were questioned on their background and individual problems.

Veterans Day was a special day for Class 8 students as they were allowed to leave the OCS area and use the recreational facilities of Fort Sill. Although the time was used by some to polish boots and do the hundreds of things that could not be accomplished on busier days, some candidates took the time to see a movie at one of Sill's three post theatres.

Friday afternoon the candidates received the first classes in what is one of the most important blocks of instruction in their 23 weeks. The class was taught the proper method of giving commands and teaching drill and physical training. They will spend many hours doing practical exercises in these subjects before their training is completed.

The usual inspection by the officers of Delta Battery took place on Saturday morning. The errors of the previous week were corrected but the well-trained eyes of the TAC staff were still able to find some flaws that the candidates had missed.

The candidates will get a look at Sill's range area as they begin their instruction in map reading next week. They will learn how to locate their position on a map by observing the area around them. This past week *they knew where they were* —IN OCS.

THIRD WEEK

"You should be able to locate your general position by observing the surrounding terrain. By taking azimuths from known points, you will be able to plot your exact position."

First Lieutenant William A. Ziegler, an instructor at Sill's Tactics and Combined Arms Department, was kept busy last Thursday afternoon instructing a section of OCS Class 8 and answering their many questions as they prepared for a final exam in map reading on Friday.

The instruction took place on Fort Sill's West Range and the candidates moved from one location to another, constantly trying to orient themselves by observing the terrain.

Using maps, coordinate scales, protractors, compasses, and grease pencils, candidates of Class 8 tried to piece together the knowledge they had been collecting all week.

The course of instruction this week also covered classes in military justice and beginning classes in survey.

On Saturday, the candidates witnessed Artillery Firepower Demonstration. They observed a display of small arms firepower and three artillery "mass missions" in which 80 pieces of tube cannonry simultaneously fired on the same target.

Following the third "mass mission," the candidates watched as H-34 Choctaw helicopters from Sill's 1st Aerial Artillery Battery demonstrated the firing capabilities of their 4.5 inch rockets.

Next week the candidates can look forward to a big Thanksgiving Day feast in the OCS messhall and the possibility of a pass for a few hours on the holiday.

Classes in Field Artillery Tactics begin next week. This block of instruction will last throughout their training. They learned this week how to read a map and travel from one location to another without getting lost. Next week the candidates will begin to learn what they should do when they get there.#

(to be continued)

At Sill's West Range, Candidate Alexander Koziolet Jr. of New Ulm, Minn., takes a compass reading on a terrain feature pointed out by 1st Lt. Ziegler, instructor in tactics and combined arms.

First Lieutenant Sharpe (left) snaps weapon from hands of Candidate Joseph W. Howze of Cleveland. Sharpe's inspection covered individual weapons and personal appearance.

BECOMING AN ARTILLERY OFFICER

By 2nd Lt. Joseph A. Rollo
Information Office
U. S. Army Artillery and Missile Center
Fort Sill, Okla.

> Second in a series of articles reporting on highlights of a typical artillery OCS class from the first week of training to completion of course and graduation as commissioned officer.

Through these portals. . .

FOURTH WEEK

Time is a very valuable but limited ingredient in the life of an officer candidate. The high standards of personal appearance and military deportment must be learned quickly for there is little free time in the candidates' daily routine.

From 5 a.m. to 10 p.m., the candidates rush through a hectic schedule fighting the clock trying to get all their duties and assignments completed.

Their day begins with a mad scramble as the candidates prepare for their physical training. Five minutes after awaking, they are in formation prepared either to run two miles around the OCS track or begin their twelve repetitions of the "daily dozen," the Army's standard physical training exercise.

The candidates have 15 minutes after PT for showers and a change to their duty uniform. At 5:45 they are in formation for Reveille.

After breakfast, the candidates have 45 minutes to prepare their barracks and the battery area for inspection. At 7:15 the academic schedule begins with classes until 11:30.

After an hour-long break for lunch, classes resume. The candidates' lunch hour includes traveling to and from classes and waiting in formation to enter the mess hall.

Study Hall begins at 7 p.m. and lasts for two hours. The candidates must have completed their duties prior to Study Hall because they have only an hour after it is over before "lights out" at 10 p.m.

When candidates fall below the standards expected of them,

demerits are imposed by Tactical Officers. In addition to being restricted to the OCS area during off-duty hours of the weekend, too many demerits will "qualify" a candidate for participation in the weekend "Jark March."

Named for Lt. Gen. Carl H. Jark, retired, former Fourth Army Commander and the director of OCS in 1941, the march is a four mile struggle up Medicine Bluff 4, a steep hill on Sill's West Range.

Several members of Class 8 took their first "Jark" this weekend. Upon returning, all candidates had a little more incentive to budget their time more carefully, accomplish all required tasks, get fewer demerits, and avoid the next "Jark March."

FIFTH WEEK

Azimuth, traverse, triangulation, orienting angle, and declination constant became familiar terms to members of Officer Candidate Class 8 this week as they prepared for their final exam in Survey.

Working with logarithms, aiming circles, theodolites, steel tapes and plumb bobs, the candidates checked their measurements of Fort Sill's terrain against the accurate data held by the Survey instructors of Sill's Target Acquisition Department.

Accurate survey is necessary if the artillery is to provide effective fire support. The exact position of the artillery piece must be known if fire is to be directed accurately on a target.

Wednesday and Thursday mornings and all day Friday the

Candidates John A. Kilcoyne (left) and Barton H. Ishizaki take tape measurement during survey exercise

Candidate Wayne Smoot of Newcastle, Calif., sights through the eyepiece of a theodolite (a survey instrument)

Candidate Michael S. Moseley, Moorestown, N. J., receives green felt and eagle, symbols of middle class

candidates worked in the field putting their classroom instruction to practical use. On Wednesday the candidates performed a *position* area survey problem. A *target* area survey problem faced them on Thursday, and Friday's problem was an artillery battalion survey requirement.

Friday's exercise afforded the candidates the pleasure of spending the duty day away from the pressures of Robinson Barracks. The candidates performed the various duties of a survey party. Some candidates acted as tapemen, some as recorders and others as instrument operators.

The final survey exam is scheduled for 7:30 Monday morning. After completion of the exam the candidates immediately begin another block of instruction in communications. They will be taught the techniques and the importance of being able to send and receive messages by wire and by radio.

This Saturday morning the class will view a helicopter demonstration. The show will feature a 105mm howitzer battery being airlifted into position. An indirect fire demonstration by the 4.5-inch rockets mounted on the H34 helicopters of Sill's 1st Aerial Artillery Battery will also take place during the show.

Next Wednesday morning the physical fitness of members of Class 8 will be determined as they take the Army's Combat Proficiency Test.

SIXTH WEEK

Crawling, swinging, running, dodging, jumping, throwing and more running highlighted this week for Class 8 at Sill's Officer Candidate School.

The candidates put away their classroom notes Wednesday morning and struggled through the five physical fitness exercises which make up the Army's Physical Combat Proficiency test. A maximum of 100 points can be made in each event.

Starting the test is a 40-yard crawl over a dirt course. Two candidates in the class completed the course in 21 seconds.

The second problem is the dodge, run and jump, a test of agility and speed. The test includes several obstacles, a six-foot wide ditch and four gates to hinder the runner. A time of 22 seconds is necessary for a maximum score. Five candidates mastered this test with perfect scores.

The horizontal ladder or "monkey bars" is next on the list. Strong hands and arms and good timing are essential in this exercise. Candidate Francis J. Sloan, Chicago, Ill., led the class in this event.

Candidates limbered up their arms in the next test, the

grenade throw. A man must throw five disarmed grenades 90 feet to the center of a bulls-eye. Three candidates scored the maximum points for the test.

A mile run is the finale of the 500-point-possible PT test. The run must be completed in six minutes and 2 seconds to score the maximum points. Candidate Andres O. Ortiz, San Antonio, Tex., excelled with a time of five minutes and 40 seconds.

Candidate Bobby L. Parker, Thomasville, Ga., led the class with a score of 497. Passing score for the test is 300. Forty-nine of the 109 candidates who took the test scored more than 425 points. Average score for the class was 411.

SEVENTH WEEK

Green felt and golden eagles became a part of the uniform of the officer candidates of Class 8 last week when they advanced to middle class status.

In a ceremony Thursday, the class received the symbols of middle class from the members of Class 3-66, who are in upper class. The green felt is worn under the OCS brass on the collar. The eagles are worn on the helmet liners which the candidates wear when in fatigues or marching in a parade.

As members of the middle class, the candidates are now charged with supervisory responsibility. They hold positions in OCS which would normally be alloted to noncommissioned officers in other units.

In the various details which must be accomplished in the OCS area Class 8 will now supervise the lower classes.

Prior to Thursday's ceremony, the candidates took a written exam in communications. On Friday, they spent the day in the field putting to use the knowledge gained in the classroom as they were required to set up a battalion communications system.

Next week the candidates take a six-day break for Christmas leave. They will return to their classes on Dec. 29.

On Saturday, the class entered into the spirit of the holidays by sponsoring a Christmas party for the Lawton Boy's Club. The candidates obtained a cartoon movie for the party and took up a collection for prizes, fruit, and candy. About 400 boys attended.

When the candidates attained middle class status, a milestone was passed; the next hurdle is advancement to upper class. With an abundance of fine training, hard work and intense study behind them, members of Class 8 may confidently anticipate upper class standing eight busy weeks from now.#

(To be continued)

Aiming circle procedure

105mm breech block mechanism

becoming an artillery officer

By 2nd Lt. Joseph A. Rollo
Information Office
U. S. Army Artillery and Missile Center
Fort Sill, Okla.

> Third in a series of articles reporting on highlights of a typical artillery OCS class from the first week of training to completion of course and graduation as commissioned officer.

EIGHTH WEEK

The complexity of accurate artillery fire support was demonstrated to the aspiring officers of Class 8 this week as they began courses in gunnery, the largest block of instruction.

Instruction began with an orientation class in fire direction. Candidates learned the general duties of personnel in the fire direction center and became familiar with the various equipment used to convert fire missions to fire commands to the howitzers.

On Tuesday, the class was taught correct terms used in preparing an artillery piece to fire. These fire commands give all the necessary information to the gun crew for commencement, conduct and cessation of fire.

Beginning on Thursday, the candidates were allowed to take time for Christmas leave.

Returning to OCS, the class began practical exercises in laying an artillery piece by use of an aiming circle. The tubes of the howitzers must be pointed in a known direction if accurate fire is to be achieved.

Rounding out the general introduction to gunnery, instruction in observed fire was given to the class members. They learned the proper method of requesting a fire mission as a forward observer. Other classes will soon follow in the adjustment of fire.

More specific courses in the art of gunnery begin next week when the candidates receive instruction in the artillery weapons. Each major weapon will be covered in detail and classes conducted in the maintenance, care, and firing characteristics of the piece.

Next week Class 8 will have a new battery commander. First Lieutenant Gerald W. Sharpe, former platoon leader, will replace Capt. Floyd D. Whitehead as battery commander. Captain Whitehead is leaving OCS for duty in Vietnam.

Settling down following Christmas, Class 8 looked forward to a new course of instruction, gunnery; a new battery commander, Lieutenant Sharpe; and a new year. One of the classes' New Year resolutions was sure to be the determination to walk across the stage in April with a second lieutenant's commission.

NINTH WEEK

Answers to questions which might save a lieutenant's life were provided for the candidates this week as they took classes that ran the gamut from first aid to insurgency to night patroling to 105mm howitzers.

CH-37 'copter loading demonstration

Fixed-wing aircraft explained

All courses of instruction had one thing in common—preservation.

Beginning Monday, the candidates were instructed in the Army's system of medical support and the medical aspects of nuclear warfare. They learned how the Army cares for its wounded and how to survive and continue their mission in the event of nuclear warfare.

First aid classes were given on Thursday and the candidates learned how to treat themselves if wounded and no other help was immediately available.

Gunnery instruction this week concentrated on the weapons of the Artillery and the care, maintenance and functions of each part of the 105mm and 155mm howitzers. Learning to keep their weapons combat-ready by careful and knowledgeable maintenance was emphasized.

The gunnery classes featured instruction in small groups, with approximately eight candidates assigned to a howitzer with one instructor from Sill's Gunnery Department.

Friday afternoon and evening the class was given instruction and a practical exercise in night patroling techniques. This session included how to locate the enemy and gather necessary information about his position and activities. The candidates experienced the difficulties of negotiating unfamiliar terrain and moving undetected during darkness.

Friday's class was only a preliminary to the instruction which will follow next week. Classes in escape and evasion and counterinsurgency operations will further test the skills acquired during Friday night's practical exercise.

The material presented this week may well save their lives in combat. And, more immediately, if properly learned, it will help them with their cannon exam next week as well as the escape and evasion practical exercise later on Sill's West Range.

10TH-11TH WEEKS

Standing at a rigid brace, the candidate knocked three times and was told to enter the room. Approaching the desk, he saluted and reported to his platoon leader. He was asked to sit down and the lieutenant read from a piece of paper as the candidate listened attentively.

Thus began another counseling session at Sill's Officer Candidate School. Each candidate is regularly rated by the other candidates in his platoon and by the tactical staff in his battery.

The tactical staff is charged with the responsibility of developing, instructing, and evaluating the candidates in their charge.

Each candidate receives a daily inspection by his platoon leader and every candidate in a leadership position receives a written observation report on his performance. Leadership positions are rotated weekly and a candidate may hold varied offices from candidate battery commander to squad leader.

Daily inspections and ratings on each leadership assignment keep the tactical staff busy learning more about each candidate every day. The candidates frequently are counseled by the staff on their progress and special problems.

Facing difficult assignments each day and under constant evaluation of their performance, the candidates learn more about themselves and develop confidence in their ability to lead.

A well-timed snow storm provided the candidates of Class 8 with physical proof of the problems of mobility. The Tactics/Combined Arms Department of Sill's Artillery and Missile School provided part of the Army's answer—movement by air.

On Wednesday the officer candidates were instructed in the role of Army Aviation and the characteristics of various fixed-wing and helicopter aircraft. After a classroom presentation on the Army's concept of air-ground operations, the class traveled to a hangar at Fort Sill's airport to view several aircraft and learn their capabilities.

A medium transport helicopter, the CH-37 Mojave; a fixed-wing utility aircraft, the U-6 Beaver; and an experimental fighter-escort helicopter, the Bell Huey Cobra, were among the aircraft shown to the candidates. The Huey Cobra drew the most attention because of its sleek design, speed, and firepower.

The blanketing snow storm forced some classes at the Officer Candidate School to move indoors. Drill instruction and physical training exercises took place in the gymnasium and inside the barracks.

Instructed mostly within the warm, cozy confines of Snow Hall this week, Class 8 journeys outside one day next week to put into practice techniques learned in the classroom.

Included in next week's schedule is a "split shoot" which will test the candidates' ability to fire the 105mm howitzer and adjust the rounds to a target on Sill's impact area.

Fixed-wing aircraft, helicopters, and 105mm howitzers are all a part of the Army's new airmobile division. Secretary of Defense Robert S. McNamara praised the division as a "different approach to the solution of tactical problems." To the candidates of Class 8, the instruction was another important phase of learning how to be an officer.#

(To be continued)

Preparing to load 105mm

Firing practice round

becoming an artillery officer

By 1st Lt. Joseph A. Rollo
Information Office
U. S. Army Artillery and Missile Center
Fort Sill, Okla.

Fourth in a series of articles reporting on highlights of a typical artillery OCS class from the first week of training to completion of course and graduation as commissioned officer.

12TH-14TH WEEKS

Having completed 11 weeks at Artillery Officer Candidate School, which included five weeks of gunnery instruction, candidates of Class 8 finally begin practicing their chosen military profession—*firing artillery weapons.*

Although as officers they will probably never fire the weapons, the candidates must learn the duties of each cannoneer to insure that they can train and supervise personnel in efficient operation of the firing battery they will someday command.

After an exhaustive, practical exercise and "dry run" beginning on Monday, the class traveled to West Range for an all-day actual firing exercise. The class was split into two groups as one section fired the 105mm howitzers while the other section observed the rounds and adjusted the fire to targets in the impact area.

Classroom instruction in the duties of the executive officer of a firing battery and a practical exercise in the operation of the fire direction center were additional classes in gunnery given this week.

The candidates will look at Sill's impact area from a different angle next week as they take a class in aerial observation. From helicopters and fixed-wing aircraft, they will observe and adjust artillery rounds.

The skills of an artillery battalion's forward observer, the most hazardous artillery job, were practiced by the candidates after many hours of preparatory indoor classes before actually expending ammunition on West Range.

Peering through their binoculars, the students took turns calling for fire on targets designated by their Gunnery Department instructor. After initial rounds were fired, the candidate was required to sense the deviation from the target and adjust the fire within destructive distance.

Yellow and red automobile bodies were the primary targets in the impact area, representing enemy patrols and command post headquarters. The candidates in Section 1 had already seen these targets from a different angle—observed from helicopters and fixed wing aircraft. Calling for fire from the air and adjusting the rounds to the targets presents additional problems with which the forward observer in modern warfare must cope. This aerial method is presently being used with much success in the jungle warfare of Vietnam.

When the candidates were not in the field practicing fire missions, they were in Snow Hall learning more about the operation of the fire direction center. Another important class given was in methods of military instruction. Proper teaching techniques are essential because when they become second lieutenants the candidates will be required to give many classes to members of their battery.

Observing-adjusting fire

On range with M-14

For two days next week the class will put aside artillery matters as it journeys to one of Sill's rifle ranges to practice firing M-14 rifles. A forward observer protects his battery and front lines with artillery fire but he must also be able to protect himself with small arms fire if necessary.

Switching from pulling lanyards to squeezing triggers, the candidates first fired the M-14 for practice and later for score, ably assisted by the personnel of Sill's Trainfire Branch. During the two days of firing, the skills acquired in basic training were relearned.

M-14 rifle familiarization is part of a weapons orientation which all artillery officer candidates receive during the 23 week course. Before graduation each candidate will fire the 3.5 rocket launcher, the .22 caliber pistol and the M-60 machine gun.

Returning to the classroom, an exam in Tactics was given. Observed fire exercises followed by another Tactics exam highlighted the remainder of the week's activities.

Although another exam in gunnery and several classes in tactical field operations are scheduled, the class members look forward to next week when they attain upperclass standing.

As upperclassmen, the green felt on their uniforms will change to red. They will become candidate officers with increasing responsibility, and more important supervisory tasks will be assigned to them. After 15 weeks at OCS, the last hurdle, upperclass, is a welcome sight on the horizon.

15TH-16TH WEEKS

One hundred nine members of Class 8 received their red tabs this week and were promoted to upperclass. After 14 weeks of academic training, they are now regarded as junior officers with added responsibilities—but additional privileges.

On Thursday the candidates assembled in the battery area and received their tabs from Class 4 which is in its 18th week. After pinning on the tabs, Class 8 is eligible for salutes from all lower class candidates.

Walking in the OCS area is only one of the privileges the candidates now receive; formerly they were required to double-time. The north-south walks are accessible to the Class and they may now smoke outside. Jark Marches will be easier for the new upperclassmen because they will no longer be required to carry their rifles which weigh some 11 pounds at the start but "grow heavier with every step."

Although the physical efforts required will be easier for the next 8 weeks, the academic schedule and added supervisory responsibilities will keep the candidates busy. The hardest part of gunnery is still before them, and the leadership positions in the all-day field problems are now theirs to assume.

The classroom schedule this week was mainly in preparation for the field problems coming up in the future. During the next eight weeks class 8 will have two tactical field problems which will test all the skills learned in previous training. This week's training dealt with the defense of an artillery battery against enemy attack and the proper methods of reconnaissance before occupying a position area.

Friday afternoon 64 candidates took the physical examination for Airborne training. Ranger and Special Forces tests were also given and attracted many candidates.

Washington's birthday highlights next week's schedule as OCS celebrates the holiday from the academic training schedule. An important class will be given next Friday as the candidates learn the duties of safety officer, a job which occupies much of a new second lieutenant's time.

Upon graduation from Artillery OCS, a candidate receives a gold bar. After assignment to a firing battery, he receives another item of equipment—the yellow helmet of an artillery safety officer. This week, class members learned their duties when wearing this helmet.

The safety officer is responsible for the safe firing of an artillery piece and insures that the shell lands within the prescribed safety zone. During all firings at Fort Sill, a safety officer is present to personally check that the weapon is pointed in the right direction and that the range and deflection settings on the weapon are within the safety limits.

During their first full week as OCS upperclassmen, Class 8 looked forward to the traditional "Redbird Party" on Saturday evening. The party, which is given for all new upperclassmen, took place at Fort Sill's Polo Club. In addition to the party, the candidates were free for the remainder of the weekend.

Washington's Birthday was an additional holiday but training continued as usual the rest of the week. An exam on Monday and one on Wednesday kept the candidates occupied during the weekend and the holiday. The class also continued its gunnery instruction and fired some 155mm howitzers Thursday morning.#

(To be continued)

becoming an artillery officer

By 1st Lt. Joseph A. Rollo
Information Office
U. S. Army Artillery and Missile Center
Fort Sill, Okla.

Fifth in a series of articles reporting on highlights of a typical artillery OCS class from the first week of training to completion of course and graduation as commissioned officer.

16TH-17TH WEEKS

In modern warfare, where there is no front line, it is not unusual for an artillery battery to have the enemy "at its front door." Consequently, an effective perimeter defense must be employed if safety for the firing battery is to be insured.

In previous weeks, the candidates learned from the instructors of the Tactics Department of the Artillery and Missile School about the defense of a firing battery.

Last week instructors of the Officer Candidate School and the tactical officers from Delta Battery showed the candidates how to emplace their available weapons to defend the perimeter.

Class 8 combined with Class 4 (upper class of the battalion) in a firing exercise which involved selecting, occupying and firing from a position area.

After darkness had fallen, the candidates moved into another location and resumed firing. OCS instructors gave them a tour of the area perimeter and demonstrated the weapon emplacement which would protect the battery.

A rotating guard was maintained, and at 3:45 a.m., the candidates returned to the garrison area and began cleaning the equipment used on the exercise. Later that day an inspection of all individual field gear was conducted.

Next week two all-day field problems are scheduled with Class 4. Duties will be rotated so that all candidates can become familiar with the problems in communications, survey and gunnery. Candidates of Class 4 will be assigned the leadership positions and Class 8 will perform the jobs normally handled by the lower grade noncommissioned officers and privates in a battalion.

UPPER CLASS STANDING

With the graduation of Class 4, Class 8 assumed the senior role in the 2nd Battalion, the school having two battalions, 1st and 2nd, of seven batteries each.

Of the 108 candidates now in Class 8, six now hold the senior command and staff assignments in the 2nd battalion. The chain of command fluctuates as the candidates rotate through the leadership positions during the 23-week course.

Wednesday the following new staff-command assignments were announced:

Candidates Kenneth W. Simpson, battalion commander; James P. Dower, executive officer; James G. Branden, S3; Jon C. Henderson, S1; and Barton H. Ishizaki, S2.

CLASS LEADERS

Another significant event occurred last week; Class leaders were announced. After 16 weeks at OCS, the leading four candidates, from four different states, range from 22 to 27 years of age. Their prior military service ranges from nine months to nine years. While the average educational level for all Artillery OCS students in Fiscal Year 1966 is 13 and a half years, the average schooling for Class 8's top four is 17 years. Army schools and extension courses helped in achieving this high academic average for the four.

The leader in the class, Candidate Simpson (battalion commander), never went to college but has attended many Army schools. "The more service a man has, the easier OCS is," he stated. He also enrolled in many extension courses offered by the Army before coming to OCS.

Simpson is 22 years old, married, and the father of two children. He entered the active Army in January 1963 after a tour in the National Guard. His home is San Diego, Calif.

Candidate Dower (battalion executive), Class 8's number two candidate, is 27 years old, married and the father of four. He entered the Army in November 1956 and reached the rank of staff sergeant before entering OCS. When questioned about his educational background, he mentioned that he had taught Survey in the U. S. Army Artillery and Missile School at Fort Sill before becoming a candidate. He never attended a college and feels that "prior service outweighed prior education in preparing me for OCS"—Candidate Dower comes from Burlington, Vt.

The S3 and third ranking member of Class 8, Candidate Branden, has completed his course work for his masters degree in Sociology. A self-styled "citizen-soldier," he plans on attaining a PhD and becoming a college teacher.

Born in Albany, Ga., Candidate Branden went to Georgia Southern College and did his graduate work at Florida State University. He is 25 years old and single. "My academic background has been very helpful at OCS. I have had more to learn but my years in school have given me a better ability to learn and a broader outlook," he explained.

Candidate Henderson (S1) was a warrant officer before becoming a candidate. He has served in the Army for eight years and decided to become a commissioned officer because it would "allow more variety in future assignments." His previous assignment in the Army was as a calibration technican.

Although he previously attended three Army schools, Candidate Henderson stated that he "never knew there was so much involved in firing an artillery weapon."

He cited his previous military service and education as very beneficial in preparation for OCS. Henderson attended Kalamazoo College in Michigan for one and a half years. Married and the father of one child, he is 27 years old and grew up in Michigan.

18TH-19TH WEEKS

Physical training classes serve a twofold purpose at Artillery OCS, (1) achieving physical stamina and (2) learning proper instructional techniques by teaching the exercises to his classmates.

Class 8 members will have spent some 76 hours in formal physical training by the time they complete the 23-week course.

Each candidate is rated by the tactical officers on his ability to teach and his composure before a group. The candidate is judged on his voice, knowledge of the material, and ability to present the instruction clearly.

The physical training drills range from the regular Army "daily dozen" to the various rifle exercises, grass drills and guerilla exercises.

In addition to the physical training classes, each class receives 60 hours of instruction on dismounted drill. Field Manual 22-5, the bible of dismounted drill, is digested and taught by the candidates paragraph by paragraph.

Practical opportunities for marching and PT at OCS are abundant. The classes march in formation wherever they go and spend at least 30 minutes each day doing PT.

A candidate's rating as an instructor is an important ingredient in his leadership rating. As a second lieutenant, he may be required to give his battery instruction in the techniques of drill and PT and his methods of presenting the material must be faultless.

This week's training schedule included classes in artillery fire planning, an observed fire examination, and an orientation on the Honest John and Little John rockets.

On Saturday, the Class spent the morning in the field operating a fire direction center in preparation for next week's gunnery examination.

Speed and accuracy via computer was the topic this week for OCS Class 8.

The M18 Gun Direction Computer (or FADAC) replaced coordinate squares, range deflection protractors, and graphical site and firing tables in the minds of the officer candidates.

Fire direction, perhaps the most difficult part of gunnery instruction, involves the use of many items of equipment to determine the proper position of the howitzer tube in order to place a round in a specific location.

The computer solves the fire direction problem with greater speed and accuracy than is possible with human calculations.

The FADAC can be programmed to solve fire control problems for five firing batteries, meteorological computations and survey operations. Although not in general use, it is currently being tested and used in several types of situations.

The candidates operated the computer to test its speed and were given instruction in the maintenance requirements of the system.

No automation has yet been developed to replace the coordination between the human eye and brain so the candidates had to be content to settle down to the old method of observed fire practice.

Peering through binoculars, the candidates adjusted artillery rounds to a target in the impact area. Some candidates were given the opportunity to observe the rounds from the air as they hovered over the target in helicopters.

On Wednesday afternoon, class members received an orientation on reporting for their first duty assignment as second lieutenants. For the next four weeks before graduation, the candidates will attend several classes geared to acclimate themselves to their future status as officers.

Class 5 graduates next Tuesday and Class 8 will become the senior class in the entire school.#

To be continued

Candidate John R. Naas leads members of his battery in physical training on which they will be graded.

From left to right: Candidates Simpson, Branden, Henderson, Dower, Class 8's most outstanding.

Direction computer explained to Candidate Paul L. Tessier by 1st Lt. Ralph V. Nichols of Gunnery Department.

becoming an artillery officer

By 1st Lt. Joseph A. Rollo
Information Office
U. S. Army Artillery and Missile Center
Fort Sill, Okla.

Sixth and last article reporting on highlights of a typical artillery OCS class from the first week of training to completion of course and graduation as commissioned officer.

Honor Graduate Simpson has bars pinned on by Fort Sill's Colonel Cullen.

With the graduation of Class No. 8 on April 15, 1966, the U. S. Army gained 99 new second lieutenants of artillery.

THE BEGINNING

On entering Artillery OCS in November 1965, and being faced with 23 long weeks of intensified study, comprehensive examinations and inspections, and exhaustive-exhausting field and physical exercises, the members of Class 8 almost immediately adopted the same attitude toward the several months ahead. They found it would be more reassuring to regard each successfully completed day as "one more behind me" rather than count the days remaining to be successfully completed— days sure to be filled with increasingly difficult challenges and requirements, to say nothing of the ever-present chance of being eliminated or dropped back to another class.

For many of Class 8's future officers, the month before graduation had all characteristics of a first-class paradox—the passage of time was slow as it rapidly sped into the past. It seemed time went by at such a snail-like pace that each minute took an hour. On the other hand, days were dissipated with such great speed that a week was gone before it was missed.

THE ENDING

Regardless of the rate of speed at which time passed, or seemed to pass, it is believed safe to say that at this point the candidates had abandoned the attitude of "one more day behind me" and that they are now confidently counting the days ahead.

Final examinations began with a comprehensive gunnery examination. The two-hour test covered all previous instruction given over a 14-week period. Next came a tactics examination which included fire-planning and organization for combat. This was followed by two demonstrations conducted by the Tactics Department with a rifle company performing in the defense and the attack phases of operation.

Familiarization firing with the .45 caliber pistol, the artillery officer's basic weapon, preceded a 24-hour field problem on the West Range. The candidates performed normal artillery operations until close to midnight and then established an overnight bivouac.

As graduation drew nearer and nearer, the agenda included a panel discussion for Class 8. Heading the panel was Col.

Charles E. Howard, commandant of the Officer Candidate School, assisted by Lt. Col. Donald H. Richardson, assistant commandant, and Capts. James R. Heldman and William R. Kulik, battery commanders.

Having undergone week after week of learning their future duty responsibilities as commissioned officers, the candidates did not ask questions regarding what an officer does. Instead, they asked, and were encouraged to ask, questions concerning educational opportunities, advanced Army schooling possibilities, and the many personal benefits accruing to an officer and his family.

Fielding questions on personnel matters was CWO John M. Futch, personnel officer. The chief of Sill's educational division, Mr. Russell Croach, was also on hand to answer queries about civilian schooling through the Army's several bootstrap programs.

The schedule also included two all-day tactical artillery field problems. While these exercises were nothing new to the candidates, they did provide another chance to sharpen their artillery skills before reporting to their first assignments as officers.

However, something new was provided in the form of instruction and practical work in rappelling and river crossing.

Descents were made from three cliffs of different heights on Sill's Medicine Bluffs mountain. Using rope and hip harness, the candidates lowered themselves down the cliffs. On reaching ground level, they were required to reach the far side of an expanse of water on ropes strung across for that purpose. Two methods of crossing were used—first with two ropes, then with one—and all candidates tried both.

Before graduation next week, the class will take part in an escape and evasion problem. It will begin with an afternoon orientation. As darkness approaches, the candidates will set out in groups of five and move to a prescribed area nine miles away. They will attempt to make the move without being spotted and upon arriving in the designated area set up a bivouac and spend the night.

COMMENCEMENT

Traditional events leading to issuance of certificates of graduation and commissions as officers are the graduation parade and the reception-dance.

Held on the OCS parade ground, the graduation parade of Class 8 was reviewed by Col. Paul S. Cullen, CO, 1st Field Artillery Brigade. That night at the reception-dance, attended by many of Fort Sill's high-ranking officers with their wives, the Gunner's Trophy was presented to Lawrence E. Soper for highest gunnery average, and the Shooter's Award went to Wayne Smoot for highest grades in observed fire.

With Colonel Cullen as guest speaker, the graduation ceremony was conducted at Snow Hall Auditorium.

Class 8's honor graduate was Kenneth W. Simpson. Distinguished military graduates with Simpson included Jon Henderson, James P. Dower, James G. Brandon, Marvin C. Williams, Thomas E. Konkle, Barton H. Ishizaki, James H. Parsley and Michael F. McCardle.

Second lieutenants all, the 99 graduates of Class 8 have passed an important milestone on their road to accomplishment—a milestone they can look back on with personal pride and satisfaction which will enable them to look the future squarely in the face with justifiable confidence, keen understanding, and sincere dedication.#

Going down

Familiarization firing

Down

www.ingramcontent.com/pod-product-compliance
Lightning Source LLC
Chambersburg PA
CBHW041532120626
46551CB00019B/2672